Physics in Food Manufacturing

Case studies in fundamental and applied research

Physics in Food Manufacturing

Case studies in fundamental and applied research

Edited by
Megan J Povey, Melvin J Holmes, Sameera Rafiq, Elena Simone, Michael Rappolt and Mat Francis
School of Food Science and Nutrition, University of Leeds, Leeds, UK

IOP Publishing, Bristol, UK

ISBN 978-0-7503-2596-7 (ebook)
ISBN 978-0-7503-2594-3 (print)
ISBN 978-0-7503-2597-4 (myPrint)
ISBN 978-0-7503-2595-0 (mobi)

DOI 10.1088/978-0-7503-2596-7

Version: 20200601

IOP ebooks

British Library Cataloguing-in-Publication Data: A catalogue record for this book is available from the British Library.

Published by IOP Publishing, wholly owned by The Institute of Physics, London

IOP Publishing, Temple Circus, Temple Way, Bristol, BS1 6HG, UK

US Office: IOP Publishing, Inc., 190 North Independence Mall West, Suite 601, Philadelphia, PA 19106, USA

This book is dedicated to all my students and colleagues who have over very many years created and developed Food Science to the point where it is indispensable to the production of food for billions of people.

—Megan J Povey

Contents

Acknowledgements

I would like to thank the editorial team who helped prepare this book—my colleagues Mat Francis, Mel Holmes, Sameera Rafiq, Elena Simone, Michael Rappolt… Without their help this book would not exist. Secondly, I would like to thank all the contributors for the great efforts they have each made which together makes this book unique. I personally have learnt a great deal from reading and editing all the contributions. The standard of physics here is very high and impressive and as a result, this work will be a standard reference for the subject.

Editor biographies

Megan J Povey

Megan J Povey leads the Food Physics group in The School of Food Science and Nutrition at The University of Leeds. She was the first person in the world to hold a university chair in Food Physics being appointed in 2002 and has 50 years of experience in the use of ultrasound for the characterization of soft materials. Her group is internationally leading in the area of ultrasound characterization of nano-particulate material, primarily but not exclusively food particles such as the self-assembling protein casein.

Current publications and research:
https://environment.leeds.ac.uk/food-nutrition/staff/7140/professor-megan-povey
https://scholar.google.co.uk/citations?user=qF-f3UkAAAAJ&hl=en

Melvin J Holmes

Dr Melvin J Holmes has been involved in research and teaching since 1997 in mathematics and food science. He has primary research interests in the use of partial and ordinary differential equations to model physical systems, in particular in visco-inertial and thermal interactions of emulsions with incident sound fields. These models in conjunction with experimental results allow a valuable insight into the complex interactions that occur in real systems.

Sameera Rafiq

Sameera Rafiq is a Research Officer in the School of Food Science and Nutrition at the University of Leeds. She has experience in editing and presenting in various contexts including television and has organised a number of large successful conferences, including the Physics in Food Manufacturing conference held in Leeds in January 2020. She specialises in promoting Food Science on social media and publications.

Elena Simone

Elena Simone joined the School of Food Science and Nutrition at the University of Leeds in September 2016 as Lecturer in Food Crystal Engineering. Her research interests include crystal engineering, understanding the phenomena of polymorphism, nucleation and crystal growth using on-line, *in situ*, analytical techniques as well as off-line solid state characterisation; and exploring novel technologies and non-invasive techniques for food safety and reassurance.

Michael Rappolt

Michael Rappolt holds a Leadership Chair of Lipid Biophysics at the School of Food Science and Nutrition at the University of Leeds. He is a physicist and a leading authority on the use of various small and wide angle scattering techniques on investigating the structure, shape and flexibility of model membranes with various, ground-breaking achievements concerning their nanostructural and mechanical analysis. Recent activities have been concentrated on the study of drug/membrane interactions with applications to drug delivery and food. His research vision is to enforce building bridges between physics (nanostructure of self-assembled soft matter), biology (biointerfaces), chemistry (response of membranes to environmental changes) and applied research (food, health and bioenergy sectors).

Mat Francis

Dr Mat Francis uses physical and chemical principles to improve engineering processes in the food, water and energy sectors. Dr Francis has undertaken research into different applications of 3D printing using inputs such as extrudable food and flexible plastics. He has provided public demonstrations of 3D printing and recently created a free FutureLearn online digital module which included a section on how food 3D printing could be integrated with existing unit operations

List of contributors

Megan J Povey
School of Food Science and Nutrition, University of Leeds, Leeds, UK

D Flick
AgroParisTech

Matt D Sinnott
CSIRO Data61, Melbourne, Australia

Paul W Cleary
CSIRO Data61, Melbourne, Australia

Sharen J Cummins
CSIRO Data61, Melbourne, Australia

James E Hilton
CSIRO Data61, Melbourne, Australia

Gerald G Pereira
CSIRO Data61, Melbourne, Australia

Simon M Harrison
CSIRO Data61, Melbourne, Australia

Sergey V Lishchuk
Materials and Engineering Research Institute, Sheffield Hallam University, Sheffield, UK

John R Melrose
Nottingham University, Nottingham, UK

Elena Simone
School of Food Science and Nutrition, University of Leeds, Leeds, UK

R S Ward-Smith
Malvern Panalytical Ltd, Enigma Business Park, Grovewood Rd, Malvern, UK

R Taylor
Malvern Panalytical Ltd, Enigma Business Park, Grovewood Rd, Malvern, UK

S Cairns
Malvern Panalytical Ltd, Enigma Business Park, Grovewood Rd, Malvern, UK

M Kaszuba
Malvern Panalytical Ltd, Enigma Business Park, Grovewood Rd, Malvern, UK

D Fernandes
Malvern Panalytical Ltd, Enigma Business Park, Grovewood Rd, Malvern, UK

M Zoontjes
Malvern Panalytical Ltd, Enigma Business Park, Grovewood Rd, Malvern, UK

T Rowney
Malvern Panalytical Ltd, Enigma Business Park, Grovewood Rd, Malvern, UK

S Marsh
Malvern Panalytical Ltd, Enigma Business Park, Grovewood Rd, Malvern, UK

L Newey-Keane
Malvern Panalytical Ltd, Enigma Business Park, Grovewood Rd, Malvern, UK

Zachary J Glover
School of Food Science and Nutrition, University of Leeds, Leeds, UK

Melvin J Holmes
School of Food Science and Nutrition, University of Leeds, Leeds, UK

Xiaobo Zou
Agricultural Product Processing and Storage Lab, School of Food and Biological Engineering, Jiangsu University, Zhenjiang, Jiangsu, China

Xiaowei Huang
Agricultural Product Processing and Storage Lab, School of Food and Biological Engineering, Jiangsu University, Zhenjiang, Jiangsu, China

Xiaodong Zhai
Agricultural Product Processing and Storage Lab, School of Food and Biological Engineering, Jiangsu University, Zhenjiang, Jiangsu, China

Jiong Shi
Agricultural Product Processing and Storage Lab, School of Food and Biological Engineering, Jiangsu University, Zhenjiang, Jiangsu, China

Amin Sadeghpour
School of Food Science and Nutrition, University of Leeds, Leeds, UK
Empa, Swiss Federal Laboratories for Materials Science and Technology, Centre for X-ray Analytics, St Gallen, Switzerland

Filippo Ceola
School of Food Science and Nutrition, University of Leeds, Leeds, UK
University of Padova, Department of Agronomy, Food, Natural Resources, Animals and the Environment DAFNAE, Padova, Italy

Richard Smith-Uchotski
School of Food Science and Nutrition, University of Leeds, Leeds, UK

Arwen I I Tyler
School of Food Science and Nutrition, University of Leeds, Leeds, UK

Michael Rappolt
School of Food Science and Nutrition, University of Leeds, Leeds, UK

Mat Francis
School of Food Science and Nutrition, University of Leeds, Leeds, UK

Chapter 1

Physics in food manufacturing: case studies in fundamental and applied research—an introduction

Megan J Povey

As I began to write this introduction, Australia was on fire and Indonesia suffering appalling floods. Now we are amid a coronavirus pandemic in which global agribusiness has undoubtedly played a role. We face an existential threat from global warming and disease; the scientific contribution of physicists will be needed in order to meet the challenges we all face. It could be argued that our existential challenges require political solutions as much as scientific ones. Nevertheless, we will be obliged to change one way or another.

The scope of this book comprises invited contributions against three core themes:

- rheology and complex fluids;
- process engineering and manufacturing;
- measurement science and physical measurement of foods.

The coverage reflects the role that physicists play in solving the inherently multidisciplinary science and technology challenges in food manufacturing—from designing safe, nutritious and great-tasting foods to the process technology and manufacturing know-how needed to deliver compelling product innovation. This book is the first authoritative text on the subject, providing a foundational resource for the transformation of engineering and materials characterisation in the food and pharmaceuticals industries.

The editors' aim is to create a 'go-to' reference source for the global PiFM community—interdisciplinary physical scientists, food/nutrition scientists and engineers working in academic research, government labs and industry. The book will also be a valuable resource for R&D staff and product engineers working for suppliers of specialist instrumentation and equipment to the food processing industry. The book is augmented by complementary presentations from the *Fourth*

doi:10.1088/978-0-7503-2596-7ch1

IOP Physics in Food Manufacturing (PiFM) Conference which was held in Leeds, UK, in January 2020 (http://foodphysics2020.iopconfs.org/presentations).

Beyond research, technology and application, this book will have a role to play in promoting food-sector careers and professional development. Food manufacturing companies need a pipeline of talented scientific and technical staff, so we are keen to raise awareness among physicists that there are interesting careers to be had in the food industry.

Chapter 2 by Denis Flick from Agro Paris Tech is a comprehensive, clearly written and up-to-date account of the underlying physics and current techniques available for modelling heat transfer in foods. Dr Flick's group is internationally renowned for their work, which is reflected in the chapter bibliography.

Chapter 3 (Matt Sinnott and colleagues at CSIRO, Australia) contains some very impressive examples of the achievements of particle based modelling as applied to complex food processing problems such as mixing and heat transfer.

In *Models of surface viscosities of particle-laden fluid interfaces* (chapter 4) Sergey Lishchuk outlines computational and mathematical models for the rheology of the complex heterogeneous materials that make up food.

Coffee brewing (chapter 5—John Melrose) is a fascinating subject for all coffee lovers and the incredible complexity of the process arising from the complex microstructure, chemistry and physical chemistry of the raw material is detailed here, with lessons for other extractive processes. Chapter 6 is an extension of chapter 5 in which the extractive process is modelled.

A pioneer of crystal engineering approaches to crystallisation in food systems, Elena Simone (chapter 7) describes novel and potentially far-reaching engineering and modelling approaches to the growth of food crystals. Detailing amongst other things (lactose, succinic acid), the surprising behaviour of a highly polar polyphenol, quercetin, is described, which nevertheless crystallises in aqueous solution. The chapter begins with a valuable introduction to the fundamentals of crystal engineering.

In chapter 8, Steven Ward-Smith provides a comprehensive description of contemporary measurement techniques for the characterisation of foods. A number of case studies (chocolate, wheat and gluten-free flours, milk casein, KHT crystallisation in wine and coffee) really bring the application of measurement techniques to life.

The dispersion of dairy powders is the topic of chapter 9, authored by Zac Glover and Mel Holmes. Powder dispersion in liquids is problematic in many processes and new insights are offered into the dispersion process which have wider applicability than simply dairy powders. In particular, novel ultrasound and optical techniques for determining the process and end point of powder dissolution are described.

Xiaobo Zou and co-authors discuss non-invasive sensing for food reassurance in chapter 10. This comprehensive account of the techniques available for testing food by non-contact means could scarcely be more important in this era of global food production as one vector for the transmission of disease.

Amin Sadeghpor and colleagues describe novel methods and systems for encapsulation of food components in nanoparticles which can be used, for example, in the delivery of natural polyphenolic compounds, Vitamin K and polyunsaturated fatty acids (chapter 11).

Finally, in chapter 12, Mat Francis presents the latest 3D printing methods for foods. Better described as additive manufacturing, 3D printing has found a place in producing novel foods, albeit on a relatively small scale, such as personalised food products.

Chapter 2

Modelling heat transfer in foods

D Flick

2.1 Introduction

Heat transfer is involved in many food processes: cooking, baking, pasteurization, sterilization, drying refrigeration, freezing etc. Heat transfer occurs inside the food material, which can be solid or liquid, and around the food products in different equipment: oven, pressure cooker, autoclave, heat exchangers, cold room, display cabinet, refrigerator [1–4].

Heat transfer in food is of major importance because thermal treatments transform food:

- in terms of composition, due to various reactions (e.g. lipid oxidation, vitamin degradation, Maillard reaction, protein denaturation, formation of acrylamide…);
- in terms of structure, due to phase change (e.g. water or fat crystallization, water or ice vaporization) and due to modification of structural elements (e.g. starch gelatinization, cellulose softening, protein aggregation, bubble fragmentation…).

Heat transfer has also a large effect on water. Due to vaporization or crystallization the amount of free water can decrease, which leads to a lower water activity, lower microbial growth rate and stronger texture.

Of course, composition, structure and state of water have a great impact on the thermo-physical properties and therefore on fluid flow and heat transfer. Thus, two-way coupling between heat/mass/momentum transfer and food transformation appears in many modelling problems.

So far, modelling of heat transfer (most often associated with transfer of water), allowed prediction of:

- temperature in continuous liquid food processes (generally at steady state): tubular, plate and scraped surface heat exchangers, ohmic heating…;

doi:10.1088/978-0-7503-2596-7ch2

- temperature and water content in solid food products versus time and position in batch processes assuming known external (boundary) conditions: baking, frying, freezing…;
- temperature and water content in both food and surrounding air (coupled heat and mass transfer) in different equipment: oven, solids dryer, spray dryer, cold room….

From temperature and water content evolution, mechanistic or empirical models were sometimes used to predict microbial growth or destruction [5], different reaction rates, weight loss, oil intake, texture change [6].

Today, there are difficulties in predicting:

- thermo-physical properties from composition and structure. Thermal conductivity is rather well predicted from available correlations except for highly porous media where the structure has an important impact [7]. Water diffusivity associated with Fick's law corresponds often to a lumped parameter. In fact water often migrates by capillarity and/or osmotic effects (e.g. intracellular and extracellular water being not at equilibrium). Rheology of food is often complex (non-Newtonian, visco-elastic), it depends greatly and in a non-linear way on the composition of macromolecules and on the amount/state of structural elements (starch granules, protein micelles/aggregates, gas bubbles, ice crystals, fat globules…).
- structure modification (which influences physical properties). Many models are able to predict the local amount of water vaporization or crystallization but they rarely predict the size distribution of the created bubbles or crystals. The fragmentation of solid particles, liquid droplets (e.g. fat globules) or bubbles is often treated by empirical (dimensionless) correlations at the scale of specific equipment. It should ideally be predicted by mechanistic models as a function of local shear and elongation rates. Aggregation (e.g. of milk proteins) is also a difficult task; it depends on temperature (Brownian agitation) and shear (which increases collision probability) and on many physico-chemical interactions (e.g. electrostatic repulsion). Since the aggregates have generally a fractal structure, their volume fraction highly depends on their size distribution. Therefore, the suspension rheology changes significantly during aggregation: if aggregation is not well-predicted, fluid flow and heat transfer cannot be accurately simulated.
- deformation of solid-like food products. Deformation can be due to expansion (e.g. ice crystallization), to bubble inflation (e.g. bread baking [8]) or temperature/water content gradients (freezing). Deformation of course influences heat transfer; for example, a rise of porosity lowers the conductivity and increases the distances, which tends to decrease the heat flux. Deformation also greatly impacts the contact heating. But deformation is difficult to predict because it needs to couple heat/mass transfer with mechanics of complex and evolving rheology materials.
- turbulent flow. Air flow around food products is often turbulent. The available and easy-to-use turbulence models (k-ε, k-ω SST, RSM) are efficient

for a large number of problems but some cases are still problematic (for reasonable computation cost). For example for not fully developed turbulence, when large unsteady structures are present the LES models can be used but they are very time-consuming. Simultaneously, laminar and turbulent zones can be present, for example, before heat treatment a starch suspension has almost the same viscosity as water but suddenly it becomes thousands of times more consistent in a heat exchanger. Difficulties arise also in the interface between a porous medium (e.g. stack of food products) and free region (air flow around the products) [9].

More generally, simulation becomes problematic for large complex geometries (computing limitation, geometric data acquisition). For example, it is impossible to simulate a truck loaded with 10 tons of apples by meshing each apple or to simulate one apple by meshing each cell. Another possibility of progression is to take into account the different variabilities (product properties, product geometry, external ambiance) and to assess the uncertainties (input data, mesh and convergence sensibility…) [10].

2.2 Background

This section introduces some commonly used modelling approaches for solid-like and liquid-like food products.

2.2.1 Processing solid-like food products

Usually continuum mechanics is used considering the food as single-phase or multiphase. This is generally acceptable since the molecules and even the structural elements (cells, micelles, aggregates, bubbles, crystals) are small compared to the scale of a product item or of the equipment (e.g. diameter of tubular exchanger). When food is multiphasic, generally, local thermodynamic equilibrium is assumed: the temperature and the component chemical potential are the same in all the phases. For example during freezing the product follows the liquidus curve or during drying the partial water vapor pressure is the saturated one (at the local temperature) multiplied by the local water activity.

The fluid around the food products is most often a gas. In the case of freeze-drying, very low pressure is encountered and the standard continuum mechanics cannot be applied, the Knudsen effect has to be taken into account [11]. Sometimes fogging is used (e.g. refrigeration of vegetables) then two-phase flow with water droplets impact and evaporation has to be considered [12].

The earliest models considered 1D simple shapes with analytical solving. Now, 3D (or 2D axisymmetric) complex geometries are easily treated numerically with the finite volume method (e.g. Ansys Fluent) or the finite element method (e.g. Comsol).

2.2.1.1 *Transfer inside one product item*

Heat conduction without water evaporation

Basically, Fourier's law with a convection/radiation boundary condition is applied in many cases when water evaporation can be neglected (figure 2.1(a)).

$$\rho C_p \frac{\partial T}{\partial t} = \vec{\nabla} \cdot (\lambda \vec{\nabla} T) \text{ at boundary: } -\lambda \vec{\nabla} T \cdot \vec{n} = h(T - T_a) + \sigma\varepsilon(T^4 - T_w^4)$$

The radiation term is often omitted or integrated in the convective term. The heat capacity can be considered constant or depending on temperature. Phase change can be taken into account by using an apparent heat capacity (derivative of the enthalpy versus temperature). For freezing, C_p becomes then very high some degrees below the initial freezing temperature. For water vaporization, a very simplified (most often inaccurate) approach is to neglect evaporation and consider only boiling near 100 °C. C_p then becomes very high some degrees below and above 100 °C.

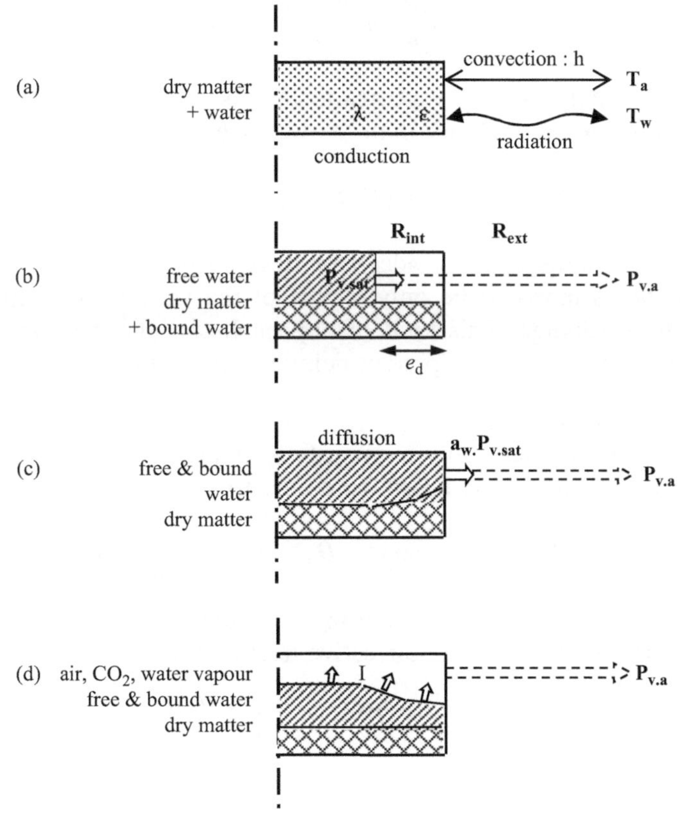

Figure 2.1. Heat transfer in solid-like products: (a) without evaporation; (b) vaporization at a front; (c) evaporation at surface; and (d) porous medium approach.

When the fluid around the product is not a simple gas or liquid, the boundary condition has to be adapted, for example when there is steam condensation in a pressure cooker [13] or during refrigerating with flake ice [14].

Water vaporization at an evaporation/boiling front

One basic model considers that water is free or bounded (water activity is a Heaviside function versus water content) and does not migrate. An evaporation/boiling front progresses from surface to core (figure 2.1(b)). Only free water is vaporized. At the front, the partial water vapor pressure is the saturated one at the local temperature. The evaporation rate is proportional to the difference of partial water pressure between the front and the external ambient and inversely proportional to mass transfer resistance. This resistance is the sum of an internal one (diffusion in the pores of the dry zone) and an external one (convection/diffusion in the boundary layer). The volume occupied by evaporated free water becomes the volume of the pores. When the partial water vapor pressure at the front reaches the total external pressure (boiling condition), pure water vapor flow occurs in the pores according to Darcy's law. This eventuality is not always considered (e.g. low temperature drying). The thickness of the dry zone follows an ordinary differential equation.

$$\rho_w \varepsilon_w \frac{de_d}{dt} = \frac{p_{v.\mathrm{sat}(T_f)} - p_{v.a}}{R_{\mathrm{int}} + R_{\mathrm{ext}}}$$

At the front, the latent heat of vaporization is taken into account. The dry zone generally has a lower conductivity than the wet one.

$$-(\lambda \vec{\nabla} T \cdot \vec{n})_{f.\mathrm{wet}} + (\lambda \vec{\nabla} T \cdot \vec{n})_{f.\mathrm{dry}} = L_v \rho_w \varepsilon_w \frac{de_d}{dt}$$

Of course, some variants of this model were developed; for example the volume of the pores can be assumed to be only a part of the volume of evaporated water allowing drying shrinkage. This kind of approach can also be applied for freeze drying (sublimation of ice) or to predict dehydration during freezing [15].

Water evaporation at the surface

Another approach considers only evaporation at product surface (figure 2.1(c)). Fickian diffusion of water is assumed from core to surface.

$$\frac{\partial W}{\partial t} = \vec{\nabla} \cdot (D_w \vec{\nabla} W) \text{ at boundary: } -D_w \vec{\nabla} W \cdot \vec{n} = \frac{kM_w}{RT}(a_w p_{v.\mathrm{sat}(T)} - p_{v.a})$$

The latent heat of vaporization is taken into account at the surface. Generally the Lewis analogy is used to link the convective heat and mass transfer coefficients.

$$-\lambda \vec{\nabla} T \cdot \vec{n} = h(T - T_a) + \sigma\varepsilon(T^4 - T_w^4) + L_w \frac{kM_w}{RT}(a_w p_{v.\mathrm{sat}(T)} - p_{v.a})$$

Porous medium approach

In fact, evaporation occurs often first at the surface and then progressively a more or less dry/porous zone appears and progresses toward the core (figure 2.1(d)). The product can be initially free of gas phase or not. The gas phase may contain water vapor, air and/or CO_2. This approach was notably developed by Datta [16]. The product is considered as a porous medium with three phases (solid: dry matter), liquid (water: concentration W) and gas (water vapor: concentration V). The main equations of such a porous medium model with diffusion and Darcian flow become:

$$\text{liquid water: } \frac{\partial W}{\partial t} + \vec{\nabla} \cdot (-D_{\text{eff}} \vec{\nabla} W) = -\dot{I} \tag{2.1}$$

$$\begin{gathered}\text{water vapor: } \frac{\partial V}{\partial t} + \vec{\nabla} \cdot \left(\frac{V}{V+C} \rho_g \vec{u}_D - \rho_g D_g \vec{\nabla} \left(\frac{V}{V+C} \right) \right) = \dot{I} \\ \text{with } \vec{u}_D = -\frac{k_g}{\mu_g} \vec{\nabla} p \end{gathered} \tag{2.2}$$

$$\text{energy: } \rho C_p \frac{\partial T}{\partial t} + \vec{\nabla} \cdot (\rho_g C_{p.g} T \cdot \vec{u}_D - \lambda_{\text{eff}} \vec{\nabla} T) = -L_w \dot{I}$$

where $\dot{I}$ is the rate of local water vaporization (kg s^{-1} m^{-3}). If local hydric equilibrium is assumed, this unknown parameter can be eliminated by summing equations (2.1) and (2.2) and locally the water activity relates the liquid water concentration and the partial water vapor pressure.

$$p_v = a_{w(W)} p_{v.\text{sat}(T)}$$

In fact, local hydric equilibrium is not always achieved. During bread baking, for example, each gas bubble is surrounded by a film of dough of typically 100 microns containing starch, gluten and water. The characteristic diffusion time of liquid water in this film is not negligible compared to the baking time. The partial water vapor pressure in the bubble corresponds to the liquid water concentration at the gas/dough interface but not to the space-averaged concentration in the film (W) which appears in equation (2.1). In this case, the local evaporation rate can be simply considered proportional to the equilibrium deviation:

$$\dot{I} = \beta(a_{w(W)} p_{v.\text{sat}(T)} - p_v)$$

where β is an adjusted parameter.

A more sophisticated, multi-scale, approach of local non-equilibrium using memory functions was recently proposed by [17].

The Darcian velocity ($\vec{u}_D$) is proportional to the total pressure gradient and to the gas permeability (k_g) which depends on the porosity, on the water content (liquid saturation), and eventually on the food transformation.

This formulation takes into account the phenomenon of evaporation–condensation: at a warm and wet part of the product, water evaporates (endothermic), then the water vapor can migrate to a colder part where it condenses (exothermic). This enhances significantly the heat transfer. The model has to be modified to take into account evaporation/condensation in non-permeable media ($k_g = 0$, nonconnected bubbles) [18].

Porous medium approaches were also used to simulate frying. In this case, liquid water can be expelled and oil can flow inward through the pores [19, 20]. A model using an unsaturated porous medium approach including gravity (Richards equations) was also used to predict drip loss during ice cream melting [21].

Deformation

As mentioned previously, deformation impacts the heat transfer. In the porous medium approach, a momentum balance can be added involving the pressure forces in the gas phase (bubbles) and the viscous or visco-elastic forces in the solid/liquid phase. This needs to know how temperature, water content, porosity, and food transformation affect the rheology. These aspects will not be developed here [22]. For simpler models, deformation is often ignored or empirically related to water content evolution.

Coupling with electromagnetic phenomena

To simulate heating or thawing using microwaves, the heat (and eventually mass) transfer equations have to be coupled with electromagnetic equations. Most often the Maxwell equations are not solved directly but some simplifying assumptions are made. The main difficulty is to obtain the dielectric properties as a function of temperature and water/ice content of the product as well as realistic boundary conditions [23]. Electric and magnetic fields can also modify ice crystallization [24, 25].

2.2.1.2 Flow around solid food products

In the previous section, the boundary conditions (fluid temperature and humidity, convective transfer coefficients) at the product/fluid interface were supposed to be known. Most often the fluid surrounding the product items is air (oven, cold room…) but it can also be saturated or superheated steam (autoclave, steam dryer), water (boiling) or oil (deep frying). In order to determine these conditions, it is necessary to simulate the fluid flow. The flow can be classified as laminar or turbulent and can occur in free or porous zones

Laminar flow (figure 2.2(a))

The Navier–Stokes (NS) and energy equations with the Boussinesq approximation are most often applied. Fluid density is considered constant except in the gravity term to take into account free convection (e.g. domestic refrigerator [26], flow of liquid in a can containing some tomatoes [27]).

$$\vec{\nabla} \cdot \vec{v} = 0 \tag{2.3}$$

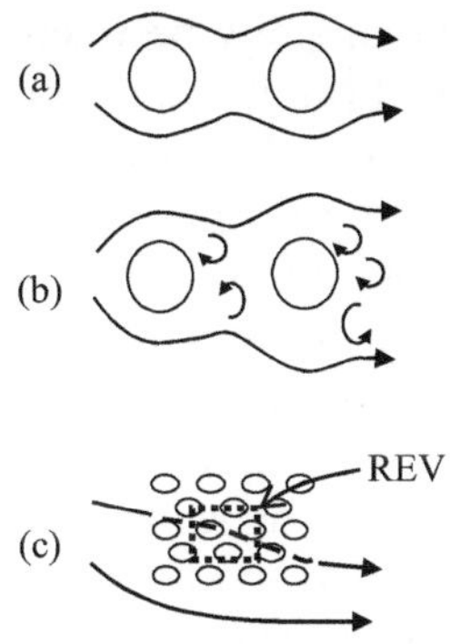

Figure 2.2. Flow around solid products: (a) laminar flow; (b) turbulent flow; and (c) flow in porous medium.

$$\frac{\partial \rho \vec{v}}{\partial t} + \vec{\nabla} \cdot (\rho \vec{v} \vec{v}) = -\vec{\nabla} p + \mu \nabla^2 \vec{v} + \rho \vec{g} \; (\mu \text{ cst}) \tag{2.4}$$

$$\frac{\partial \rho C_p T}{\partial t} + \vec{\nabla} \cdot (\rho C_p T \vec{v}) = -\lambda \nabla^2 T \; (\lambda \text{ and } C_p \text{ cst, viscous dissipation neglected}) \tag{2.5}$$

Turbulent flow (figure 2.2(b))

The NS equations are theoretically still applicable in turbulent flow but they should be solved in transient state with a time step lower than the quickest fluctuation and with cells/elements smaller than the smallest eddy. Therefore, the direct Navier–Stokes (DNS) approach is not applicable in practice. The Reynolds averaged Navier–Stokes equations (RANS), which involve not the instantaneous velocity but the time averaged one, are then solved usually. Because of the non-linearity of the NS equations, the RANS equations contain additional terms called Reynolds stresses. Physically they represent the friction between fluid layers of different velocities due the presence of eddies (vortices). They have to be modelled to close the set of equations. The RANS equations give explicitly the production of turbulent kinetic energy (k): large eddies are created due to the mean velocity gradients. Large eddies degenerate into smaller ones, the smallest are slowed down by viscous forces and their energy is dissipated into internal energy (molecular agitation) with a rate called ε. The most usual turbulence models (k-ε and k-ω SST where $\omega = \varepsilon/k$) use a kind of analogy between molecular agitation and turbulent agitation (eddies). Just as molecular agitation is represented only by the internal energy (which contains the kinetic energy of molecular agitation), turbulence is only represented by the turbulent kinetic energy. Just as the friction due to molecular agitation is taken into account by the (molecular) viscosity, the friction due the eddies (Reynolds stresses) is taken into account by a turbulent viscosity depending empirically on k and ε. Finally, the RANS equations are completed by balance equations for k and ε for which some terms are in turn empirically modelled. Heat transfer is also enhanced by turbulence, a simple analogy between heat and momentum transfer is often assumed (empiric turbulent Prandtl number).

The first limitation of this approach is that the RANS equations assume that the velocity is the sum of a time-averaged one and a random fluctuation. But in the transition to turbulence, some organized (periodic) flow structures may appear. The second limitation is that unlike molecular agitation, the eddies are not isotropic: the turbulent viscosity approach does not strictly apply. The RSM model can overcome this limitation. The third limitation lies in the modelled terms. The fourth limitation is that, whereas the characteristic length of molecular agitation (mean free path) is small compared to the macroscopic scale, the largest eddies have a scale comparable to the macroscopic one. The very expensive LES approach can overcome this limitation. The fifth limitation is that, since all the equations (mass, time-average momentum, k and ε) are coupled, numerical convergence is difficult to obtain for fine enough meshes. All this makes that standard turbulence simulations should always be validated by experiment [28].

Porous medium flow (figure 2.2(c))

Obviously to simulate a cold room or a refrigerated truck loaded with fruit, like apples, it is not possible to mesh in detail each apple. The same is true for a can of peas. Then the stack of food product is often considered as a porous medium [29, 30]. In the porous zone, the momentum equation does not involve the local velocity (in the detail of the pores) but the space-average over a representative elementary volume (REV) also called Darcy's velocity or superficial velocity. Two additional terms take into account the force exerted by the solid matrix on the flowing fluid: one is proportional to viscosity and velocity (Darcy's term) the other is proportional to fluid density and square of velocity (Forchheimer term). The latter becomes significant when inertial effects (contraction, direction change…) become important and predominant when turbulence occurs in the pores [31]. For turbulent porous flow, time-average and space-average are applied, additional production and dissipation terms of turbulent kinetic energy also appear. This approach is only valid if the product items are small compared to the macroscopic scale (e.g. apple/pallet).

2.2.1.3 Conjugated transfers of heat/mass and momentum

In some cases (forced convection) the flow field can be determined independently of the heat transfer, but for free or mixed convection it depends on temperature field. In almost all cases, heat transfer in the fluid is conjugated (coupled) with the heat transfer inside the product items. The same is true for water transfer [32].

Direct meshing of the product items

The equations of fluid flow, conduction/convection in the fluid and conduction inside the product, have to be solved together. Infrared radiation often plays a significant role even at low temperature if the convection is weak [11]. The absorption by the fluid can generally be ignored, only surface-to-surface radiation is then coupled with the other equations. Water transfer equations have eventually to be integrated (e.g. condensation/dehydration in a domestic refrigerator [26])

Porous medium approach
When a stack of food product is considered as a porous medium, generally locally the product temperature differs from the fluid one. A two-temperature model (one for each phase) is then often proposed. This approach can be adapted to simulate frost formation during storage of frozen food for example [33]. But the characteristic conduction time inside the product can be non-negligible compared to the process one, then the temperature in the product items cannot be considered uniform (core and surface temperature are different). Then a multi-scale model (macro-porous approach) with a 1D conduction equation is added, leading to a partial differential equation with five variables (x, y, z, r, t) where x, y, z are the coordinates in the stack and r is the radial direction (r) in a product item [34].

Product as a dispersed phase
In the special case of spray drying, the product (first liquid, then solid) is dispersed in hot air. The air flow and the dispersed droplets are coupled for momentum (drag force), heat and mass (water) transfer. Fluid flow and transfers are solved by the Eulerian approach (with fixed cells), whereas the droplet movement, temperature and water content evolutions are solved by the Lagrangian approach [35] (following representative trajectories). Sometimes sophisticated models are used for the transfers inside the droplets (including 1D in their radial direction).

2.2.2 Processing liquid-like food products

For engineering purposes, the chemical engineering approach considering plug flow (eventually with axial dispersion) or combination (in series/parallel) of perfectly stirred tanks with dimensionless heat transfer correlations is often satisfactory for heat exchanger design. Correlations are available for tubular, plate and scrapped surface heat exchangers for Newtonian and non-Newtonian fluids. To better understand some phenomena (e.g. overheating near the walls, heterogeneity in food transformation, fouling…), 2D or 3D continuum mechanics can be applied including for complex geometries with moving elements. The problems can be classified as steady state (continuous process) or unsteady (batch process), laminar or turbulent, single-phase or multiphase.

Laminar flow
For single-phase laminar problems, the NS and energy equations apply for viscous food products (equations (2.3)–(2.5)). Often the liquid food products are non-Newtonian; for Ostwald fluids, the apparent viscosity is then a function of local shear rate:

$$\mu_{\text{app}} = K\dot{\gamma}^n$$

where K is the consistency index and n the flow behaviour index.

Of course these parameters depend on the temperature and the transformation stage of the product. For example, once the temperature exceeds the gelatinization one, a starch suspension becomes very much consistent. In this case, there is a strong two-way coupling between fluid flow (depending on apparent viscosity), heat transfer (depending on velocity) and food transformation (depending on temperature) which can become challenging to solve [36]. For yield stress or visco-elastic fluid, the mechanical model becomes more complex, specific software (e.g. Polyflow) were developed in this domain.

In the specific case of ohmic heating, the electro-kinetic equations have to be integrated to the model. The electric conductivity is highly dependent on temperature and composition.

An additional difficulty occurs when a free surface (of unknown shape) is present like in stirred tanks with a high viscosity product.

Turbulent flow

Turbulent flow occurs for heat-treatment of low viscosity liquids like milk. The previously mentioned precautions apply when using the RANS approach.

Multiphase flow

Many liquid-like food products are multiphasic, they contain solid particles, droplets of non-miscible liquid, bubbles…. [37]. Most often the models consider that the dispersed phase has the same velocity and temperature as the continuous phase. This is acceptable if the dispersed elements are small (ice crystals, starch granules…) so that the drag forces for small velocity decays are high compared to buoyancy forces (due to density difference). So the problem is the same as for a single phase.

For particles which are small compared to the macroscopic scale (e.g. exchanger diameter) having a different velocity (and/or temperature) than the liquid, a two-phase approach (Euler/Euler or Euler/Lagrange) could be used. But for large particles moving in a liquid, each particle should be meshed and the interaction with the surrounding liquid could be treated as for fluid/structure interactions. Since a very complex moving mesh would be necessary, the numerical problem would be tricky.

2.3 Current directions

As mentioned previously, one challenge is to integrate the knowledge of the food structure and its evolution in the heat and mass transfer models. Here are some examples of recent work in this direction.

$$\text{Structure} \rightarrow \text{Properties} \rightarrow \text{Transfers (heat/mass/momentum)}$$
$$\rightarrow \text{Temperature–Shear} \rightarrow \text{Structure}$$

2.3.1 Using detailed structure data to assess heat/mass/momentum transfer

When modelling the refrigeration of carcasses or meat pieces, often uniform thermo-physical properties are assumed. In fact, muscles, bones, fat, connective tissues have different properties. The exact geometrical structure of a pork leg can be obtained by IRM or x-ray scan and then transcript in a finite element mesh. This approach was used by Harkouss *et al* [38] to simulate the heat and mass transfers (water, salt) during the production of dry-cured ham.

X-ray microtomography allows determining precisely the microstructure of food. From geometric reconstruction, some macroscopic properties like permeability or conductivity can be estimated and further used in continuum mechanics models. This approach was, for example, used for apple tissues [39]. More generally, the research team of B Nicolaï has a multi-scale approach of postharvest problems of fruit and vegetable [40].

An apple puree has an in-mouth texture which is related to its rheology which in turn depends notably on concentration and geometry of the apple cells. Confocal microscopy can be used to determine this geometry. For a suspension of rigid spheres (at low concentration) Einstein's law predicts the relative increase of viscosity versus the solid volume, the proportionality coefficient is 2.5. To determine the equivalent coefficient for apple cells, Plana-Fattori *et al* [41] simulated a creeping flow in several directions around different apple cells. This allows prediction of rheology for apple juice. Further work is planned to apply this approach for higher concentrated products.

2.3.2 Coupling a Lagrangian approach of food structure transformation and an Eulerian approach of heat and fluid flow

During thermal treatment of liquid-like foods, the product transformation can often be represented by a simple kinetics like a chemical reaction. For example, in a first approach, starch gelatinization can be represented by the evolution of the mean granule diameter. This equation can easily be integrated in the Eulerian approach of heat and fluid flow:

$$\frac{DS}{Dt} = \alpha(1 - S)^{\beta}, \text{ where } S = \frac{d - d_{\text{ini}}}{d_{\text{max}} - d_{\text{ini}}}$$

The rheology of a suspension greatly depends on the particle volume fraction which can be obtained from the mean granule diameter. The problem is two-way coupled but can be solved by the finite element method for example.

But sometimes more complex structure modifications occur. This is the case for milk protein aggregation, notably β-lactoglobulin. The first step is a reversible unfolding reaction, then depending on collision probability (and other factors), two unfolded monomers can form a dimer and small polymers can aggregate forming larger ones containing up to billions of monomers. Due to the fractal structure of the aggregates, the larger the aggregates the more they incorporate water. Thus the volume fraction of aggregates increases and thus also the apparent viscosity. Collision probability depends on temperature (Brownian agitation) and on shear rate. A simple way to predict the evolution of the size distribution of the aggregates is to use the Smoluchowski population balance equation (PBE).

$$\frac{DN_i}{Dt} = \frac{1}{2}\sum_{j=1}^{i-1} k_{i-j,j}N_{i-j}N_j - N_i\sum_{j=1}^{j_{\max}} k_{i,j}N_j$$

where N is the number of aggregates containing i monomers (of size L_i) and $k_{i,j}$ are the aggregation kernel parameters which depend on the sizes of the interacting aggregates, on the temperature and on the shear rate [42] .

For a perfectly stirred tank, the pivot method (kind of size discretization) is generally used to solve this equation; about a hundred pivots are typically necessary. But in a heat exchanger, temperature and shear rate are not uniform and plug flow is unrealistic. An Eulerian approach would need typically a hundred additional partial derivative equations (one for each pivot). An alternative is to use the Lagrangian approach for a number of representative fluid parcels trajectories.

The fluid flow and heat transfer problem is first solved with uniform viscosity (value at inlet). A set of representative trajectories is determined. For each one, the temperature and shear rate evolution versus time being known, the PBE allows determining the evolution of the aggregate size distribution and of the corresponding apparent viscosity. Then fluid flow and heat transfer is solved with the new apparent viscosity field and alternative iterations with PBE are carried out until convergence (under-relaxation is used if necessary) [43].

This Euler/Lagrange approach was used with success for the heat treatment of β-lactoglobulin in a tubular heat exchanger [44]. It could be applied even for very different types of food structure evolution models (e.g. neural network) [45].

2.3.3 Non-equilibrium crystallization modelling

Many models are available to predict the amount of ice crystallization during freezing of solid food, or during production of ice cream or sorbet [46]. But they rarely predict ice crystal size distribution which is a major parameter for in-mouth texture and influences the rheology and therefore fluid flow or deformation.

Nucleation and growth of ice crystal depends in fact on the deviation to equilibrium. Therefore, when using continuum mechanics (REV containing a large number of crystals) a non-equilibrium model of heat transfer and nucleation/growth is necessary. In a perfectly stirred tank, a population balance equation can be written

$$\frac{D\phi_{(L)}}{Dt} + \frac{\partial}{\partial L}(G\phi_{(L)}) = \delta(L - L_N)N$$

where $\varphi_{(L)}$ is the number of ice crystals of size L by unit volume.

This equation can be discretized by size classes (about a hundred classes) or approximated by the method of moments (3 or 4 moments). In a scraped surface heat exchanger (SSHE, ice cream freezer) temperature is not uniform, coupling with heat and fluid flow leads to a large number of variables. To simplify the complex flow field in an SSHE, the radial mixing effect of the blades can be represented by a radial diffusion (like the effect of eddies in turbulence) allowing coupling with nucleation/growth PBE [47].

The texture of an aerated sorbet also greatly depends on the bubble size distribution. Fragmentation of the bubbles depends on shear and elongation rates which are very heterogeneous in an SSHE. A first approach using CFD and empirical fragmentation correlations was proposed recently [48].

In order to predict the size and also the shape of the crystals, another possibility is to simulate the heat transfer and the diffusion processes (water/sucrose) at the scale of the crystals. This was done, for a motionless sucrose solution, by a finite volume method (analogue to the volume of fluid method) for hundreds of crystals [49]. Recently, Van der Sman [50] used the phase field method to simulate the growth 'competition' of some ice crystals. More generally, his team shows remarkable activity on thermodynamics in food.

2.4 Outlook/conclusion

In the next decade some improvements are expected in terms of computation: the number of computation cores in a PC will certainly become much larger, allowing direct meshing of large complex geometries and solving problems with large particles moving in a fluid. CFD software could become 'smart': deploying mesh sensitivity analysis, verifying the applicability of models and estimating uncertainties. Software at different scales: molecular dynamics [51], coarse grain, particles, continuum mechanics, unit operation and logistic chain will be better connected. Discrete element methods, smooth particles approaches, lattice Boltzmann simulations, CFD, chemical engineering tools could be used together to solve problems, coupling phenomena which cannot be treated by a single tool [52, 53]. Also, we can hope that the software used for experimental data acquisition at different scales (x-ray tomography, IRM, AFM…) will be more easily interfaced with the simulation tools.

Many more multi-scale applications of transport phenomena modelling in food will certainly be developed [54–56]. What happens during the harvest of an orange, the juice production and pasteurization, the storage in a plastic bottle, during drinking, and digestion…, at the scale of a cell inside the orange, for some polymer molecules of the bottle…, for the whole logistic chain [57]? What happens for a slice of ham, in the slaughterhouse, in the factory, in the display cabinet and in the refrigerator in terms of temperature history, quality evolution and bacterial growth taking into account various variabilities [58, 59]? What happens for a starch granule, in the plant, during separation/refining, mixing with many other components, heat treatment…, and during digestion [60]? Figure 2.3 illustrates the possible interaction at different scales and indicates some of the modelling questions.

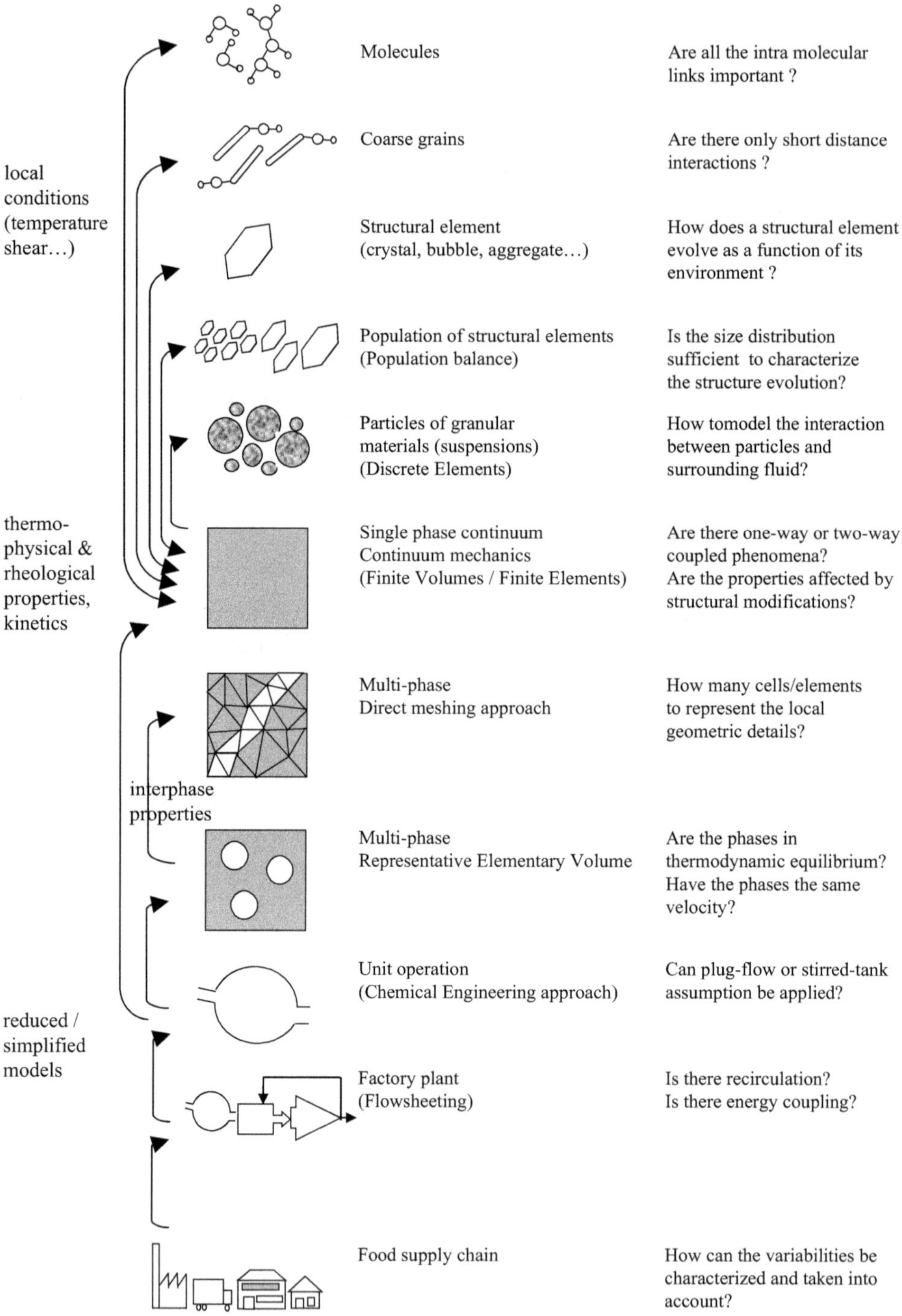

Figure 2.3. Interactions between phenomena at different scales.

References

[1] Bakalis S, Knoerzer K and Fryer P 2015 *Modelling Food Processing Operations* (Cambridge: Woodhead Publishing)

[2] Erdogdu F, Karatas O and Sarghin F 2018 A short update on heat transfer modelling for computational food processing in conventional and innovative processing *Curr. Opin. Food Sci.* **23** 113–9

[3] Castro A M, Mayorga E Y and Moreno F L 2018 Mathematical modelling of convective drying of fruits: A review *J. Food Eng.* **223** 152–67

[4] Pham Q T 2006 Modelling heat and mass transfer in frozen foods: a review *Int. J. Refrig.* **29** 876–88

[5] Cepeda J F, Weller C L, Negahban M, Subbiah J and Thippareddi H 2013 Heat and mass transfer modeling for microbial food safety applications in the meat industry: a review *Food Eng. Rev.* **5** 57–76

[6] Rocca-Polimeni R and Flick D 2019 Experimental and numerical study of the evolution of carrot texture during cooking in a pressure cooker *Food Bioproc. Tech.* **12** 1284–92

[7] Carson J K, Wang J, North M F and Cleland D J 2016 Effective thermal conductivity prediction of foods using composition and temperature data *J. Food Eng.* **175** 65–73

[8] Zhang L, Doursat C, Vanin F M, Flick D and Lucas T 2017 Water loss and crust formation during bread baking, Part I: Interpretation aided by mathematical models with highlights on the role of local porosity *Drying Technol.* **35** 1506

[9] Defraeye T, Blocken B, Derome D, Nicolai B and Carmeliet J 2012 Convective heat and mass transfer modelling at air–porous material interfaces: Overview of existing methods and relevance *Chem. Eng. Sci.* **74** 49–58

[10] Rabeler F and Feyissa A H 2019 Modelling of food processes under uncertainty: Mechanistic 3D model of chicken meat roasting *J. Food Eng.* **262** 49–59

[11] Scutella B, Plana-Fattori A, Passot S, Bourlès E, Fonseca F, Flick D and Tréléa I C 2017 3D mathematical modelling to understand atypical heat transfer observed in vial freeze-drying *Appl. Therm. Eng.* **126** 226–36

[12] Allais I, Alvarez G and Flick D 2006 Modelling cooling kinetics of a stack of spheres during mist-chilling *J. Food Eng.* **72** 197–209

[13] Rocca-Poliméni R, Flick D and Vasseur J 2011 A model of heat and mass transfer inside a pressure cooker *J. Food Eng.* **107** 393–404

[14] Laguerre O, Derens E and Flick D 2018 Modelling of fish refrigeration using flake ice *Int. J. Refrig.* **85** 97–108

[15] Mulot V, Benkhelifa H, Pathier D, Ndoye F T and Flick D 2019 Experimental and numerical characterization of food dehydration during freezing *J. Food Eng.* **263** 13–24

[16] Datta A K 2007 Porous media approaches to studying simultaneous heat and mass transfer in food processes. I: Problem formulations *J. Food Eng.* **80** 80–95

[17] Perré P 2019 Coupled heat and mass transfer in biosourced porous media without local equilibrium A macroscopic formulation tailored to computational simulation *Int. J. Heat Mass Transf.* **140** 717–30

[18] Lucas T, Doursat C, Grenier D, Wagner M, Trystram G and Flick D 2015 Modeling of bread baking with a new, multi-scale formulation of evaporation-condensation-diffusion and evidence of compression in the outskirts of the crumb *J. Food Eng.* **149** 24–37

[19] Halder A, Dhall A and Datta A 2007 An improved, easily implementable, porous media based model for deep-fat frying *Food Bioprod. Process.* **85** 220–30

[20] Vauvre J M, Patsioura A, Vitrac O and Kesteloot R 2015 Multiscale modeling of oil uptake in fried products *AIChE J.* **61** 2329–53

[21] Alvarez G, Warren D L, Hartel R and Flick D 2016 Numerical model of melting, liquid migration and deformation during ice cream meltdown test *Fourth IIR Conf. on Sustainability and the Cold Chain (Auckland, New-Zealand, 7-9/4/2016)*

[22] Nicolas V, Vanin F M, Grenier D, Lucas T, Doursat C and Flick D 2016 Modelling bread baking with focus on overall deformation and local porosity evolution *AIChE J.* **62** 3847–63

[23] Gulati T, Zhu H and Datta A K 2016 Coupled electromagnetics, multiphase transport and large deformation model for microwave drying *Chem. Eng. Sci.* **156** 206–28

[24] Dalvi-Isfahan M, Hamdami N, Xanthakis E and Lebail A 2017 Review on the control of ice nucleation by ultrasound waves, electric and magnetic fields *J. Food Eng.* **195** 222–34

[25] Sado M, Curet S, Rouaud O, Lebail A and Havet M 2017 Numerical modelling of an innovative microwave assisted freezing process *Int. J. Refrig.* **80** 66–76

[26] Laguerre O, Ben Amara S and Flick D 2010 Numerical simulation of simultaneous heat and moisture transfers in a domestic refrigerator *Int. J. Refrig.* **33** 1425–33

[27] Rabiey L, Flick D and Duquenoy A 2007 3D simulations of heat transfer and liquid flow during sterilisation of large particles in a cylindrical vertical can *J. Food Eng.* **82** 409–17

[28] Moureh J, Tapsoba M S and Flick D 2009 Airflow in slot-ventilated enclosure partially filled with porous boxes. Part II : measurements and simulations within porous boxes *Comput. Fluids* **38** 206–20

[29] Verboven P, Flick D, Nicolaï B M and Alvarez G 2006 Modelling transport phenomena in refrigerated food bulks, packages and stacks: basics and advances *Int. J. Refrig.* **29** 985–97

[30] Laguerre O, Ben Amara S, Alvarez G and Flick D 2008 Transient heat transfer by free convection in a packed bed of spheres: Comparison between two modelling approaches and experimental results *Appl. Therm. Eng.* **28** 14–24

[31] Hoang H M, Duret S, Flick D and Laguerre O 2015 Preliminary study of airflow and heat transfer in a cold room filled with apple pallets: comparison between two modelling approaches and experimental results *Appl. Therm. Eng.* **76** 367–81

[32] Caccavale P, De Bonis M V and Ruocco G 2016 Conjugate heat and mass transfer in drying: a modeling review *J. Food Eng.* **176** 28–35

[33] Urquiola A, Alvarez G and Flick D 2017 Frost formation modelling during the storage of frozen vegetables exposed to temperature fluctuations *J. Food Eng.* **214** 16–28

[34] Alvarez G and Flick D 2007 Modelling turbulent flow and heat transfer using macro-porous media approach used to predict cooling kinetics of stack of food products *J. Food Eng.* **80** 391–401

[35] Gianfrancesco A, Turchiuli C, Flick D and Dumoulin E 2010 CFD modeling and simulation of maltodextrin solutions spray drying to control stickiness *Food Bioproc. Tech.* **3** 946–55

[36] Plana-Fattori A, Chantoiseau E, Doursat C and Flick D 2013 Two-way coupling of fluid-flow, heat-transfer and product transformation during the heat treatment of a starch suspension inside a tubular exchanger *Eng. Appl. Comput. Fluid Mech.* **7** 334–45

[37] Barigou M, Mankad S and Fryer P J 1998 Heat transfer in two-phase solid-liquid food flows: a review *Food Bioprod. Process.* **76** 3–29

[38] Harkouss R, Chevarin C, Daudin J D, Sicard J and Mirade P S 2018 Development of a multi-physical finite element-based model that predicts water and salt transfers, proteolysis

and water activity during the salting and post-salting stages of the dry-cured ham process *J. Food Eng.* **218** 69–79

[39] Warning A, Verboven P, Nicolaï B, van Dalen G and Datta A K 2014 Computation of mass transport properties of apple and rice from X-ray microtomography images *Innov. Food Sci. Emerg. Technol.* **24** 14–27

[40] Ho Q T, Verboven P, Verlinden B E, Herremans E, Wevers M, Carmeliet J and Nicolaî B 2011 A three-dimensional multiscale model for gas exchange in fruit *Plant Physiol.* **155** 1158–68

[41] Plana-Fattori A, Doursat C, Almeida G, Moulin G, Leverrier C, Ou E and Flick D 2018 Estimation of intrinsic viscosity of apple cells by numerical simulation *FoodSim Conf. (Ghent, Belgium, 8-12/4/2018)*

[42] Erabit N, Ndoye F T, Alvarez G and Flick D 2015 A Population balance model integrating some specificities of the β-lactoglobulin thermally-induced aggregation *J. Food Eng.* **144** 66–76

[43] Plana-Fattori A, Chantoiseau E, Doursat C and Flick D 2013 An Eulerian–Lagrangian approach for coupling fluid flow, heat transfer, and liquid food product transformation *Comput. Chem. Eng.* **52** 286–98

[44] Chantoiseau E, Plana-Fattori C, Doursat D and Flick 2012 Coupling fluid flow, heat transfer and thermal denaturation-aggregation of beta-lactoglobulin using an Eulerian/Lagrangian approach *J. Food Eng.* **113** 234–44

[45] Foucquier J, Chantoiseau E, Le Feunteun S, Flick D, Gaucel S and Perrot N 2012 Toward an integrated modeling of the dairy products transformations, a review of the existing mathematical models *Food Hydrocoll.* **27** 1–13

[46] Hernandez-Parra O, Plana-Fattori A, Alvarez G, Ndoye F T, Benkhelifa H and Flick D 2018 Modeling flow and heat transfer in a scraped surface heat exchanger during the production of sorbet *J. Food Eng.* **221** 54–69

[47] Benkhelifa H, Amamou A H, Alvarez G and Flick D 2008 Modelling fluid flow, heat transfer and crystallization in a scraped surface heat exchanger *Acta Hortic.* **802** 163–70

[48] Hernandez-Parra O D, Plana-Fattori A, Alvarez G, Ndoye F T, Benkhelifa H and Flick D 2017 Prediction of bubble fragmentation during sorbet production in a scraped surface heat exchanger *Int. Conf. on Food and Biosystems Engineering (Rhodes Island, Greece, 14–17,6, 2017)*

[49] Mulot V, Doursat C, Benkhelifa H, Ndoye F T, Pathier D and Flick D 2019 Modelling freezing of food products at ice crystal scales *IIR Int. Congress of Refrigeration (Montreal, 26/30-8/2019)*

[50] Van der Sman R G M 2016 Phase field simulations of ice crystal growth in sugar solutions *Int. J. Heat Mass Transf.* **95** 153–61

[51] Nguyen P M, Guiga W and Vitrac O 2016 Molecular thermodynamics for food science and engineering *Food Res. Int.* **88** 91–104

[52] Azmir J, Hou Q and Yu A 2019 CFD-DEM simulation of drying of food grains with particle shrinkage *Powder Technol.* **343** 792–802

[53] Blais B, Lassaigne M, Goniva C, Fradette L and Bertrand F 2016 Development of an unresolved CFD-DEM model for the flow of viscous suspensions and its application to solid–liquid mixing *J. Comput. Phys.* **318** 201–21

[54] Ho Q T *et al* 2013 Multiscale modeling in food engineering *J. Food Eng.* **114** 279–91

[55] Trystram G 2012 Modelling of food and food processes *J. Food Eng.* **110** 269–77

[56] Vitrac O and Touffet M 2019 Food process modeling *Encycl. Food Sec. Sust.* **1** 434–54
[57] Nguyen P M, Goujon A, Sauvegrain P and Vitrac O 2013 A computer-aided methodology to design safe food packaging and related systems *AIChE J.* **59** 1183–212
[58] Flick D, Hoang M H, Alvarez G and Laguerre O 2012 Combined deterministic and stochastic approaches for modelling the evolution of food products along the cold chain. Part I: Methodology *Int. J. Refrig.* **35** 907–14
[59] Duret S *et al* 2015 Identification of the significant factors in food quality using global sensitivity analysis and the accept and reject algorithm, Part II: Application to the cold chain of cooked ham *J. Food Eng.* **148** 58–65
[60] Ferrua M J and Singh R P 2010 Modeling the fluid dynamics in a human stomach to gain insight of food digestion *J. Food Sci.* **75** 151–62

IOP Publishing

Physics in Food Manufacturing
Case studies in fundamental and applied research
Megan J Povey

Chapter 3

Particle based modelling in industrial processing

Matt D Sinnott, Paul W Cleary, Sharen J Cummins, James E Hilton, Gerald G Pereira and Simon M Harrison

3.1 Introduction

Commercial food production is becoming more complex for manufacturers with substantial process engineering challenges. Balancing consumer desires and an evolving regulatory environment for local and international markets constrains reformulation of existing products and creation of new foods to combat obesity and poor health (which is increasingly attributed to the heavily processed foods in western diets [1]). Together with the emerging understanding of requirements for gut health [2], these are driving demand to develop new food structures [3] with improved in-body performance and targeted health benefits while maintaining enjoyable sensory experience for the consumer. Being able to do so will increasingly be linked to ongoing commercial viability.

Industrial processing typically involves multiple sequential unit operations (either batch or continuous) that turn broad ranges of raw ingredients into edible products for a consumer market. These processes may be purely mechanical such as conveying, mixing/blending, sieving, loading/unloading from storage, or involve property changes such as occur in milling/size reduction, extrusion, fluidization, agglomeration, and spray coating; or involve additional physics such as chemical reactions, phase changes, rheological changes (e.g. thickening), or temperature treatment (e.g. drying, freezing, and melting). Process inputs can consist of granular bulk materials such as powders, grains, agglomerates, pellets, or assemblies of discrete bodies (such as fruit and vegetable components); or simple liquids; but more commonly later-stage processing involves more complex soft condensed matter that can be multi-phase and multi-material with complex rheological behaviour.

Processing may occur over a broad range of length scales from protein interactions at the colloidal scale to factory scale (metres) for mechanical conveying, sorting and mixing of bulk materials. Similarly, timescales can vary from sub-milliseconds for chemical reactions and inter-particle bonding to minutes for mixing

doi:10.1088/978-0-7503-2596-7ch3

processes to years for storage of long-life products on supermarket shelves. Direct measurement in all these circumstances and for all these processes is both physically and economically impossible. So understanding of processes and how to modify them to meet commercial goals is increasingly dependent on the use of sophisticated computer modelling in which equations representing the controlling physics and chemistry are solved on digital computers.

3.2 Background

Broadly speaking, simulation methods for process operations can be classified as Eulerian (grid/mesh based) or Lagrangian (with a common subset termed particle methods). Particle-based methods involve physics-based numerical simulation that resolves the system of interest at a 'particle' scale where the 'particle' may be a well-defined discrete rigid entity in a granular assembly or a small spatial sub-volume of a fluid or solid material. These particles may interact with each other or with an environment such as the geometry of a piece of processing equipment. These methods are particularly well suited for processing operations where materials are either discrete (and experiencing collisional interactions) or continuous (but with complex free surface, rheological, phase change or history dependent behaviour).

We will focus on two particle-based methods, which are discrete element method (DEM), used for modelling granular materials, and smoothed particle hydrodynamics (SPH) for modelling continuum materials.

3.2.1 Discrete element method

DEM is now a widely adopted approach across many industries. It is described in many papers so the details of the methodology will not be repeated here. For a review of the method see [4]. Particles are defined as volumes with shape, position, orientation, velocity and spin attributes. Calculations are efficiently performed to determine which particles are in contact. A contact model is then used to calculate contact forces from particle state information. In all the examples shown here a linear spring-dashpot model is used [5]. The amount of geometric overlap of two bodies Δx, and normal v_n and tangential v_t relative velocities determine the normal force:

$$F_n = -k_n \Delta x + C_n v_n. \tag{3.1}$$

This consists of a linear spring to provide the repulsive force and a dashpot to dissipate a proportion of the relative motion. It is restricted to being positive to prevent unphysical attractive forces at the end of collisions. The normal damping coefficient C_n is chosen to give the required coefficient of restitution ε (defined as the ratio of the post-collision to pre-collision normal component of the relative velocity). The tangential force is given by:

$$F_t = \min\{\mu F_n, \sum k_t v_t \Delta t + C_t v_t\}, \tag{3.2}$$

where the vector force $\boldsymbol{F}_t$ and velocity v_t are defined in the tangent plane at the contact point. The summation term represents an incremental spring that stores energy from the relative tangential motion and corresponds to the elastic tangential deformation of the contacting surfaces. The total tangential force $\boldsymbol{F}_t$ is limited to the Coulomb friction μF_n after which sliding occurs at the contact.

Shape is critical to flow behaviour, material yield stress, slope failure, porosity and permeability [6] and needs to be included in the geometric representation of the particle volumes. Following [7] particles are represented as super-quadrics, which in their canonical frame, are:

$$\left(\frac{x}{a}\right)^m + \left(\frac{y}{b}\right)^m + \left(\frac{z}{c}\right)^m = 1. \tag{3.3}$$

The particle aspect ratios are b/a and c/a. The exponent m determines the roundness or angularity of the particle. This is a very flexible continuous class of particle shapes.

3.2.2 Smoothed particle hydrodynamics

SPH is a powerful particle-based method for modelling complex flow and phase change behaviour of continuous materials (see [8–10] for details and application examples).

The SPH continuity equation, from [8], is:

$$\frac{d\rho_a}{dt} = \sum_b m_b(\mathbf{v}_a - \mathbf{v}_b) \cdot \nabla W_{ab}, \tag{3.4}$$

where ρ_a is the density of particle a with velocity v_a and m_b is the mass of particle b. The relative position vector from particle b to particle a is $\mathbf{r}_{ab} = \mathbf{r}_a - \mathbf{r}_b$. The relative velocity between fluid particles is $\mathbf{v}_{ab} = \mathbf{v}_a - \mathbf{v}_b$. A smoothing kernel with spatial size h is used in all interpolation evaluations. The kernel evaluated for the distance $|\mathbf{r}_{ab}|$ between particles a and b is $W_{ab} = W(\mathbf{r}_{ab}, h)$. This form of the continuity equation has good numerical conservation properties and is valid for free surfaces and for density discontinuities.

The SPH form of the momentum equation [9] gives the acceleration for each particle a:

$$\frac{d\mathbf{v}_a}{dt} = \mathbf{g} - \sum_b m_b\left[\left(\frac{P_b}{\rho_b^2} + \frac{P_a}{\rho_a^2}\right) - \frac{\xi}{\rho_a\rho_b}\frac{4\mu_a\mu_b}{(\mu_a + \mu_b)}\frac{\mathbf{v}_{ab}\mathbf{r}_{ab}}{\mathbf{r}_{ab}^2 + \eta^2}\right]\nabla_a W_{ab}, \tag{3.5}$$

where P_a is the pressure and μ_a is the viscosity at particle a. ξ is a calibration factor for the viscous term and $\mathbf{g}$ is the gravity vector. The parameter η is small and is added to regularize the singularity at $r_{ab} = 0$.

A quasi-compressible approach is used to calculate the pressure directly from the density using an equation of state:

$$P = P_0\left[\left(\frac{\rho}{\rho_0}\right)^{\gamma} - 1\right] \tag{3.6}$$

where ρ_0 is the reference density. Here we use $\gamma = 7$ which is suitable for water. The pressure scale P_0 is given by:

$$\frac{\gamma P_0}{\rho_0} = 100\ V^2 = c_s^2 \tag{3.7}$$

where V is a characteristic fluid speed. This together with the equation of state (3.6) ensures that the density variations are less than 1% and that the flow can be considered incompressible.

3.2.3 DEM-SPH coupling

If particles are small in size or large in number, then an unresolved coupling method is needed. Particles in a DEM phase are averaged to give a continuum solid fraction and velocity distribution which is then used in a multiphase SPH formulation of the fluid equations [11, 12].

The drag force exerted on a DEM particle in such a particle fluid system with solid fraction corrections due to [13] is:

$$\mathbf{F}_D = \frac{1}{2}c_d\rho\ |\mathbf{u}_r|^2\varepsilon^{-\chi}A_{\perp}\mathbf{u}_r, \tag{3.8}$$

where

$$\chi = 3.7 - 0.65\exp\left\{-\frac{1}{2}(1.5 - \log \mathrm{Re})^2\right\}, \tag{3.9}$$

$\mathbf{u}_r$ the local relative flow velocity between phases, ε the local fluid fraction and the particle Reynolds number is $\mathrm{Re} = 2r\rho|\mathbf{u}_r|/\mu$, where r is the equivalent spherical radius of the particle and μ is the fluid dynamic viscosity. The drag coefficient c_d by [14] is used and is dependent on the projected cross-sectional area of the particle in the direction of the local fluid flow which is important for rotating non-spherical particle flows.

3.2.4 Gas-DEM coupling

This is a broadly adopted method for coupling of computational fluid dynamics (CFD) to DEM and so will not be described in detail. This type of coupling was pioneered by [15] and was followed by significant contributions by multiple groups including [16–18]. Constitutive equations for gas flow through a porous bed for modelling the system, as given by [18] are used. The gas density is assumed constant so the governing equations can be formulated in terms of the superficial gas velocity $\mathbf{u}' = \varepsilon\mathbf{u}$:

$$\frac{\partial \varepsilon}{\partial t} = -\nabla \cdot \mathbf{u}', \tag{3.10}$$

$$\frac{\partial \mathbf{u}'}{\partial t} + \frac{1}{\varepsilon}(\mathbf{u}' \cdot \nabla)\mathbf{u}' + \mathbf{u}'\nabla \cdot \frac{\mathbf{u}'}{\varepsilon} = -\frac{1}{\rho}[\varepsilon \nabla p + \mathbf{f}_{fp} + \nabla \cdot (\varepsilon \boldsymbol{\tau})] + \varepsilon \mathbf{g} \tag{3.11}$$

where ε is the fluid fraction at each location, p is the fluid pressure, the stress tensor is $\boldsymbol{\tau}$ and the volume averaged fluid particle interaction body force is $\mathbf{f}_{fp}$. The stress tensor is:

$$\boldsymbol{\tau} = -\eta\left[\left(\nabla\frac{\mathbf{u}'}{\varepsilon} + \nabla\frac{\mathbf{u}'^{T}}{\varepsilon}\right) - \frac{2}{3}\left(\nabla \cdot \frac{\mathbf{u}'}{\varepsilon}\right)\mathbf{I}\right] \tag{3.12}$$

where $\mathbf{I}$ is the identity tensor.

These equations are discretised in finite difference form and solved on a Cartesian grid. The solid porosity is explicitly calculated from the DEM simulation using a recursive Octree approach [19]. The same drag laws for the coupling of the fluid on the particles are used here as were used for the SPH coupling in equations (3.8) and (3.9).

3.3 Current directions

There are a very wide range of unit operations and processes used in food manufacture. Particle-based modelling is suited to simulating and understanding many of these. Following is a selection of such models for a representative range of processing operations that highlight the utility of this type of technology for addressing knowledge shortfalls for these high value, complex processes (which are more challenging to solve via other approaches).

Unit operation and process examples are organised into two sections based on whether they are purely mechanical primary processing (often of raw materials) with no modification of the material state; and secondary processing that involves more complex physics such as thermal and bubbly flows. These are then sub-divided by type of operation to represent classes of processes with like purpose and modelling challenges (such as bulk materials handling of grains and powders).

3.3.1 Primary mechanical processing

DEM is well suited to modelling the flow of powders or grains in a variety of industrial operations. These include unloading of storage bins [20], vibrating screens [21, 22], screw conveyors [23], conveyor transfer chutes [24], bucket elevators [25], and grain [26] and fruit [27, 28] handling.

Figure 3.1(a) shows an example of particle size separation in batch operation of a vibrating screen [6] where a bed of non-round particles of varying sizes are vibrated on a sinusoidally oscillating screen with square holes. The finer particles (dark blue to light green) can percolate through the bed and then pass through holes in the screen. Coarser particles are retained on the top deck where the bed becomes increasingly dominated by red to yellow coloured particles that are too large to pass

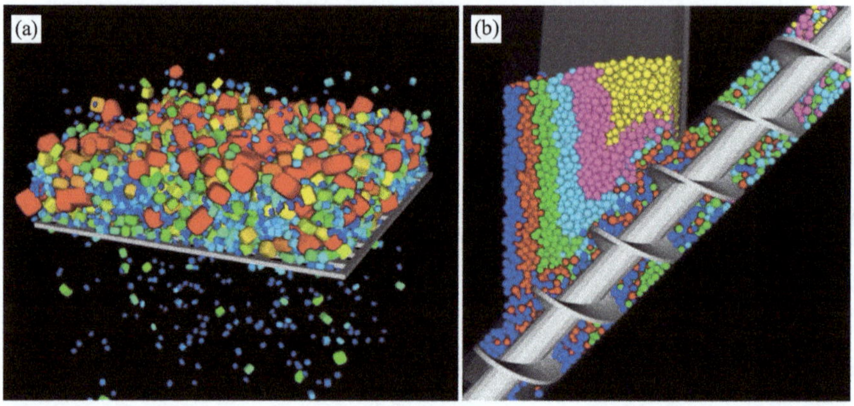

Figure 3.1. (a) Sieving of a granular raw material and (b) drawdown of grains from a bin by an inclined screw conveyor.

through the screen apertures. This type of model can be used to assess the effect of changes to screen vibration, deck aperture size and shape, and to customise these better to suit the material being separated. Figure 3.1(b) shows draw down of spherical material by an inclined screw conveyor located at the bottom of a hopper [22]. When the screw design is uniform there is strong preferential flow into the screw from the back of the hopper (red and dark blue particles) which leads to uneven draw down, non-uniform residence times, recirculation in the hopper and excess wear on screw and lower sections of the hopper. The effect of changes in screw design (adding a core, variable pitch or flight or a combination) can be evaluated using such DEM models to make more optimal selections of screw design for specific materials and hopper conditions.

Coupling gas flow to DEM allows prediction of a range of processes where air flow is important. Figure 3.2(a) shows an example of dust generation from fine particulate material flowing through a conveyor system including a transfer chute [12]. Figure 3.2(b) shows the discharge of particles from a bucket elevator. The buckets rise on the right lifting material and then throw this material from the buckets as they change direction at the top. The trajectories of these particles determine whether they properly exit the conveyor system on the left or instead recirculate (potentially causing contamination to any following products and reducing process efficiency). The discharge behaviour of the finer particles is strongly influenced by the complex gas flow generated by the passing buckets whose motion is shown by the velocity arrows on the regular grid used for the gas solution [24]. Generation and transport of powders (which can often be explosive if allowed to build up) throughout a factory's infrastructure can be a major fire hazard. Computational tools such as these can help predict not only the gas flow, but also the amount of dust generated for a given granular flow through the conveying system under study.

Figure 3.2(c) shows that particle shape has a very strong effect on gas fluidisation of granular materials. In this example, gas is injected at the base and passes up

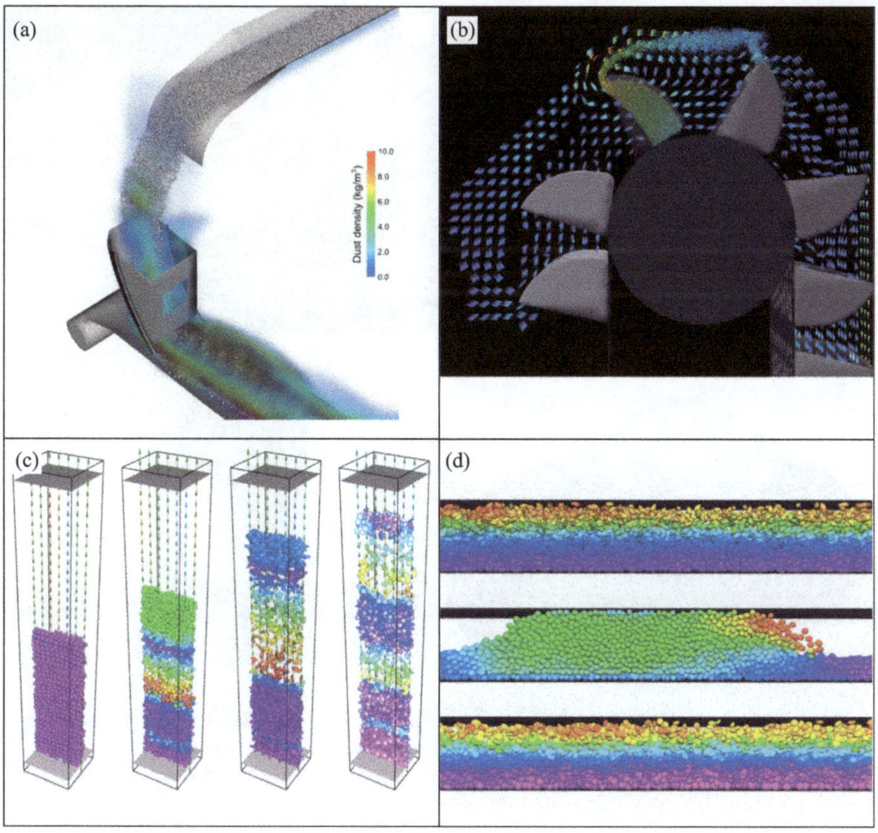

Figure 3.2. (a) Dust generation from granular flow through a conveyor transfer chute; (b) gas drag on grain ejected from a bucket elevator vertical conveying system; (c) fluidization of different grain shapes (spherical; cuboid; oblate spheroid; prolate spheroid); and (d) pneumatic conveying of different grain shapes (prolate spheroid; spherical; oblate spheroid).

through an initially packed bed. The superficial gas velocity is just smaller than that necessary to fluidise spherical particles (left). The same gas flow produces significant fluidisation with increasingly strong interaction with the gas for cuboidal, prolate spheroidal and oblate spheroidal particles, as shown in the three images on the right of figure 3.2(c) (and as discussed in [19]). Similarly, stability of slugs in dense phase pneumatic conveying [29] is also dependent on particle shape. The central image in figure 3.2(d) shows a stable slug of spherical particles which is pushed to the right by high pressure air on the left. For other particle shapes (images above and below) the slugs are not stable and a less efficient stratified flow with axially uniform particle distributions is observed. The ability to include particle shape and to predict its effect on gas–particle flows is useful in refining process designs for specific materials.

Mixing is an important and very common unit process required for much food manufacturing. It involves combining two or more sets of ingredients into a mixture. These may include bulk particulates, fluids, or both. The action of spreading and intermingling the constituent components often proceeds through mechanical

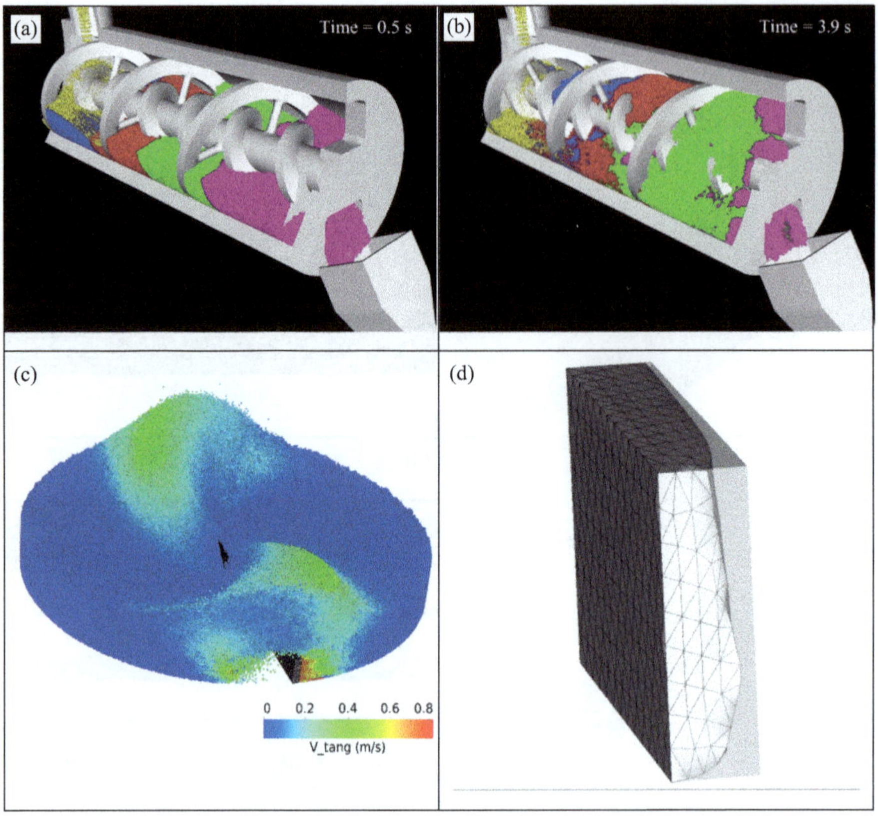

Figure 3.3. (a) Continuous processing of a cohesive food powder in an inline ribbon blender showing mixing/extrusion early in the mixing process; (b) mixing state in the continuous mixer once the process is progressed; (c) food powder mixing in a cylindrical paddle mixer; and (d) prediction of incremental abrasive wearing of the paddle superimposed over the unworn transparent blade.

agitation via tumbling or stirring using an impeller. These are typically batch processes which are slow and have low productivity or can be newer continuous processing. Design and operation of the mixer together with the material properties of the components control the quality/uniformity of the final mixture and the mixing timescale. Inefficient mixing that leads to poor product control and impractical mixing times can be a major cause of bottlenecks for product lines.

The mixing of powders has been extensively studied using DEM [30, 31]. Figure 3.3(a) and (b) show an example of DEM prediction of flow and mixing in a continuous inline blender at two times. It uses outer forward and smaller central reverse helical ribbons to agitate a cohesive powder which is then extruded from the end. Particles are coloured in axial bands so that the nature of the mixing can be understood and related to the impeller design. Such models allow for virtual prototyping of different impeller designs and studies of scale-up. Figure 3.3(c) shows batch processing using a cylindrical paddle mixer [32]. The rectangular paddle creates a heap in front of the blade and material then flows up and over the blade.

Again, the nature of the mixing process can be understood, and design and operating variations can be easily investigated. Particulate materials are inherently abrasive which can lead to progressive wear on the exposed machine surfaces. Figure 3.3(d) shows a prediction of the worn shape of the impeller by the DEM method. Changes in the shape of the impeller then changes the nature of the particle flow and mixing.

Figure 3.4 shows an example of fluid mixing. The impeller has a small radius, long flight length, inner helical blade and a shorter flight length, larger radius, external helical blade [33]. This is conceptually similar to the powder mixer shown in figure 3.3. The cylindrical vessel is fully filled with liquid but only the initial lower quadrant is shown. This allows the transport and mixing action to be visualised. The helical blades generate a strong upward axial flow which stretches the fluid in this direction. The folding action, which is required for mixing, is provided by the passage of the structural bars (attached to the outer helical blade) through the liquid.

The twin screw extruder is a workhorse of the food industry allowing mixing, gelatinisation, cooking, extrusion and other processes to be performed in one machine. The various requirements can lead to very complex impeller designs (figure 3.5(a)) with regions designed to densify feed material, break-up natural grains, heat, compress, then mix (with specific kneading elements) and finally extrude rheologically complex materials. Figure 3.5(b) shows mixing in an industrial scale Clextral twin screw extruder [34] under partial fill conditions.

Combining particulate materials with liquids can be performed using coupled DEM-SPH or DEM-CFD methods. In all cases, the uniformity of mixtures produced is important for quality control of the resulting food products.

3.3.2 More complex secondary processing

Temperature is a very important controlling factor both for preparation of food products and storage to maximize shelf life. Thermal changes during cooking and other processes can change chemical reactivity of ingredients; modify material state and rheology; influence mixing and extrusion of products; affect microbial agents present (e.g. damage to probiotics); create convective currents in fluids (e.g. important for beer fermentation); influence fouling of equipment surfaces; and

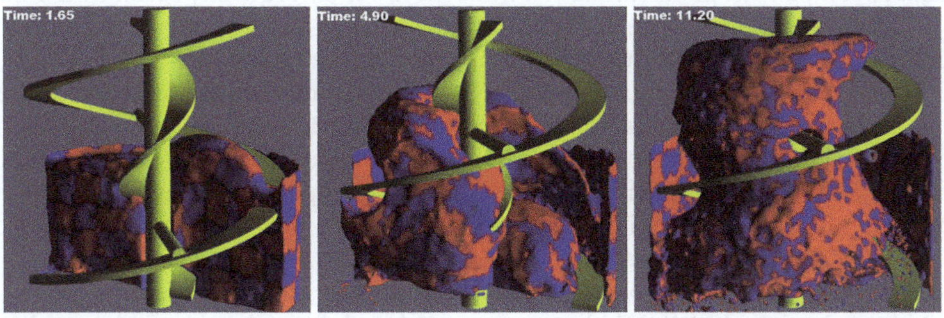

Figure 3.4. Progress of fluid mixing in a double helical blade mixer at three times. Only the fluid initially in the lower quadrant of fluid is shown.

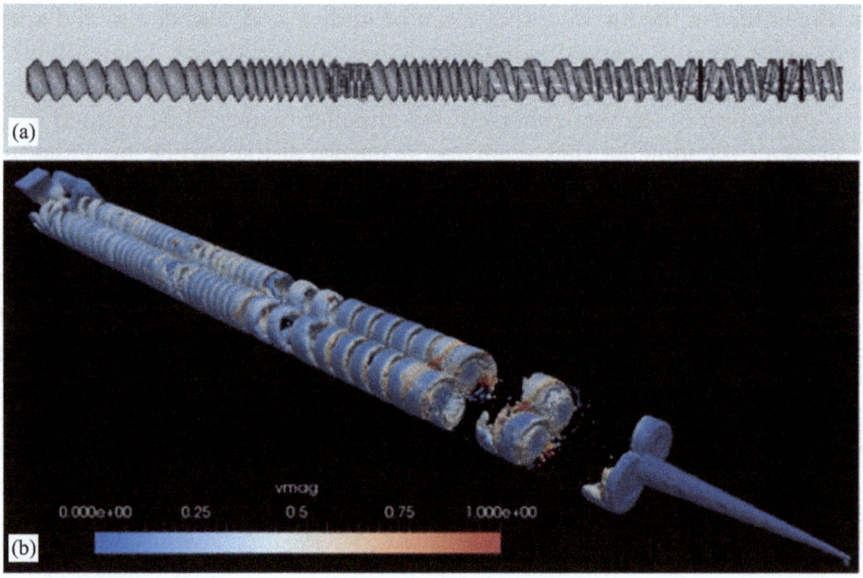

Figure 3.5. Viscous fluid flow and mixing in an industrial scale Clextral twin screw extruder, (a) impeller, (b) flow under partially filled operating conditions.

control the rate of cooling. Poor control of temperature during storage of raw ingredients and final products can lead to significant spoilage and reduced shelf-life.

Figure 3.6 shows food related examples of SPH models of boiling and melting. Boiling of eggs in a saucepan where heat is provided below creating a superheat (temperature above the boiling point) in the water at the bottom of the saucepan leading to boiling (figure 3.6(a)). The strong thermal gradients in the water produce strong convection currents that agitate the free surface and push the rigid model eggs around [35]. Figure 3.6(b) shows the melting of a solid chocolate Easter bunny driven by heat conduction from the thermal plate below. The transport of heat through the structure softens the chocolate material producing a spatially varying material behaviour. This provides a realistic representation of melting with progressive collapse of the locally weakened structure under its own weight. The upper part of the structure remains solid till late in the simulation while the lower parts quickly become fluid-like. This demonstrates one of the key benefits of the particle-based approach to modelling highly transient flows with spatially varying material behaviour.

Many food liquids contain bubbles, particularly carbonated beverages, beer, foams, and aerated liquids. These can be modelled using coupled DEM-SPH models. The SPH component allows prediction of the fluid component while the discrete bubble motion is predicted by the DEM component. Figure 3.7 shows an example of pouring of a glass of beer [36]. The incoming beer is a liquid which contains dissolved CO_2 gas. The beer glass surface contains a range of microscale defects which lead to bubble generation and growth which are represented by a bubble nucleation model. Bubbles are formed, detach from the glass and form dense rising plumes. The resulting sloshing bubbly flow has strong two-way momentum

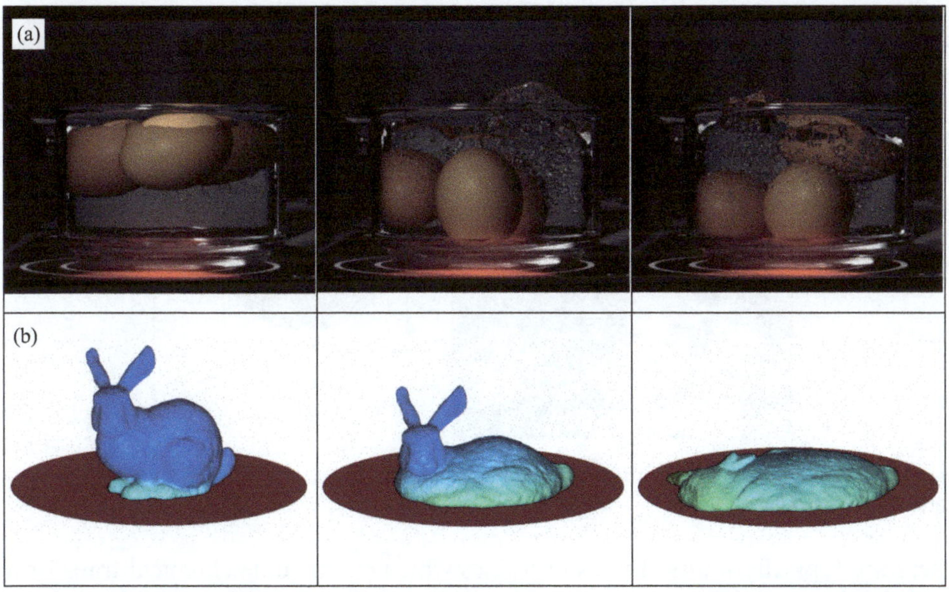

Figure 3.6. Examples of thermal processing of food materials showing; (a) convective boiling of eggs, and (b) melting of a 'chocolate bunny' via thermal conduction on a hot plate.

Figure 3.7. Pouring of a beer into a glass (at three times) demonstrating a free surface jet, bubble nucleation and foam generation.

exchange between the liquid and bubble phases. The bubbles finally rise to the surface where they form a dense foam (represented by the DEM model with a cohesion force between bubbles).

Many food processes transform the structure of the food. Examples include:

- milling (in which grains are converted to flour by grinding action in mills) [37];
- gelatinisation;
- cooking and softening;
- drying [38, 39];
- surface modification; and many more.

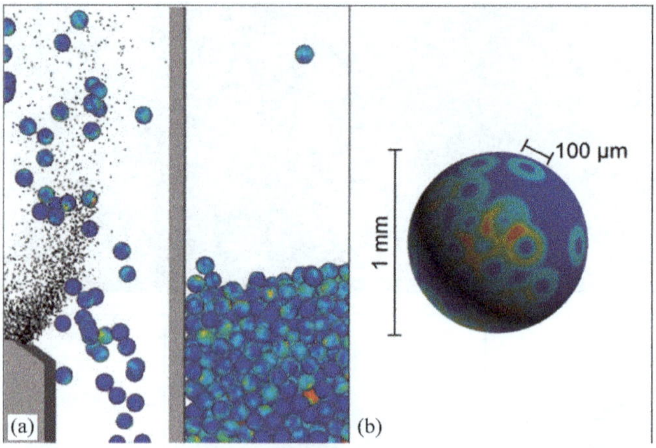

Figure 3.8. (a) Spray coating of particles inside a spouted bed reactor and (b) colouring on each particle represents coating height.

These are typically multi-physics processes which can include chemical transformation of some material components. One example where particle modelling has been useful in understanding the process behaviour is in spray coating of particles which is a surface modification method. It can be used in food and pharmaceutical applications. Figure 3.8 shows a DEM-CFD model of such a process being used for surface modification [40]. Small particles (mm) are placed into a spouted bed and high-speed gas is injected from a central nozzle. This fluidises the particles and generates flow circulations in which particles are drawn into the gas stream and are lifted up and returned to the bed. Coating liquid is introduced to the gas jet which breaks this up into a dense spray of fine droplets. These collide with and spatter onto the surfaces of the entrained particles. The model is able to track the build-up of the coating on each particle which provides useful information on the coating uniformity for each particle and the uniformity across the set of particles being processed.

3.4 Outlook

Adoption of computational modelling as a key research and engineering tool is driven by its cost effectiveness and its ability to provide illuminating insight into complex processes that are difficult or impossible to analyse in other ways. The trend to increasing intensity of model usage and to increasing model complexity will continue into the future driven by a combination of engineering demand and ever-increasing computer capacity. Key areas of model development will include:

- increasing size of the computational models allowing simulation of larger engineering scales while resolving increasing amounts of the fine scale physics and chemistry that drives the emergent behaviour of these systems;
- broader ranges of physics and chemistry will be included, particularly by coupling increasing numbers of methods to create complex hybrid solvers that will allow more complex problems to be tackled or existing problems to be better solved;

- for processes where the increases in model scale are still insufficient to resolve the enabling microscale processes then multiscale models will be required. Different solvers can be used at different scales with models linked hierarchically so as to allow representation of the microscale but avoiding the cost of resolving this at all spatial locations [41];
- inclusion of better characterised empirical and computationally derived closures for sub-resolution physical processes that are not resolved in the numerical solutions. This is particularly relevant for multiphase rheological behaviours and for phase change.

The growth in the emerging wellness market and the adoption of radical new technologies (such as additive printing [42] and microencapsulation [43]) are fundamentally changing both the requirements for food structure and the potential for improving nutrient/probiotic delivery into the body. Modelling capabilities of the type detailed in the previous section can also be applied to simulating the consumption of food and potentially for predicting the sensory experience of eating [44, 45]. They can also be used to provide insight into the progressive digestion and absorption along the GI tract. Both existing and hypothetical food structures can be represented in such models which can replicate the anatomical movements of the mouth [46] (figure 3.9(a)), stomach [47] (figure 3.9(b)), or intestines [48] (figure 3.9(c)). The modelling framework can predict:

1. mechanical and chemical breakdown of the food;
2. the flow of comminuted food and its mixing with saliva, gastric juices and other liquid and solid foods;
3. the growth and decay of microbial species due to the presence (or lack of) of food sources and chemical interactions with other species;
4. the physiological response of the body to the presence of the digesting food.

(a) (b) (c)

Figure 3.9. (a) Example of mastication of an elastic-brittle-plastic solid (a cream filled biscuit) fracturing under mechanical load from the teeth and tongue; (b) flow and mixing of multiphase gastric contents in the stomach; and (c) peristaltic wave propagation showing movements of flexible walled colon and the fluid digesta contained.

Historically, the usage of modelling in food processing unit operations has been to understand and improve the performance of machines for existing processes. The type of digestion modelling above aims to connect food structure to health and nutrition outcomes, ideally enabling creation of new high-performance food structures. The challenge will then be to manufacture them. A key emerging use of modelling technology will be about supporting choice, design, and operation of unit processes and sequences of such processes to help identify how to manufacture these new structures rather than to improve efficiency or quality of existing structures. This will remain a need going forward but the development of new structures is an important emerging trend for the future which will have large impacts on the way modelling is used and will demand higher fidelity and realism in the models, particularly as these respond to demands for personalisation of health and nutrition outcomes.

Acknowledgements

Development of many of the examples shown here involved contributions from Phil Owen, Martin Robinson, Soon-Hyoung Pyo, Charles Cooney and Matthew Abel which the authors acknowledge.

References

[1] Cordain L, Eaton S B, Sebastian A, Mann N, Lindeberg S, Watkins B A, O'Keefe J H and Brand-Miller J 2005 Origins and evolution of the Western diet: health implications for the 21st century *Am. J. Clin. Nutr.* **81** 341–54

[2] Conlon M A and Bird A R 2015 The impact of diet and lifestyle on gut microbiota and human health *Nutrients* **7** 17–44

[3] Campbell C L, Wagoner T B and Foegeding E A 2017 Designing foods for satiety: the roles of food structure and oral processing in satiation and satiety *Food Struct.* **13** 1–12

[4] Guo Y and Curtis J S 2015 Discrete element method simulations for complex granular flows *Ann. Rev. Fluid Mech.* **47** 21–46

[5] Thornton C, Cummins S J and Cleary P W 2013 An investigation of the comparative behaviour of alternative contact force models during inelastic collisions *Powder Technol.* **233** 30–46

[6] Cleary P W 2009 Industrial particle flow modelling using DEM *Eng. Comp.* **26** 698–743

[7] Cleary P W 2004 Large scale industrial DEM modelling *Eng. Comput.* **21** 169–204

[8] Monaghan J J 1994 Simulating free surface flows with SPH *J. Comput. Phys.* **110** 399–406

[9] Cleary P W 1998 Modelling confined multi-material heat and mass flows using SPH *App. Math. Model* **22** 981–93

[10] Cleary P W, Prakash M, Ha J, Stokes N and Scott C 2007 Smooth Particle Hydrodynamics: status and future potential *Prog. Comput. Fluid Dyn.* **7** 70–90

[11] Cleary P W 2015 Prediction of coupled particle and fluid flows using DEM and SPH *Minerals Eng.* **73** 85–99

[12] Cleary P W, Hilton J E and Sinnott M D 2017 Modelling of industrial particle and multiphase flows *Powder Technol.* **314** 232–52

[13] Di Felice R 1994 The voidage function for fluid-particle interaction systems *Int. J. Multiph. Flow* **20** 153–9

[14] Hölzer A and Sommerfeld M 2008 New simple correlation formula for the drag coefficient of non-spherical particles *Powder Technol.* **184** 361–5

[15] Tsuji Y, Tanaka T and Ishida T 1992 Lagrangian numerical simulation of plug flow of cohesionless particles in a horizontal pipe *Powder Technol.* **71** 239–50

[16] Xu B H and Yu A B 1997 Numerical simulation of the gas-solid flow in a fluidized bed by combining discrete particle method with computational fluid dynamics *Chem. Eng. Sci.* **52** 2785–809

[17] Hoomans B P B, Kuipers J A M, Briels W J and Van Swaaij P M 1996 Discrete particle simulation of bubble and slug formation in a two-dimensional gas-fluidised bed: a hard-sphere approach *Chem. Eng. Sci.* **51** 99–118

[18] Kafui K D, Thornton C and Adams M J 2002 Discrete particle-continuum fluid modelling of gas–solid fluidised beds *Chem. Eng. Sci.* **57** 2395–410

[19] Hilton J E, Mason L R and Cleary P W 2010 Dynamics of gas–solid fluidised beds with non-spherical particle geometry *Chem. Eng. Sci.* **65** 1584–96

[20] González-Montellano C, Gallego E, Ramírez-Gómez Á and Ayuga F 2012 Three dimensional discrete element models for simulating the filling and emptying of silos: analysis of numerical results *Comp Chem. Eng.* **40** 22–32

[21] Elskamp F and Kruggel-Emden H 2015 Review and benchmarking of process models for batch screening based on discrete element simulations *Adv. Powder Technol.* **26** 679–97

[22] Cleary P W, Sinnott M D and Morrison R D 2009 Separation performance of double deck banana screens–Part 1: Flow and separation for different accelerations *Minerals Eng.* **22** 1218–29

[23] Fernandez J W, Cleary P W and McBride W 2011 Effect of screw design on hopper drawdown of spherical particles in a horizontal screw feeder *Chem. Eng. Sci.* **66** 5585–601

[24] Cleary P W 2010 DEM prediction of industrial and geophysical particle flows *Particuology* **8** 106–18

[25] Sinnott M D, Hilton J E, McBride W and Cleary P W 2017 Coupled gas-particulate discharge from a bucket elevator *Powder Technol.* **314** 203–17

[26] Boac J M, Ambrose R K, Casada M E, Maghirang R G and Maier D E 2014 Applications of discrete element method in modeling of grain postharvest operations *Food Eng. Rev.* **6** 128–49

[27] Coetzee C J and Lombard S G 2013 The destemming of grapes: Experiments and discrete element modelling *Biosys. Eng.* **114** 232–48

[28] Van Zeebroeck M, Lombaert G, Dintwa E, Ramon H, Degrande G and Tijskens E 2008 The simulation of the impact damage to fruit during the passage of a truck over a speed bump by means of the discrete element method *Biosyst. Eng.* **101** 58–68

[29] Hilton J E and Cleary P W 2011 The influence of particle shape on flow modes in pneumatic conveying *Chem. Eng. Sci.* **66** 231–40

[30] Huang A N and Kuo H P 2014 Developments in the tools for the investigation of mixing in particulate systems—a review *Adv. Powder Technol.* **25** 163–73

[31] Cleary P W and Sinnott M D 2008 Assessing mixing characteristics of particle-mixing and granulation devices *Particuology* **6** 419–44

[32] Sinnott M D, Cummins S J and Cleary P W 2015 DEM modelling of wear in high shear mixers *11th Int. Conf. on CFD in the Minerals and Process Industries CSIRO* (Melbourne, Australia, 7–9 December)

[33] Robinson M and Cleary P W 2012 Flow and mixing performance in helical ribbon mixers *Chem. Eng. Sci.* **84** 382–98

[34] Robinson M and Cleary P W 2019 Effect of geometry and fill level on the transport and mixing behaviour of a co-rotating twin screw extruder *Comput. Part. Mech.* **6** 227–47

[35] Prakash M, Cleary P W, Pyo S H and Woolard F 2015 A new approach to boiling simulation using a discrete particle based method *Comput. Graph.* **53** 118–26

[36] Cleary P W, Pyo S H, Prakash M and Koo B K 2007 Bubbling and frothing liquids *ACM Trans. Graph. (TOG)* **26** 97

[37] Patwa A, Ambrose R K and Casada M E 2016 Discrete element method as an approach to model the wheat milling process *Powder Technol.* **302** 350–6

[38] Azmir J, Hou Q and Yu A 2018 Discrete particle simulation of food grain drying in a fluidised bed *Powder Technol.* **323** 238–49

[39] Karunasena H C P, Brown R J, Gu Y and Senadeera W 2015 Application of meshfree methods to numerically simulate microscale deformations of different plant food materials during drying *J. Food Eng.* **146** 209–26

[40] Hilton J E, Ying D Y and Cleary P W 2013 Modelling spray coating using a combined CFD–DEM and spherical harmonic formulation *Chem. Eng. Sci.* **99** 141–60

[41] Dankar I, Haddarah A, Omar F E, Sepulcre F and Pujola M 2018 3D printing technology: the new era for food customization and elaboration *Trends Food Sci. Technol.* **75** 231–42

[42] Kavitake D, Kandasamy S, Devi P B and Shetty P H 2018 Recent developments on encapsulation of lactic acid bacteria as potential starter culture in fermented foods—a review *Food Biosci.* **21** 34–44

[43] Ho Q T *et al* 2013 Multiscale modeling in food engineering *J. Food Eng.* **114** 279–91

[44] Harrison S M, Cleary P W, Eyres G, Sinnott M D and Lundin L 2014 Challenges in computational modelling of food breakdown and flavour release *Food Func.* **5** 2792–805

[45] Cleary P W, Sinnott M D, Hari B, Bakalis S and Harrison S M 2015 *Modeling Food Processing Operations* (Cambridge: Woodhead Publishing) pp 255–305

[46] Harrison S M, Eyres G, Cleary P W, Sinnott M D, Delahunty C and Lundin L 2014 Computational modeling of food oral breakdown using smoothed particle hydrodynamics *J. Text. Stud.* **45** 97–109

[47] Harrison S M, Cleary P W and Sinnott M D 2018 Investigating mixing and emptying for aqueous liquid content from the stomach using a coupled biomechanical-SPH model *Food Funct.* **9** 3202–19

[48] Sinnott M D, Cleary P W and Harrison S M 2017 Peristaltic transport of a particulate suspension in the small intestine *Appl. Math. Mod.* **44** 143–59

IOP Publishing

Physics in Food Manufacturing
Case studies in fundamental and applied research
Megan J Povey

Chapter 4

Models of surface viscosities of particle-laden fluid interfaces

Sergey V Lishchuk

4.1 Introduction

From the point of view of physics, many foods and beverages are soft interface-dominated materials. A particular class of such materials are those which contain food-grade particles adsorbed at fluid interfaces. Particle types typically used in foods include organic crystal, prolamin-based, cellulose-based, chitin-based, starch-based and microgel particles [1]. Recent works have reviewed different types of food-grade particles [1–3] and the properties of particle-laden fluid interfaces in food and drinks [1, 3, 4].

Surface rheology of particle-laden fluid interfaces plays an important role in formation, stability, and dynamics of food emulsions and foams [5] as well as in sensory physics [6]. A vast amount of experimental studies of rheology of particle-laden interfaces in different conditions has been performed [7–11]. Nevertheless, the theoretical description of the rheology behind these studies lags behind. Many studies tend to use general phenomenological constitutive models, which are fitted to experimental data and have little link to the structure and processes which occur in these complex systems [12]. There is a need in powerful constitutive laws for particle-laden fluid interfaces capable to predict their rheological properties in a range of conditions relevant to food manufacturing. Good models of surface rheology are also important for predicting large-scale properties of the multicomponent fluid systems which involve particle-laden interfaces [13, 14].

This chapter focuses on the viscous part of viscoelastic rheological response of particle-laden fluid interfaces. The selected existing theoretical models of surface viscosities, both phenomenological and microscopically based, are reviewed.

doi:10.1088/978-0-7503-2596-7ch4

4.2 Background

It has been known since the beginning of the previous century [15, 16] that colloidal particles may self-assemble at the interface between two fluids [17]. At the length scales large compared to the distance between particles, particle-laden fluid interfaces can often be regarded as continuous and infinitely thin. In this case they can be characterised by some *effective* properties in a way similar to fluid membranes. Surface tension [18], bending [19] and saddle-splay [20] rigidities of particle-laden interfaces are examples of their static effective properties.

Dynamic properties of particle-laden fluid interfaces include those which describe their rheological response [7]. Particle-laden fluid interfaces are highly viscoelastic [9]. The elastic component of the rheological response depends on the details of interparticle interactions which determine the structure formed by the particles and its response to deformation [21, 22]. The models of the viscous component of the rheological response constitute the topic of the present review.

Viscous rheology of the isotropic interfaces is described by the Boussinesq–Scriven constitutive model [23–26]. According to this model, the surface viscous stress tensor in the case of the flat interface can be represented as

$$\sigma'^{(s)}_{ij} = [(\zeta_s - \eta_s)\nabla_s \cdot \mathbf{v}_s]\mathbf{I}_s + 2\eta_s\mathbf{D}_s, \tag{4.1}$$

where ∇_s is the surface gradient operator, $\mathbf{I}_s$ is the surface projection tensor,

$$\mathbf{D}_s = \frac{1}{2}[\nabla_s\mathbf{v}_s + (\nabla_s\mathbf{v}_s)^{\mathrm{T}}] \tag{4.2}$$

is the surface rate-of-deformation tensor, and, $\mathbf{v}_s$ is the surface velocity vector.

The Boussinesq–Scriven model contains two independent material parameters which describe viscous dissipation of energy at the interface: surface shear viscosity η_s and surface dilatational viscosity ζ_s. More complicated models of surface rheology may include additional material parameters, such as bending and surface torsion-twist viscosities [27], anisotropic viscous coefficients [28], or the transport coefficients which describe various types of exchange with the adjoining bulk fluids [7].

The models of surface viscosities reviewed in the present text aim to predict the behaviour of surface shear and dilatational viscosities for different well-defined systems under different flow conditions.

4.3 Current directions

The three principal mechanisms of viscous dissipation of energy will be identified here. First, the presence of particles at a fluid interface modifies the flow of bulk fluids at both sides of the interface which, in turn, results in additional viscous dissipation of energy. This idea was put forward by Albert Einstein who used it to derive the famous formula for the effective shear viscosity of a dilute suspension of solid spherical particles [29, 30]. In the case of particles adsorbed at an interface this mechanism generates a contribution to the effective surface viscosities. Dissipation of energy due to possible contact line dynamics fundamentally also reduces to the hydrodynamic mechanism [31].

Another source of dissipation is the interfacial flow. Fluid interface can be characterised by excess viscous properties characterised by surface shear and dilatational viscosities. The presence of the adsorbed particles modifies interfacial flow and results in additional dissipation of energy. The relative magnitude of viscous energy dissipation in bulk fluids and fluid interface can be characterised by the Saffman–Delbrück length [32], which can be defined as [33]

$$l_0 = \frac{\eta_s}{\eta_1 + \eta_2}, \tag{4.3}$$

and its dilatational analogue,

$$\bar{l}_0 = \frac{\zeta_s}{\eta_1 + \eta_2}, \tag{4.4}$$

where η_1 and η_2 are shear viscosities of two adjoining bulk fluids. When the particle size is large compared to Saffman–Delbrück length, dissipation of energy occurs predominantly in the bulk fluids adjoining the interface. Conversely, if Saffman–Delbrück length is large compared to the particle size then the viscous dissipation at the fluid interface prevails [33]. Therefore, the excess viscous dissipation of energy at fluid interfaces should be taken into account in the models of the effective viscosities in the cases when Saffman–Delbrück length is not small.

Two mechanisms of energy dissipation in the particle laden interfaces described above originate from viscous fluid flow. These can be directly modelled by the methods of fluid dynamics. However, a large number of particles adsorbed at the interface can be considered as a separate subsystem in which energy can be dissipated. This mechanism of energy dissipation, which can be modelled by the methods of irreversible thermodynamics, becomes especially important for modelling interfaces with complex microstructure.

It is not surprising that only the systems that are either simple themselves or allow for certain simplifying assumptions can be described by rigorously solvable models. More complicated cases require phenomenological approaches. Currently used phenomenological approaches range from simple extrapolative formulas to powerful models which provide insight into the physics of the system. One particular constitutive model, known as soft glassy rheology model, is briefly described in this section.

Finally, continuous rise in available computational power and progress in numerical algorithms increases the role of computer simulations in the study of surface rheology.

4.3.1 Hydrodynamic models

4.3.1.1 Surface shear viscosity

The hydrodynamic approach to the modelling of surface viscosities is based on the idea put forward by Einstein [29, 30]. Following Einstein's method, the fluid dynamics equations are solved with appropriate boundary conditions corresponding to the externally applied flow. Particles adsorbed at the fluid interface modify the fluid flow, thus leading to an increase in the rate of viscous energy dissipation. This increase can be represented in terms of the effective surface viscosities.

Analytical formulas for the effective surface viscosities can be obtained for some simple model setups. For example, one can consider impenetrable spherical or disk-shaped non-Brownian particles with no interparticle interactions apart from hard-core and, possibly, hydrodynamic. Particles are assumed to be confined to a plane, either due to strong surface tension of the interface between two fluids, or by introducing the constraint that confines particles to a monolayer within a homogeneous surrounding fluid. In the latter case the difference between the properties of two fluids is neglected.

Direct generalisation of Einstein's approach to the case of a dilute particle-laden interface between two fluids yields the formula for the effective surface shear viscosity linear in area concentration of particles

$$\phi = \frac{\pi R^2 N}{A}, \tag{4.5}$$

where N is the number of spherical particles of radius R per interfacial area A. The result, illustrated in figure 4.1, was first obtained by Brady and Bossis [34] for the excess shear viscosity of a monolayer of spherical particles, normalised by the width of the monolayer $2R$. Casting the result by Brady and Bossis in terms of the surface shear viscosity defined by Boussinesq–Scriven model (4.1) and taking into account

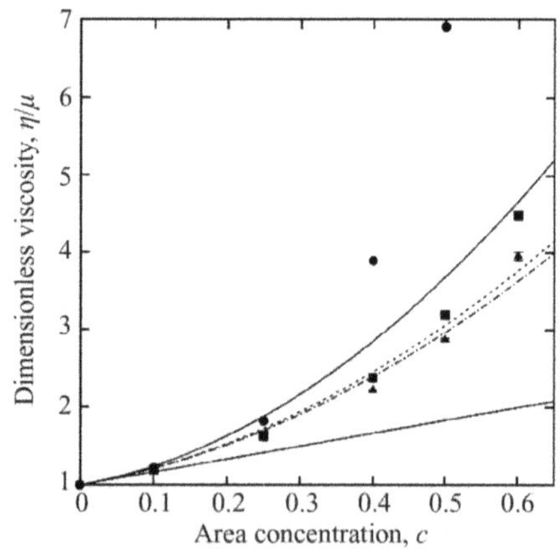

Figure 4.1. Plots of the suspension viscosity η (made dimensionless with the fluid viscosity μ) against area concentration of particles, c [36]. Points represent data from the Stokesian dynamics simulation, and curves the dilute theory. The straight line is the Einstein value $\eta = 1 + 5c/3$, and the solid curve the $O(c^2)$ result for smooth spheres based on the assumption about the probability distribution discussed in [36]. The circles represent the data for smooth spheres from Brady and Bossis [34]. The dashed line and squares are for roughness height $\zeta = 10^{-3}$ and the dash-dot line and triangles for roughness height $\zeta = 10^{-2}$. No tangential friction. Reprinted from [36] with permission from Cambridge University Press.

that two adjoining bulk fluids generally have different viscosities η_1 and η_2, the effective surface shear viscosity can be represented in the form [35]

$$\eta_s^{\text{eff}} = \frac{5}{3}\phi(\eta_1 + \eta_2)R. \tag{4.6}$$

Note, the effective surface shear viscosity depends on the size of the particles, in contrast with Einstein's result for 3D suspensions.

The results, obtained for the monolayer of particles in a homogeneous fluid, can be adapted to the case of the particles adsorbed at the interface between two different fluids by replacing the viscosity of the fluid by the arithmetic mean of the viscosities of the fluids adjacent to the interface. This straightforward generalisation is possible if (i) the contact angle between the particle surface and fluid interface is 90 degrees, and (ii) the shear rate is small enough so that the interface remains flat. The latter condition can be written as [35]

$$D_{ij} \ll \frac{\tau}{|\eta_1 - \eta_2|R}, \tag{4.7}$$

where D_{ij} are the components of the surface rate-of-deformation tensor (4.2), τ is surface tension.

Wilson and Davis [36] extended the result given by equation (4.6) to a quadratic order in ϕ under a certain assumption about the closed trajectories of pairs of particles. In terms of the surface shear viscosity defined by the Boussinesq–Scriven model, their result can be represented in the form

$$\eta_s^{\text{eff}} = \left(\frac{5}{3}\phi + 7.4\phi^2\right)(\eta_1 + \eta_2)R, \tag{4.8}$$

where, again, the presence of two fluids with different viscosities has been taken into account. Non-zero values of the first and second stress differences appear in this approximation [37].

Considering a dilute particle-laden interface between two fluids with vanishing viscosity ratio, $\eta_1/\eta_2 \to 0$, Eigenbrod and Hardt [38] obtained the expansion of the effective surface shear viscosity up to the second order in the cosine of the contact angle α formed by the fluid interface at the particle surface:

$$\eta_s^{\text{eff}} = \frac{5}{3}\phi\eta_1 R\left(1 + \frac{5}{4}\cos\alpha - 0.1089\cos^2\alpha\right). \tag{4.9}$$

Investigating the effect of surface roughness of particles, Wilson and Davis [36] found that the viscosity decreases with increase in surface roughness height and increases with increase in friction coefficient. The second normal stress difference was found to be negative while the first normal stress difference was observed to change sign from negative to positive with increase in concentration of particles [37]. The results are shown in figure 4.1.

Edwards and Wasan [39] considered the monolayer of solid spheres arranged on a square lattice and interacting by the instantaneous hydrodynamics forces. The authors derived a linear constitutive relation for this model and calculated numerically the values of surface shear and dilatational viscosities at different area concentrations of particles.

The models described above focus on the viscous dissipation of energy due to bulk fluid flow. Application of these models is therefore limited to the cases when the particles are large compared to Saffman–Delbrück length l_0, given by equation (4.3). The role of the Saffman–Delbrück length scale in particle-laden interfacial flows, described in the beginning of this section, has been exposed by Henle and Levine [33]. Following Einstein's approach, they calculated the effective shear viscosity of an incompressible fluid membrane laden with rigid disk-shaped inclusions at different sizes of the inclusions compared to l_0. The paper by Henle and Levine contains explicit formulas for the effective viscosity of disk-laden fluid interface in the limits of low and high values of l_0. The authors also propose an interpolation function for the whole range of l_0. They find that the effective shear viscosity depends on the ratio of the particles' size to Saffman–Delbrück length when the ratio of these two quantities is not large. Oppenheimer and Diamant [40] independently studied the viscosity of membranes with embedded inclusions and came to similar conclusions regarding the role of Saffman–Delbrück length.

Some works consider fluid flow in purely two-dimensional suspensions. Belzons *et al* [41] directly generalised Einstein's formula to the 2D case. Doyeux *et al* [42] investigated the effect of confinement of two-dimensional suspensions upon their effective shear viscosity and found that the latter increases with the degree of confinement. In the extreme case of large Saffman–Delbrück length, which corresponds to the large values of surface viscosities of the fluid interface, the energy is predominantly dissipated at the interface, making it tempting to model three-dimensional system with a particle-laden fluid interface as a two-dimensional suspension. However, several authors [33, 40] point out that the results for two-dimensional systems are not directly applicable to three-dimensional systems even in the case of large Saffman–Delbrück length because of the difference in the pressure field in both systems (the momentum at the 2D interface is not conserved in the 3D system). Berlyand *et al* [43] emphasise another difference between two- and three-dimensional systems: for topological reasons, the viscous dissipation rate in purely two-dimensional suspensions contains singular terms which may lead to anomalous increase of dissipation at high concentration of particles.

Non-spherical particles pose additional mathematical challenges. Camley and Brown [44] present a numerical scheme for modelling the motion of arbitrary shaped solid inclusions embedded in a fluid membrane. The authors apply this method to calculate the effective surface viscosity of a suspension of rigid linear oligomers composed of circular monomers. They observe that the effective surface viscosity increases with the increase in the number of monomers, and the rate of this increase depends on the ratio of the monomer radius to Saffman–Delbrück length.

If the concentration of adsorbed species, such as surfactant molecules, is small then the excess viscosity and elasticity of a fluid interface can be neglected. Nevertheless, this system does not always reduce to the case of a surfactant-free interface. As pointed out by Fischer [45, 46], in the presence of even a small amount of adsorbed species the surface flow becomes practically incompressible. Incompressibility of the interfacial flow modifies the dynamics of the adsorbed particles. For example, the translational drag on a sphere half-immersed in an incompressible monolayer or membrane of low viscosity was found to be 25% larger than the drag on a sphere immersed in a free surface [46]. However, the effect of surface incompressibility upon surface viscosity is not as pronounced as that on the drag: the effective surface shear viscosity of a flat, low-concentration, particle-laden incompressible interface separating two immiscible fluids was found to be only 7.6% larger than the value obtained without account for surface incompressibility [47].

4.3.1.2 Surface dilatational viscosity

In the case of a 90° contact angle between fluid interface and surface of the particles, the effective surface dilatational viscosity of fluid interfaces laden with solid spherical particles can be expressed in analytical form in the limits of small and large area concentration of the particles. Both limits are illustrated in figure 4.2.

In the limit of low area concentration of particles the interaction between the particles can be neglected, and the excess dissipation due to modification by the particles of the flow in the bulk fluids can be calculated in the lines of Einstein's derivation. The effective dilatational viscosity is given by the formula [35]

$$\zeta_s^{\mathrm{eff}} = 5(\eta_1 + \eta_2)R\phi. \tag{4.10}$$

In the case of large viscosity difference between the adjacent fluids, $\eta_1 \gg \eta_2$, it is possible to represent the low-concentration limit of dilatational viscosity for arbitrary contact angles α between particle surface and fluid interface in the form [38, 48]

$$\zeta_s^{\mathrm{eff}} = 5\eta_1 R\phi K(\alpha). \tag{4.11}$$

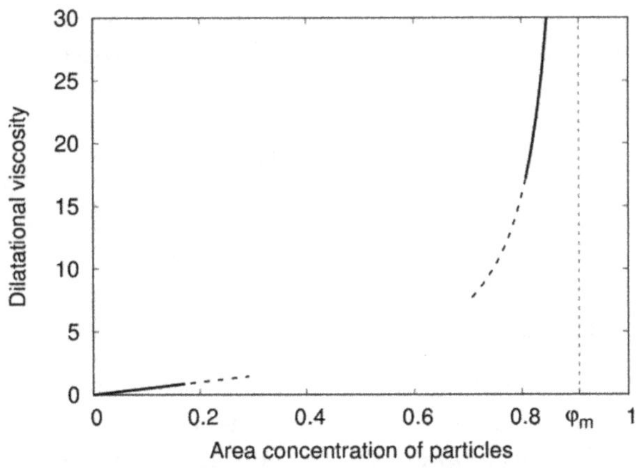

Figure 4.2. Dilatational viscosity at low (equation (4.10)) and high (equation (4.13)) area concentrations of particles.

Eigenbrod and Hardt [38] derive the following expansion of the function $K(\alpha)$ up to the second order in $\cos\alpha$:

$$K(\alpha) = 1 + \frac{15}{8}\cos\alpha - 0.9036\cos^2\alpha. \tag{4.12}$$

The authors find that the effective dilatational viscosity shows stronger dependence on the contact angle than the effective shear viscosity given by equation (4.9). The function $K(\alpha)$, calculated by numerical integration in a wider range of α, is tabulated and plotted in reference [48].

In the limit of high area concentration of particles, the effective surface dilatational viscosity can be derived on the basis of the following facts: (i) highly concentrated arrays of spherical particles in a plane form hexagonal structure, and (ii) most of viscous energy dissipation in highly concentrated suspensions takes place in the thin gaps between the neighbouring particles. The latter fact was noticed by Frankel and Acrivos [49], who used this idea to calculate the effective shear viscosity of highly concentrated three-dimensional suspensions of solid spherical particles. The derivation for the case of particle-laden interface based on the above assumptions results in the formula [50]

$$\zeta_s^{\text{eff}} = \frac{3\sqrt{3}\pi}{16}\frac{(\eta_1 + \eta_2)R}{\sqrt{\phi_m/\phi} - 1}, \tag{4.13}$$

where

$$\phi_m = \frac{\pi}{2\sqrt{3}} \tag{4.14}$$

is the maximum packing density of circles in the plane. Surface dilatational viscosity given by equation (4.13) diverges as h^{-1}, where h is the distance between the surfaces of the neighbouring particles.

In a similar manner, Timounay and Rouyer [51] studied a thin film with dilatational viscosity ζ_s laden with spherical particles. They concluded that at high area concentration of particles the effective dilatational viscosity should scale as

$$\zeta_s^{\text{eff}} \propto \zeta_s\frac{\phi}{1 - \sqrt{\phi/\phi_m}}. \tag{4.15}$$

Similarly to the case of a concentrated particle-laden fluid interface, the effective surface dilatational viscosity diverges at $\phi \to \phi_m$ as h^{-1}.

4.3.2 Thermodynamic models

The irreversible thermodynamic approach allows developing thermodynamically consistent phenomenological and structural constitutive models for a number of different systems [52]. Sagis *et al* employed the formalism of irreversible thermodynamics to construct phenomenological [53, 54] and microscopically based [55] models of surface rheology of particle-laden fluid interfaces.

The microscopically based model of the interfacial rheology by Luo *et al* [55] employs the specific non-equilibrium thermodynamic framework known as GENERIC formalism, where GENERIC stands for general equation for non-equilibrium reversible–irreversible coupling. The framework is based on a GENERIC equation proposed by Grmela and Öttinger [56, 57].

GENERIC equation describes the time evolution of an arbitrary observable of the system, A (which can be represented as a functional of state variables), in a general form

$$\frac{dA}{dt} = \{A, E\} + [A, S], \tag{4.16}$$

where E is the total energy and S is the entropy of the system. The first term on the right-hand side of equation (4.16), the Poisson bracket, represents the reversible contributions to the time evolution of A. The irreversible contributions are represented by the dissipative bracket (the second term). In order to solve the GENERIC equation, one must choose an appropriate set of state variables, the forms of the total energy and entropy of the system, and formulate the form of Poisson and dissipative brackets. Equation (4.16) then describes the time evolution of the state variables.

Applying the GENERIC framework to the case of a fluid interface stabilised by hard ellipsoids, Luo *et al* [55] demonstrate viscoelastic behaviour of this system and calculate rheological response in different shear and dilatational regimes. The calculated shear-thinning behaviour is shown in figure 4.3.

Hydrodynamics is neglected in the model by Luo *et al* [55] and difficulties in accurate numerical solution make the current form of this model difficult to apply for the analysis of experimental results [9]. However, this approach is important because it provides a basis for constructing thermodynamically consistent constitutive models. It is especially promising in highly non-equilibrium situations, where

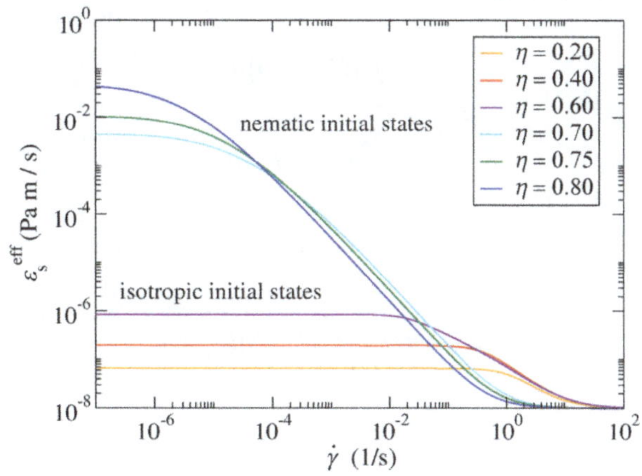

Figure 4.3. The effective shear viscosity for ellipses of aspect ratio 4, ε_s^{eff}, plotted as a function of shear rate for various packing fractions. Reproduced from [55] by permission from The Royal Society of Chemistry.

the surface rheological response is not linear [58]. Hydrodynamics and other microscopic details of the system can be included in the model, making it possible to construct and analyse realistic rheological models, even for more complex, highly non-equilibrium interfaces observed in food emulsions. This is provided difficulties stemming from mathematical complexity of the model can be overcome.

4.3.3 Soft glassy rheology model

In particle-laden interfaces subjected to shear flow, the positional rearrangement of the structure becomes hindered at high area concentrations of particles. The system has to adopt structurally disordered metastable configurations. The model of this behaviour, typical of soft glassy materials, has been proposed by Sollich *et al* [59].

In the soft glassy rheology model by Sollich *et al*, the ability of the structure to rearrange is parameterized by an effective mean-field noise temperature, x, which depends upon the properties of the system and the flow history. The effective noise temperature is the only model parameter which determines the rheological response.

The main equation describing soft glassy rheology is

$$\frac{\partial P}{\partial t} = -\dot{\gamma}\frac{\partial P}{\partial l} - \Gamma_0 e^{-(E-\frac{1}{2}kl^2)/x}P + \Gamma(t)\rho(E)\delta(l), \tag{4.17}$$

where the time evolution of $P(E,\ l;t)$ is shown, together with the joint probability of finding an element with a yield energy E and a local strain l in a system subjected to macroscopic strain γ. A total yielding rate $\Gamma(t)$, together with other parameters entering equation (4.17), is discussed by Sollich *et al* [59]. On the basis of this formulation for the soft glassy rheology model, Sollich has constructed a constitutive equation [60]. This phenomenological equation, which relates the stress to the strain history, was later found to be consistent with the behaviour of particle-laden fluid interfaces at area concentrations approaching close packing [61–67].

In a particular case of a steady shear flow ($\dot{\gamma}$ = const) the shear stress was found to be given by the formula [60]

$$\sigma(\dot{\gamma}) = \frac{\int_0^\infty lG_\rho(Z(l))dl}{\int_0^\infty G_\rho(Z(l))dl} \tag{4.18}$$

with

$$Z(l) = \frac{1}{\dot{\gamma}}\int_0^l e^{l'^2/2x}dl' \tag{4.19}$$

and

$$G_\rho(z) = \int \rho(E)\exp(-ze^{-E/x})dE. \tag{4.20}$$

With an additional assumption about the form of the density of yield energies, $\rho(E) = \exp(-E)$, the stress can be calculated numerically as a function of the shear rate. The results shown in figure 4.4 demonstrate shear thinning behaviour at large shear rates and transitions between different dynamic regimes (Newtonian fluid, power-law fluid, Herschel–Bulkley fluid) at small shear rates [59, 60].

4.3.4 Computational models

The theoretical study of particle-laden interfacial rheology at intermediate and high concentrations is difficult because of the need to account for the formation of different structures formed by the particles and to calculate many-body hydrodynamic interactions between them. Difficulty increases even more in the case of particles or systems with complex structure and interparticle interactions. Useful information about the rheological properties of the models intractable by theoretical methods can be obtained from computer simulations.

In computational models the viscous dissipation of energy can be obtained directly from the simulated velocity field under given particle properties, interparticle interactions and flow conditions. Calculated rate of viscous energy dissipation yields the values of the effective surface viscosities.

Various computer simulation methods have been applied to study the rheology of particle-laden fluid interfaces, including Stokesian dynamics [34, 36, 37], Brownian dynamics [68], lattice Boltzmann [35] and finite elements method [42]. Molecular dynamics simulations have been used in the construction of theoretical thermodynamic models of rheological response [12, 55].

4.4 Outlook

We have outlined the selected theoretical and computational models of surface viscosities of particle-laden fluid interfaces relevant for food manufacturing. We

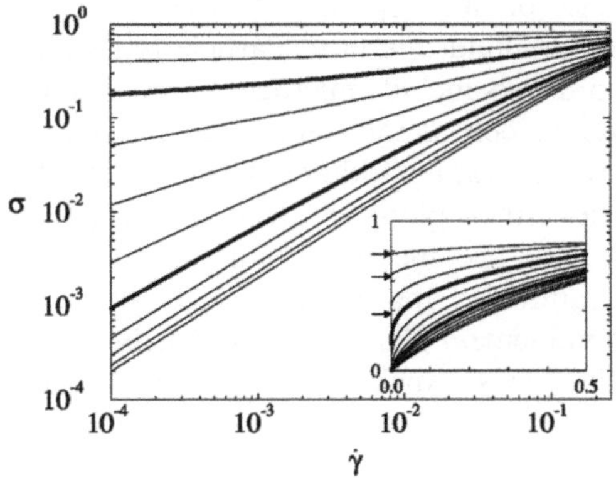

Figure 4.4. Shear stress σ versus shear rate $\dot{\gamma}$, for $x = 0.25, 0.5, \ldots, 2.5$ (top to bottom on left); $x = 1$ and 2 are shown in bold. The inset shows the behaviour on a linear scale, with yield stresses for $x < 1$ indicated by arrows. Reprinted figure with permission from [60]. Copyright (1998) by the American Physical Society.

hope the provided references will allow interested readers to begin exploring the topic on their own.

Due to the complexity of the problem, theoretical modelling currently uses either simplified models or phenomenological approaches. Rigorous models provide description of these systems in well-defined limits. The key challenge in this area is the construction of good predictive constitutive models able to describe realistic systems and conditions. We mention below some of the phenomena worth including in the models, in certain conditions.

First of all, while limits of low and high area concentrations of the adsorbed particles are easier to describe, intermediate range of concentration is the most relevant in food and other applications. Proper account of interparticle interactions of different nature in the rheological models is an important task. In particular, control and modification of interparticle forces can be used as a means to alter microstructure and viscoelasticity of particle-laden interfaces [66]. A good model should provide description of fluid interfaces with realistic concentration of the adsorbed particles having realistic interparticle interactions.

Developing detailed models using only theoretical methods is a difficult problem. Due to the continuous advances in computing power and methodological efficiency, the role of computer modelling in predicting the rheological behaviour of complex particle-laden fluid interfaces and guiding the development of theoretical models is expected to increase.

An important line of research lies in the investigation of rheology in a wide range of conditions. The rheology of complex fluid–fluid interfaces often becomes non-linear even at small deformation and deformation rates [58]. Various interesting phenomena arise at high concentration of particles, for example, jamming [69, 70], buckling [71], rearrangement of particles [72, 73] and their expulsion from the interface [71, 72]. Particle adsorption [74] and desorption [75] kinetics may also significantly affect rheology in certain situations.

Other factors which may be relevant in certain situations include the effects of particles' physical and chemical structure and properties, morphology and size distribution, the presence of other species adsorbed at the interface, the curvature and the fluctuations of the fluid interface. Particle types that can act as interfacial stabilisers in foods and drinks include microgel particles and supramolecular aggregates, which can undergo considerable internal reorganization on adsorption [1] and deform under flow, generating additional dissipation of energy. If the particles are small ('nanoparticles') then physical effects, which are not captured by simple macroscopic models, may become important and require further investigation; for example, thermal motion of the particles at the interface [76], modification of fluid viscosity due to confinement [77] or viscoelectricity [78].

Powerful constitutive laws capable of predicting structure and dynamics of particle-laden fluid interfaces in a wide range of realistic conditions are still lacking. One of the ways to construct them could lie in combining different existing approaches. For example, an approach based on irreversible thermodynamics can be strengthened by detailed microscopic models of hydrodynamic dissipation of energy. A good model, whether it is based on combining and enhancing the existing approaches or employing new ideas, should be able to agree with existing experimental data, successfully predict future experimental results, and guide engineering of foods and beverages with desired, targeted rheological properties. Successful accomplishments in this direction should lead to an increase in quality and safety of food products.

References

[1] Murray B S 2019 *Curr. Opin. Food Sci.* **27** 57–63
[2] Dickinson E 2017 *Food Hydrocoll.* **68** 219–31
[3] Linke C and Drusch S 2018 *Crit. Rev. Food Sci. Nutr.* **58** 1971–85
[4] Dickinson E 2019 *Trends Food Sci. Technol.* **83** 31
[5] Berton-Carabin C C, Sagis L and Schroën K 2018 *Annu. Rev. Food Sci. Technol.* **9** 551–87
[6] Chen J 2009 *Food Hydrocoll.* **23** 1–25
[7] Sagis L M C 2011 *Rev. Mod. Phys.* **83** 1367–403
[8] Fuller G G and Vermant J 2012 *Ann. Rev. Chem. Biomol. Eng.* **3** 519–43
[9] Mendoza A J, Guzmán E, Martínez-Pedrero F, Ritacco H, Rubio R G, Ortega F, Starov V M and Miller R 2014 *Adv. Coll. Int. Sci.* **206** 303–19
[10] Maestro A, Santini E and Guzmán E 2018 *Eur. Phys. J.* E **41** 97
[11] Thijssen J H J and Vermant J 2018 *J. Phys. Condens. Matter* **30** 023002
[12] Sagis L M C, Liu B, Li Y, Essers J, Yang J, Moghimikheirabadi A, Hinderink E, Berton-Carabin C and Schroen K 2019 *Sci. Rep.* **9** 2938
[13] Gross S and Reusken A 2011 *Numerical Methods for Two-Phase Incompressible Flows* (Berlin: Springer)
[14] van der Sman R G M and Meinders M B J 2014 *Adv. Coll. Int. Sci.* **211** 63–76
[15] Ramsden W 1903 *Proc. R. Soc. London* **72** 156–64
[16] Pickering S U 1907 *J. chem. Soc. Trans.* **91** 2001–21
[17] Binks B P 2017 *Langmuir* **33** 6947–63
[18] Vignati E and Piazza R 2003 *Langmuir* **19** 6650–6
[19] Kralchevsky P A, Ivanov I B, Ananthapadmanabhan K P and Lips A 2005 *Langmuir* **21** 50–63
[20] Lishchuk S V 2009 *EPL* **85** 56001
[21] Tambe D E and Sharma M M 1994 *J. Coll. Int. Sci.* **162** 1–10
[22] Kralchevsky P A, Denkov N D and Danov K D 2001 *Langmuir* **17** 7694–705
[23] Boussinesq J 1913 *Ann. Chim. Phys.* **29** 349–57
[24] Boussinesq J 1913 *Ann. Chim. Phys.* **29** 357–64
[25] Boussinesq J 1913 *Ann. Chim. Phys.* **29** 364–71
[26] Scriven L E 1960 *Chem. Eng. Sci.* **12** 98–108
[27] Aguilar Gutierrez O F, Herrera Valencia E E and Rey A D 2017 *J. Col. Int. Sci.* **503** 103–14
[28] Rey A D 2000 *Phys. Rev.* E **61** 1540–9
[29] Einstein A 1906 *Ann. Phys.* **19** 289–306
[30] Einstein A 1911 *Ann. Phys.* **34** 591–2
[31] Joanny J F and Robbins M O 1990 *J. Chem. Phys.* **92** 3206–12
[32] Saffman P G and Delbrück M 1975 *Proc. Nat. Acad. Sci. USA* **72** 3111–3
[33] Henle M L and Levine A J 2009 *Phys. Fluids* **21** 033106
[34] Brady J F and Bossis G 1985 *J. Fluid Mech.* **155** 105–29
[35] Lishchuk S V and Halliday I 2009 *Phys. Rev.* E **80** 016306
[36] Wilson H J and Davis R H 2002 *J. Fluid Mech.* **452** 425–41

[37] Wilson H J and Davis R H 2017 *J. Fluid Mech.* **814** 614–7
[38] Eigenbrod M and Hardt S 2019 The effective shear and dilatational viscosity of a particle-laden interface in the dilute limit arXiv:1909.03660
[39] Edwards D A and Wasan D T 1991 *Chem. Eng. Sci.* **46** 1247–57
[40] Oppenheimer N and Diamant H 2009 *Biophys. J.* **96** 3041–9
[41] Belzons M, Blanc R, Bouillot J L and Camoin C 1981 *C. R. Acad. Sci. Paris Ser. II* **292** 939–44
[42] Doyeux V, Priem S, Jibuti L, Farutin A, Ismaïl M and Peyla P 2016 *Phys. Rev. Fluids* **1** 043301
[43] Berlyand L, Gorb Y and Novikov A 2009 *Arch. Rational Mech. Anal.* **193** 585–622
[44] Camley B A and Brown F L H 2019 *J. Chem. Phys.* **151** 124104
[45] Fischer T M 2004 *Phys. Rev. Lett.* **92** 139603
[46] Fischer T M, Dhar P and Heinig P 2006 *J. Fluid Mech.* **558** 451–75
[47] Lishchuk S V 2014 *Phys. Rev.* E **89** 043003
[48] Lishchuk S V 2016 *Phys. Rev.* E **94** 063111
[49] Frankel N A and Acrivos A 1967 *Chem. Eng. Sci.* **22** 847–53
[50] Lishchuk S V 2014 *Phys. Rev.* E **90** 053005
[51] Timounay Y and Rouyer F 2017 *Soft Matter* **13** 3449–56
[52] Lebon G, Jou D and Casas-Vázquez J 1990 Extended irreversible thermodynamics versus rheology *Rheological Modelling: Thermodynamical and Statistical Approaches* (Lecture Notes in Physics vol 381) ed J Casas-Vázquez and D Jou (Berlin: Springer) pp 257–77
[53] Sagis L M C 2011 *Soft Matter* **7** 7727–36
[54] Sagis L M C 2013 *Phys. Rev.* E **88** 022150
[55] Luo A M, Sagis L M C, Öttinger H C, Michele C D and Ilg P 2015 *Soft Matter* **11** 4383–95
[56] Grmela M and Öttinger H C 1997 *Phys. Rev.* E **56** 6620–32
[57] Öttinger H C and Grmela M 1997 *Phys. Rev.* E **56** 6633–55
[58] Sagis L M C and Fischer P 2014 *Curr. Opin. Coll. Int. Sci.* **19** 520–9
[59] Sollich P, Lequeux F, Hébraud P and Cates M E 1997 *Phys. Rev.* E **78** 2020–3
[60] Sollich P 1998 *Phys. Rev.* E **58** 738–59
[61] Krishnaswamy R, Majumdar S, Ganapathy R, Agarwal V V, Sood A K and Rao C N R 2007 *Langmuir* **23** 3084–7
[62] Hooghten R V, Imperiali L, Boeckx V, Sharma R and Vermant J 2013 *Soft Matter* **9** 10791–8
[63] Samaniuk J R, Hermans E, Verwijlen T, Pauchard V and Vermant J 2015 *J. Dispersion Sci. Tech.* **36** 1444–51
[64] Pradilla D, Simon S, Sjöblom J, Samaniuk J, Skrzypiec M and Vermant J 2016 *Langmuir* **32** 2900–11
[65] Zhang H, Yu K, Cayre O J and Harbottle D 2016 *Langmuir* **32** 13472–81
[66] Rahman S E, Laal-Dehghani N, Barman S and Christopher G F 2019 *J. Coll. Int. Sci.* **536** 30–41
[67] Pitois O and Rouyer F 2019 *Curr. Opin. Coll. Int. Sci.* **43** 125–37
[68] Wijmans C M and Dickinson E 1998 *Langmuir* **14** 7278–86
[69] Stratford K, Adhikari R, Pagonabarraga I, Desplat J C and Cates M E 2005 *Science* **309** 2198–201
[70] Reichhardt C and Olson Reichhardt C J 2014 *Soft Matter* **10** 2932–44
[71] Garbin V 2019 *Curr. Opin. Coll. Int. Sci.* **39** 202–11
[72] Basavaraj M G, Fuller G G, Fransaer J and Vermant J 2006 *Langmuir* **22** 6605–12

[73] Madivala B, Vandebril S, Fransaer J and Vermant J 2009 *Soft Matter.* **5** 1717–27
[74] Hirose Y, Komura S and Kato T 2008 *Progr. Theor. Phys.* **175** 81–92
[75] Ettelaie R and Lishchuk S V 2015 *Soft Matter.* **11** 4251–65
[76] Bresme F and Oettel M 2007 *J. Phys. Condens. Matter* **19** 413101
[77] Hoang H and Galliero G 2012 *Phys. Rev.* E **86** 021202
[78] Dukhin S S 1993 *Adv. Coll. Int. Sci.* **44** 1–134

Chapter 5

Part I: Internal coffee particle phases and coffee brewing release profiles

John R Melrose

The steps in the release of flavour and aroma molecules from food matrices is a complex problem to study. The amounts released over time in total and for individual molecules are termed release profiles. In recent years, there has been significant progress in the measurement of release profiles during coffee brewing from packed beds of coffee grinds. This chapter reviews advances and reported observations on the structure and internal phases of roast and ground coffee particles, their size distributions and effects in beds of coffee particles which together influence the release profiles.

5.1 Introduction

Concentrations and *yield* in coffee brews measured over time of either the individual coffee molecules or the total mass of these molecules are referred to here as *release profiles*. At a given time, the yield is the mass released into the coffee brew normalised by the dry weight of the coffee grinds from which they were released. Whilst in the early literature profile data for brewing was mainly restricted to total mass over time, more recently aliquot collection and modern dynamic analytical techniques such as proton transfer reaction mass spectrometry (PTR-MS) have led to the reporting of release profiles for individual molecules. This paper will review this recent literature. There is a substantial literature on the chemical composition and analysis of roasted coffee and coffee brews, for example, see the books by Clarke and Macrae (1985) and Clarke and Vitzthum (2001). Roasting clearly affects chemical composition and hence release profiles; but the aim here is to focus on the effects of physical structure on release profiles, including that internal to the particles and that of the particle size distribution and the packed bed of particles.

Coffee particles have a heterogeneous internal structure including cell-pockets, nano-porous cell walls and oil phases. The molecules released from particles during

doi:10.1088/978-0-7503-2596-7ch5

brewing, transport through this structure to the outer particle surface. For some molecules the structure and internal co-solutes simply hinder their diffusion, but for others they interact with each other, the macro-molecules and oil phases within the particles. The current knowledge on this internal structure will be reviewed.

This chapter, part, I has a companion, part II, published in the same book. Part II (chapter 6) reports modelling of some selected release profiles and parameterisation of the modelling is in part based on the data reviewed below.

5.2 Background

This section gives an overall review of observations of coffee particles: the particle size distribution of coffee grinds and the internal structure and phases of coffee particles. Later sections then discuss brewing effects occurring in bed and finally effects on brewing imposed by hindered diffusion and phases internal to particles.

5.2.1 Particle size distribution (PSD)

For coffee grinds, the PSD measured by dry laser diffraction techniques is bimodal with a distribution of fines with peak dimension (plotted on a log scale) between 30 and 60 μm and a coarse distribution—flockor the finer grinds, a sum of two log-normal distributions fits well. The level of fines increases as the coarse distribution dimensions are reduced (ground finer); some data with fits taken from Melrose *et al* (2019a) are shown in figure 5.1

The peak dimension of the fines remains relatively constant as the coarse distribution is varied. The fines are sections of cell walls and are of the same length scale as the macro-pores in the particles (see section 5.2.2). If measured in wet conditions with dispersants, there is a significant increase in the level of fines

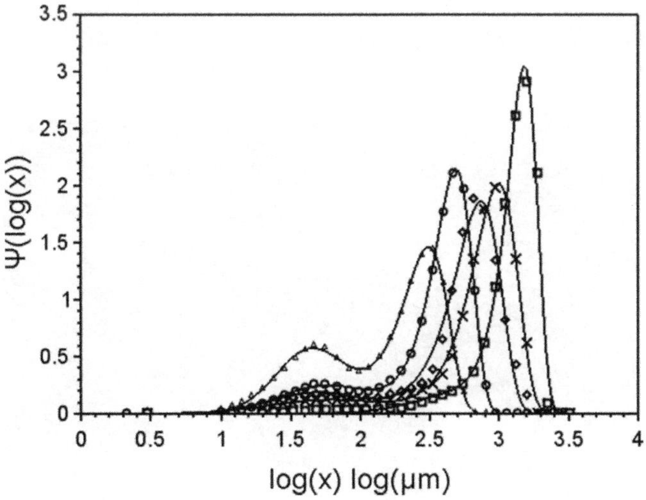

Figure 5.1. Particles size distributions, density function plotted against particle dimension, for series of grinds. The data points are fitted by a sum of log-normal distribution for the fines distribution <100 μm and a Weibull (stretched exponential) for that above, the coarse particle distribution.

(Melrose *et al* 2019a) and a third distribution of ultra-fines appears in the range 1–10 μm. Observations of electron-micro-graphs show that the coarse particles are encrusted with fines (and presumably the ultra-fines), these likely release in wet conditions increasing the overall level of fines. The situation is further complicated because many commercial coffees are densified; this process involves tumbling grounds with paddles which drives fines into the surface cell pockets of coarse particles.

The PSD is a key input into modelling coffee brewing and the resulting release profiles which will be reviewed in part II. Whilst modelling with a wide range of particle sizes is feasible with significant computing power, this may not benefit as a number of other effects such as shape and bed packing effects may have comparable impact to a more complete PSD model (see later). A simpler approach, Melrose *et al* (2012, 2018a, 2018b), is to assume that the overall physics scales with the basic two length scales of the grind and to just include this in modelling. The fines and surface features of the coarse particles (see section 5.2.2), give one length scale, 30–60 μm and including a level of fines gives estimates of the fast-release from these features. Melrose *et al* (2019a) shows that for a range of coffee grinds, the obvious candidates for the second length scale, the volume averaged dimension, $d_{4,3}$ (Co.) and the area averaged dimension $d_{3,2}$ (Co.) of the coarse part of the PSD, to a fair approximation, scale together as grind size is varied: $d_{3,2}$ (Co.) is between 0.8 and 0.9 of $d_{4,3}$ (Co.). The simplest approach is to model particles as spheres, which individual particles are clearly not (see figure 5.2). Additionally, some data on particle sphericity is given in Corrochano *et al* (2015). Commercial exploitation of shape effects has been made with flaked coffee, McSwiggin and Brandt (1969), Joffe (1970). Corrochano (2017) includes some experimental data on flaked coffee particles and brewing and has also developed modelling for this case.

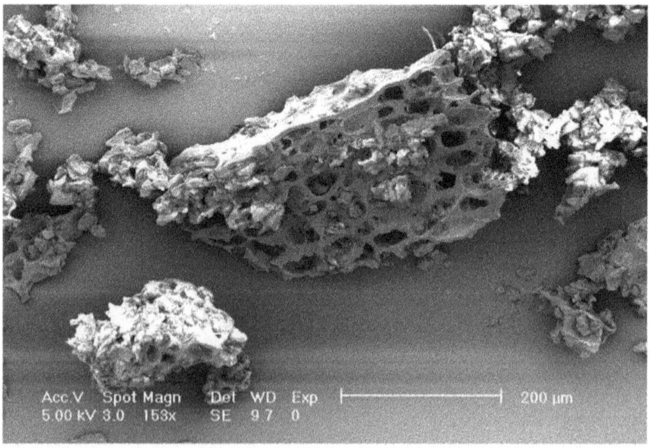

Figure 5.2. An electron micro-graph of a coffee particle (reproduced from Corrochano (2017)).

5.2.2 Internal structure of coffee particles—dry state

Figure 5.2 gives an image of a roast and ground (R&G) coffee particle. Dry R&G particles contain air and CO_2, oil, cell-walls built of macro-molecules, lipids and the molecules and colloids that will be referred collectively as 'solubles'—all can in principle be released from the particles during brewing.

There is a residual content of CO_2 generated by the roasting process and for commercial coffees left over from a degassing process following roasting. The precise state of the CO_2 is unclear, whether it is absorbed or a trapped gas. There is also fraction of residual water (typically for commercial coffee below 5% by weight). NMR (nuclear magnetic resonance) studies of the moisture in dry grinds, Mateus *et al* (2007a), showed that 90% of this is bound to the solid phases but some is partitioned inside the oil phases. Coffee is grown in two major varieties Arabic and Robusta. This review of structure is limited, references to earlier measurements of bean porosity can be found in Schenker *et al* (2000).

The gases occupy the heterogeneous pore structure of the particles: the pore structure includes large pores of dimensions 30–60 μm—which are the cell-pockets that held the biological cells before their destruction during roasting. The macro-pores sit within percolating, nano-porous cell-walls of thickness 1–10 μm and are built of macro-molecules (galactomannans, arabinogalactans, cellulose). In the green coffee bean, the cell-walls also have plasmodesmata pores (typically around 50 nm); the middle lamella between neighbour cell walls also provides larger scale pores. Wilson *et al* (1997) reported two scales of pores, 5 nm and 50 nm in the cell walls, but that the walls are partially degraded by roasting and become more porous, For roasted coffee, Schenker *et al* (2000) and Schenker (2000) reported a distribution of pore sizes over the range 10 to 100 nm, with an average size increasing with degree of roast by 12–13 nm. In addition, grinding may introduce large fractures in the cell-walls; for example, the process of flaking coffee was argued to introduce such fractures McSwiggin and Brandt (1969). Grinding also destroys some cell pockets and produces the fines which are fragments of intact cell-wall material and ultra-fines formed by breaking of the walls themselves. Schenker (2000) found ink-bottle effects in Hg-porosimetry data: large increases of volume at high pressure indicating that some cell-pockets can only be accessed through nano-scale pores in cell walls. Corrochano *et al* (2015) probed structure for a range of grind sizes with He-pycnometry and Hg-porosimetry. The particle specific volume increased with grind size and this was interpreted as an increasing fraction of closed pores (ones into which the He cannot penetrate) being included in the 'solid' volume. The very finest grind ($d_{4,3} = 90$ μm) was assumed to be all fines without closed pores and thus gave a true value of the intrinsic density of the solids. The increase in apparent solid volume with increasing grind size thus gave an estimate of the closed porosity. With Hg-porosimetry, a bi-modal distribution of pores was found with pores in the range 10–30 nm and 5–40 μm. The former was interpreted as cell wall pores, the latter in part as the cell-pockets. Integration to the maximum applied Hg-pressures was used to give an estimate of the open porosity: all regions that can be entered through pores of scale above 4 μm. There were significant hysteric, ink-bottle, effects.

The sum of this estimated open and closed porosity gave an estimate of the total porosity (volume occupied by the air phase), it was found to increase with grind size $d_{4,3} = 200$ to 1250 from 0.5 to 0. 65. The fraction of volume in closed pores increased over this size range from 0.1 to 0.28, see figure 5.3.

The average weight fraction of oil/lipid phases in beans is 10% for Robusta and 10%–15% for Arabica; in the green bean the oils are distributed in oleosomes, located typically at the surface of the biological cells and of scale 0.5–1 μm– (Wilson *et al* (1997)). It is unclear to what degree the green bean oleosomes and their surface structure of oleosins are destroyed by the roasting process

The 'solubles' include numerous small molecules, organic acids, caffeine, chlorogenic acids, aldehydes, thiols, pyarazines, and higher molecular weight molecules like carbohydrate oligomers and melanoidins. The latter are products of the Maillard reactions between proteins and reducing sugars during roasting and are present as both low and high molecular weight compounds (Bekedam 2008). Many of these are the roast reacted contents of the biological cells. Other soluble molecules are naturally present in the cell-walls including fragments of the structural carbohydrates and products of the roasting reactions of the cell-walls and small molecules whose biological function in the green bean locates them in the cell-walls. The books by Clark and Macrae (1985), Clark and Vitzthum (2001) collect together detailed data on the molecules found in coffee.

In R&G coffee particles, Schenker (2000) reported the surfaces of the cell-pockets to be coated with a layer, a few micron thick, of roast modified cell cytoplasm embedded with numerous micron scale oil droplets. Given the weight fraction and size of oil drops it can be estimated that there are a few hundred oil drops per cell-pocket. Schenker (2000) reported images of voids in the modified cytoplasm layer which were interpreted as burst oil drops. Wilson *et al* (1997) reported an oil layer on the surface of the cell pockets but also a fraction of the oil in the cell-walls after roasting. With increasing degree of roast, oil can migrate through coffee beans and appear at the surface of the roast bean.

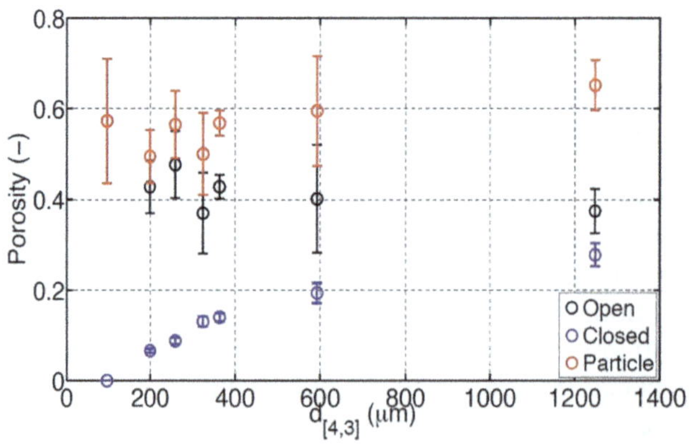

Figure 5.3. Data on porosity versus grind size, taken from Corrochano (2017); figure 4.14.

The volume fractions occupied by the various phases can be determined with equations (5.1)–(5.4). The following relationships should hold:

$$\varepsilon_{ad} = (1 - \varepsilon_c)\varepsilon_w + \varepsilon_c \tag{5.1}$$

$$\varepsilon_{ad} + \varepsilon_o + \varepsilon_{ss} + \varepsilon_m = 1 \tag{5.2}$$

$$\rho_g = \rho_{int}(1 - \varepsilon_{ad}) \tag{5.3}$$

where:

ε_{ad}: the total porosity of the particles, the volume occupied by the gas phase.
ε_w: the porosity of the cell walls
ε_c: the fraction of volume in the cell-pockets
ε_o: the volume fraction of oil in a coffee particle.
ε_{ss}: the volume fraction of 'solubles' in a coffee particles.
ρ_g: the density of coffee particles
ρ_{int} : averaged intrinsic density, ρ_{int}, is that of the non-gas phases (cell-walls, 'solubles' and presumably oils averaged together).

The residual water volume fraction will be accounted for within ε_o, ε_{ss} and ε_m. It is useful to make various estimates of these parameters from the measurements of medium roast Arabica samples reported by Corrochano (2017) and Corrochano *et al* (2015). The intrinsic density of the dry coffee particles studied was found by pycnometry on the finest grinds to be 1337 ± 2.31 kg m^{-3}. Using values $\varepsilon_{ad} = 0.5$–0.65 gives an averaged particle density of ρ_g = 670–467 kg m^{-3} for these samples, decreasing with increasing grind size. The oils and soluble solids volume fractions are given by equation (5.4).

$$\varepsilon_o = f_o\rho_g/\rho_o \text{ and } \varepsilon_{ss} = f_{ss}\rho_g/\rho_{ss} \tag{5.4}$$

where:

ε_{ss}, ρ_{ss}, f_{ss} : the volume fraction, the averaged density of the mass of 'soluble' molecules, and the fraction by weight of 'solubles' in a coffee particles.
ε_m: the volume fraction of the insoluble macro-molecules.
ε_o, ρ_o, f_o: the volume fraction, the average density of the oil phase and the fraction by weight of oil in a coffee particle.

The oil fraction, by weight, is in the range $f_o = 0.1$ for Robusta varieties and 0.1–0.15 for Arabica varieties (Wilson *et al* (1997)). Assuming the oils have a relative density around 0.95, ε_o will be about 0.06 for Robusta and 0.09 for Arabica.

Assuming that the content of solubles does not vary with grind size, the soluble solids weight fraction f_{ss} is set by the maximum mass released by a very fine grind. Larger grinds release less mass externally in brew conditions because some soluble content remains trapped or adsorbed in the larger structure. For samples from Corrochano (2017), f_{ss} = 0.32 and the soluble solids have a relative density of 1, so ε_{ss} in this case will be about 0.2. A tiny fraction of the 'solubles' will be hydrophobic molecules in the oil phase, but this ignored in this estimate. If it is further assumed that the soluble solids are predominantly (70%–80%) the roasted cell contents, then the observation by Schenker (2000) of layers on the surface of cell pockets just a few micron thick is add at odds with the above estimates of ε_o and ε_{ss}. As can be seen in figure 5.2, the cell-pockets are between 40% and 50% of the overall volume, and with dimensions between 30 and 60 μm the soluble solids and oils should occupy at least half of their volume with much thicker layers. *These estimates may be reconciled if it is assumed that whatever processes forced the soluble solids to the cell-pocket surfaces during roasting must also force much of this material into the nano-pores of the cell walls. That is in the dry state some of the soluble solids and oils are a filler within the nano- pores of the cell walls.* This hypothesis, made here, is assumed true in the discussion here but requires experimental validations testing. Note, if such a 'filler' of molecules in the cell-walls is a gas-barrier, which is likely, this has significant implications for molecular gas diffusion and aroma loss of volatiles in the dry state. It may be a mechanism by which some cell-pockets and possibly pores with the walls can hold CO_2 post roasting above atmospheric pressures and will influence how the fraction of closed pores (ones not penetrated in He-pycnometry) is observed to increase with particle size (Corrochano 2017, Corrochano *et al* 2015). Closed pores containing trapped CO_2 may also explain the observations by Anderson *et al* (2003) of slower degassing rates at later times variation in degassing rates (slower at later times)

5.2.3 Wet state

There is a lack of clarity on the time scale of wetting and the degree of swelling of wet coffee particles. The hydrophilic coffee particles are assumed to wet through capillary forces. As the particles wet, the gas phase is released—observations by the author of wetting grains under a microscope show a gas volume comparable to that of the particle pore space appearing as bubbles on the particle surfaces within a few seconds. Coffee particles may swell as they wet, but measurement of this effect is difficult. Shifts in wet PSD measurements are hard to interpret as particles may also lose surface encrusted fines and ultra-fines. Indeed the data sets of Melrose *et al* (2019a) show a decrease on wetting of the volume averaged dimension of the coarse part of the measured PSD. Conversely, Mateus *et al* (2007a) report a swelling factor of 1.23, although one of their figures seems at odds with their text. Water quality is likely to affect any swelling through the effects of ionic-strength screening electro-static interactions and divalents causing specific interactions. Wetting experiments with de-ionised water need to be performed for comparison. Inside the wetted particles, the water phase with solutes occupies what was the gas phase in the dry particles: both the macro-pores and pores within the cell wall. Adopting the assumption above that much of the soluble solids in the dry sate are a 'mortar' filling the nano-porous cell-walls, in the wet state the dissolution of these into solutes will change the porosity surrounding the cell-pockets. *It is unclear if the macro-pores all become open pores in the wet state, but the release profiles of individual molecules may give some insight into this.* Much of the cell wall 'filler' argued to exist in section 5.2.2 will solubilise; although oil components may remain.

The overall volume fraction of the water + solute phases, ε_w with its solutes will thus exceed that of the dry total porosity and can be calculated from equation (5.5)

$$\varepsilon_w = \varepsilon_{pd} + f_{ss} * (1 - \varepsilon_{pd}). \tag{5.5}$$

Equation (5.5) yields values between 0.5 and 0.7. It is argued that for hydrophilic solubles the time-scale on which they transfer into solutes in the intra-particle water phase, is fast relative to the time scales of brewing. As there are oil phases present, the poorly soluble hydrophobic molecules will partition into the oil phases; this will affect their release as will be modelled in part II. Given the maximal yields are around 30% of the weight, the initial solute concentration, before any release, will be between 250 and 350 kg m^{-3}. This will enhance the solution viscosity and hence contribute to the hinderance of the diffusion of molecules within the particles. However, as brewing proceeds the solute concentration will drop. An approach for estimating this hindrance will be discussed in section 5.2.6

5.2.4 Brewing, release profiles, and maximal yield

Coffee is brewed in many ways; heated in pots, drip filters with fluid driven by gravity and by using pressure to force fluid through a packed bed. The latter approach is used for espresso style beverages prepared at low volumes 25 (ristretto), 40 (espresso) and 80 ml (lungo). The focus of this review will be on the packed bed brewing techniques, typically used in on-demand coffee systems, such as the Nespresso™ and Tassimo™ systems, commercial café brewing and also in the manufacturing of instant coffee. In commercial café brewing, brews are typically prepared using 7–16 g of coffee grind while in-home systems typically use 5–7 g. In the manufacturing case, volatile aromas are often stripped from coffee grinds by gas fluxes prior to full brewing, they are condensed and then added back in post-brewing—this allows more extreme pressure and temperature conditions to be applied during the commercial extraction; data reported from a laboratory scale experiment on a stripping process will also be discussed and modelled in part II.

The coffee industry has discovered that total mass yield is correlated with flavour; indeed, this is the basis for the well-known barista control chart, Lockhardt (1957), which overlays various flavour regions on brew strength and brew yield. Sensory data suggests there is a region of optimal coffee flavour. On the low yield side of this, the brew is said to be under-extracted and described as sour and lacking in sweetness. On the high yield side of this region it is said to be over extracted with a bitter, smoky/burnt flavour. The claimed dominant dependence of flavour on yield shows there is something generic about the relative time scales for the release of different flavour component families, hence the need to understand the profiles of individual molecules. The barista control chart is not a control chart in the full engineering sense—it is just a correlation of outcomes. Process parameters, flow rates and brew time are not included. Melrose *et al* (2018a) proposed a modelling approach to a more complete brewing process chart with time and flow rates.

For a range of grind sizes, Melrose *et al* (2019a, 2019b) fitted multi-scale models to data of total mass yield profiles from dilute batch and brewing bed experiments by Corrochano (2017). In the case of total mass, the fitted diffusion coefficients are lumped parameters, representative of a whole collection of molecules. For batch data measured over a range of grind sizes but collapsed together, Melrose *et al* (2019a) fitted a bimodal distribution of diffusion constants D : one mode of $O(10^{-10}\ m^2\ s^{-1})$ with around 70%–80% of the mass and the other of $O(10^{-11}\ m^2\ s^{-1})$. This result was interpreted as a distribution of fast diffusing molecules and a distribution of predominately slower macro-molecules (high molecular weight melanoidins and oligomers of structural carbohydrates). The fits were argued to be consistent with literature data on composition. An alternative explanation is that the two modes indicate that a wetted grain has closed and open pores. However, the molecular weight distribution explanation is preferred as the fits are consistent with expected molecular weights and diffusion constants. Additionally, the profiles of individual molecules releasing from larger particles do not seem to show two modes e.g. caffeine (Spiro and co-workers 1984a, 1984b, 1989, 1993) and acetaldehyde stripping (Mateus *et al* 2007b).

5.2.5 Bed and grind effects on brewing

A feature of release profiles extensively discussed in the early literature was that data was better fit if an increase in yield was added a $t = 0$. Volliey and Simatos (1979) called this 'wash-off'. A first order form plus a constant was used to fit data by Spiro (1993). Stapley (2002) noted that the early time solution of the diffusion involves high fluxes and deviates from a first order form. There are however other effects influencing the early release. The surfaces of coarse particle are pitted with exposed macro-pores, see figure 5.2; fines are on the length scale of the cell-pockets (40–60 μm). These short length scale features, will give a fast-release at the start of brewing, over a few seconds. Melrose *et al* (2012) include fines in their modelling whilst Moroney *et al* (2015) include a fast mode which equated release with such surface features. Melrose *et al* (2019a) showed that release profiles for dilute batch data were fit with an expected distribution of hindered diffusion constants if a fast-release mode was included. Moreover, they used a model for the fraction of surface and fines to estimate fast-release in dilute batch conditions, which gave optimal fits when based on length scales comparable to that of the fines and surface features. Melrose *et al* (2019b) compared model fits to the total yield over time for 'brewing' the grinds in dilute batch conditions and flow through a packed bed. For the packed bed, fast-release was found to be suppressed relative to the batch case; more than the suppression estimated for the effect of bed pore space concentration alone. For espresso style grind sizes, simulations fit well with fast-release by including in the model a fraction of fine particles $\theta_{fs} = 0.15$ of the total particle volume; significantly lower than the fraction of 0.4 in the corresponding dilute batch brews at these grind sizes.

There are numerous physical effects that may occur in a packed bed of coffee particles, these are discussed in detail elsewhere (Melrose *et al* 2019b). These effects include uncertainty in the surface area the packed particles present to the bed pore space, inhomogeneous flow within the bed and the gas released from the particles at the start of brewing. It is unclear how bed packing effects the relevant diffusion path lengths from the particles into the bed pore space. Migrating fines could trap into plugs on the length scale of the coarse particles and hence act as coarse particles with higher hinderance. In the simple approach to bed modelling used in part II, an underlying assumption is that this diffusion path in the bed will scale with particle sizes measured as moments of the PSD. However, there could be mechanisms by which this scaling is broken: coffee particles are shaped with surface facets, packing facet against facet may reduce surface areas open to the pores space and if the facet size is fixed, this effect will become stronger with finer grinds. Beds do consolidate under pressure. Typical coffee bed bulk densities are in the range 400–480 kg m^{-3} which corresponds to bed (inter-particle) porosities ε_{bed} of 0.1 to 0.25 accounting for consolidation of the bed under applied brew pressures (Corrochano *et al* 2015). In capsule brews, air is observed sometimes to trap in the bed leaving part of the bed dry. All these mechanisms will introduce variability in the measured profiles. One particular issue is the time scale over which they might vary brew yield and give rise to an apparently variable maximal yield when extrapolating from short data sets measured over typical brewing times. *These effects are an area for future research but would require difficult observations inside beds*; clearly results and insights gained on this may be more widely applicable beyond coffee beds. Variability of hydrodynamic resistance is another source of variation in brewing systems. Some simulations couple a time dependent bed hydrodynamic resistance to pressure flow relationships set by pumps and allow flow to vary (Melrose *et al* (2012, 2018b)). Some measurements of the hydrodynamic resistances of coffee beds are reported in Corrochano *et al* (2015).

The gas phases, air and CO_2, release into the bed (mainly from the grains themselves in a packed bed) and appear as a foam, known as *crema*, on top off the coffee beverages. It is unclear how this is coupled to release, either via affecting the transfer between the particles and bed pore space or by acting as a 'carrier' of released species. The crema is part of the sensory experience of the beverage, especially for espresso brews. Navarini and Rivetti (2010), Rivetti *et al* (2001) made a systematic study of how water quality affects espresso brews. They report little effect on brew times, but they find significant effects on the quality (appearance and stability) of crema. They estimated the relative roles of CO_2 arising from the reaction of water carbonates and brew acids versus that contained in the grinds.

The coffee literature often asserts that an applied pressure head is directly affecting release (e.g. Illy *et al* 2005) and optimal pressures are often quoted; however, from a simple engineering perspective flow rate is the key parameter and pressure sets the flow rate dependent on the hydrodynamic resistance in turn determined by intrinsic quantities of brew viscosity, grind, PSD, bed packing and extrinsic quantities such as the bed shape. The pressure, however, may also directly affect the gas phase and this may complicate its influence on the sensory experience.

5.2.6 Hinderance

For molecules that have no partitioning, reaction or adsorption effects, the prime effect of the particle structure and contents is to hinder (slow) diffusion inside the particles; diffusion inside particles controls the release rate. Two hindrance mechanisms are discussed here. The porous structure of the particle will give rise to a *geometric hinderance* factor, H_g. Prior to a loss of mass in brewing, the total solute concentrations, c, are high in the wetted particle interior, which will enhance the viscosity of the fluid phase and slow diffusion. This will be referred to as *solute hinderance*, $H_s(c)$. The effective diffusion constant is calculated in equation (5.6)

$$D_p = D_{\text{bulk}}/(H_g H_s(c)) \tag{5.6}$$

where D_p is the effective diffusion constant of the molecule inside particles, and D_{bulk} is its diffusion constant in a dilute bulk solution at the appropriate temperature. The geometric hinderance was discussed in Melrose *et al* (2018a); The approach adopted previously was to assume the factor as shown in equation (5.7)

$$H_g = \left(\frac{\beta}{\varepsilon}\right) \tag{5.7}$$

where ε and β are porosity and tortuosity averaged over the particle pore space (Cussler 2009). From porosimetry data combined with a model of tortuosity, Corrochano (2017) estimated geometric hinderance factors using equation (5.7) were found to be 6, 7, 9, and 10 for a series of grinds increasing with grind size $d_{4,3}$ = 200, 360, 600 and 1250 μm, respectively. The geometric model can be refined to take into account the heterogeneity of the internal structure. Microstructure models such as the Maxwell Garnet approximation, model the cell-pockets as spherical inclusions in cell walls with a separate porosity and tortuosity explicitly (e.g. Beverly *et al* (2020)). However, solute effects complicate the interpretation. Alternatively, structural heterogeneity may be explicitly included in modelling, this is on-going work by the author which will be reported elsewhere.

The hinderance due to solute effects is assumed here to be just directly proportional to the ratio of the solution viscosity to that of water (the Stokes–Einstein relation for diffusivity). Sobolik *et al* (2002) reported a model for the viscosity of coffee brews. Figure 5.4 shows the predicted solute hinderance with increasing weight fraction.

Given the range of solute concentrations inside the wetted particle pore space of between 250 and 350 kg m^{-3}, this suggests an upper limit to $H_s(c)$ of 3–8 $H_s(c)$. As solute is released this will reduce in the depleted exterior regions of the particles. In addition, there are oil droplets (at about 10%–15% by particle weight) which may also contribute to the hinderance. The viscosity estimate of the effect of co-solutes may be inaccurate as macromolecular solutes may produce additional hinderance.

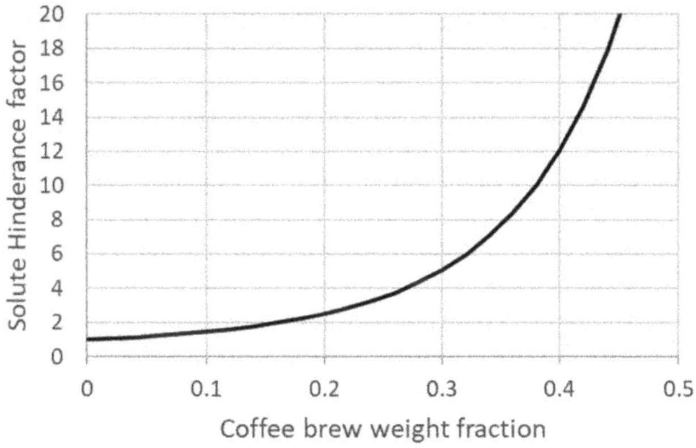

Figure 5.4. The viscosity model of Sobolik *et al* (2002) used to estimate the hinderance factor at 80 °C by normalising it by the water viscosity and 80 °C.

5.2.7 Solubility, volatility, and partitioning

The small molecules released during brewing have a wide range of solubilities, volatilities and octanol–water partition coefficients (mainly determined by the polarity of the molecules). How these physical chemical parameters affect brew profiles will be modelled in part II.

It is sometimes stated in the coffee literature that solubility limits can affect release rates. Although the overall mass concentration is high, most individual small molecules are at a few ppm concentrations and, if considered dissolved, are far below solubility limits even for molecules of moderately low solubility ($10–10^3$ mg L^{-1} - ppm). For substances such as caffeine and chlorogenic acids, which are both moderately soluble in water and are also present at relatively high concentrations, estimates of the pore space concentrations are well below their solubility at high brew temperatures around 80 °C–95 °C but can approach solubility limits at low temperatures (25 °C). Solubility effects on dissolution rates are commonly expressed in the Noyes–Whitney equation which models that the dissolution of a solid particle is slowed by diffusion into the bulk from a boundary layer at the solid surface set at the solubility of the solute. However, this is only relevant for macroscopic solid particles above colloidal length scales > 1 μm; inside coffee particles, molecular species are expected to be much more finely distributed.

If wetted particles are placed in an isolated system, molecules will reach an equilibrium partition across the matrix, oil and water phases. In the dry state, however, the distribution of molecules may be far from that of this wetted equilibrium. An example of this is hydrophobic molecules, where it is unclear if they are partitioned in the oil phase after roasting or if they are located elsewhere during this stage, and transport and partition to the oil phases on wetting. Once the particles are wetted at the start of brewing, the molecules will diffuse, and it is assumed that the majority of this transport occurs in the water phase occupying the pore volume; the oil phase is assumed non-percolating, partitioning between the oils and the water phase will modulate the concentration in the water phase and hence affect release profiles.

Volatility of molecules from the brew into a headspace will affect the aroma experience. Profiles of head-space aromas have been measured by Sánchez-López *et al* (2014, 2016). Volatile molecules will of course diffuse across dry particles. It is unclear, however, if volatility can affect release from the wetted particle into the brew. Mestdagh *et al* (2014) found no correlation between volatility and brewing release profiles. However, in the case where a coffee bed is stripped by gas flow, volatility is directly relevant to release profiles (Beverly *et al* 2020); this scenario will be modelled in part II.

Molecules may also undergo chemical reaction with each other and moieties on the macro-molecules. For example, in part II it will be argued that a large portion of the acids appearing in the brew are neutralised to a degree set by the cationic mineral content of the grains. Phenolic binding is common in food solutions (Harrison and Hills 1997) and melanoidins in coffee brews offer a wide range phenolic sites (Bekedam 2008). Beverly *et al* (2020) included a model for phenolic binding and used it to model the stripping data for pyridine reported by Mateus *et al* (2007b). If there are strong adsorption reactions, a species may only release from the surface of particles and thus appear to be fast releasing, reaching its maximal yield rapidly.

5.3 Current directions—measurement of release profiles

There have been many reports of release profiles for R&G coffee brewing. In the early literature, several authors reported profiles for the yield of total mass released, Volliey and Simatos (1979) and Zanoni *et al* (1992). Cammenga *et al* (1997) showed data for yield profiles in drip filter brewing and examined the effects of several process parameters, in particular a sharp jump in yield at temperatures between 75 °C and 85 °C; this may result from some structural change in the cell-wall such as a glass transition and deserves more detailed investigations (especially in light of the current popularity of cold-brew coffee.)

There are many papers seeking optima in brewing conditions based on target molecules and sensory studies. Albanese *et al* (2009) studied the effect of thermal profiles on quality; this reference contains many further references to early work on quality and brew conditions. Caprioli *et al* (2012) comparing both Robusta and Arabica varieties used sensory and measured odorants to conclude for one café espresso brewer that the optimal conditions were 92 °C and 9 bar. As noted in section 5.2.5, however, the principle effect of pressure on release is via the relationship of flow rate, grind size (bed permeability) and brew time (beverages prepared to target volumes) This makes interpretation difficult, furthermore, as in some home on-demand brewers, the pump characteristics also plays a role (Melrose *et al* 2018b). Severini *et al* (2015) concluded that grinding is the most significant factor in the variation of commercial café style espresso brewing. Cordoba *et al* (2019) found that bean type, grind and contact time are all key factors in cold brewing (20 °C), but that grind was the most important. Others sought to compare more broadly across different brewing methods. Parenti *et al* (2013) compared café espresso brews with a capsule system. Using the same coffee grind and roast they measured physical parameter (density, yields etc), lipids, caffeine and a range of volatiles, they show that a principal components analysis was able to distinguish between the brewing methods. Caporaso *et al* (2014) compared the chemical profiles of a range of preparation techniques (flow and grind) they varied and identified chemical signatures for each technique. Sanz *et al* (2002) and Ribeiro *et al* (2009, 2011) are examples where correlations between a large number of measured volatiles and sensory analysis are attempted. Lopes *et al* (2019) reported the used of response surface methods to show the relationship between brewing parameters and compositions, with emphasis on the carbohydrate content.

This review focuses on measurements of physical properties of R&G coffee and their effects on release profiles of individual molecules and their modelling. Measurements of total mass over time for a series of grind sizes in both batch and bed brewing were made by Corrochano (2017); as discussed above these were fit by the models used in part II (Melrose *et al* 2019a, 2019b). Kuhn *et al* (2017) also show some total mass data. Navarini *et al* (2004) give profiles for viscosity, density and yield in the context of studying the body of espresso brews. Spiro and co-workers (1984a, 1984b, 1989, 1993) reported profiles for caffeine in dilute batch conditions; Jaganyi and Madlala (2000) report data for minerals and caffeine release in dilute batch conditions over 90 min. Lee *et al* (1992) reported profiles for a range of molecules for brewing through a packed bed. They characterised the profiles by half-lives identifying slow and fast extractors. Caprioli *et al* (2014) measured caffeine, trigonelline and nicotinic acid in espresso coffee brews over seven time points; they compared bean origins, temperatures and machine pressure, however, they varied grind at each set of conditions to maintain a fixed flow rate which complicates interpretation. Wang *et al* (2016) measured brews for a single serve Keurig™ brewer, they reported data for total mass yield, pH, TA and caffeine but over a limited set of three time points. Kuhn *et al* (2017) collected aliquots over some 15 time points which they analysed for caffeine and trigonelline by GCMS (gas chromatography mass spectrometry); notably they collected three time points in the first 5 s. Sánchez-López *et al* (2014, 2016) collected data in real time using dynamic PTR-MS (proton transfer reaction mass spectrometry), measuring volatiles in the brew headspace above espresso brews. Mateus *et al* (2007b) used PTR-MS to analyse molecules extracted by gas stripping through a bed. Mestdagh *et al* (2014) reported profiles for several volatiles in an espresso style brewing, showing significant effects of polarity; they fitted data with a hyperbolic equation. The profiles of molecules believed to have anti-oxidant properties have been of recent interest. Bartel *et al* (2015) compare the profiles of chlorogenic acids and melanoidins for several espresso brew systems, despite having filtered for high MW components they note that melanoidins are extracted rapidly; they also fitted data with a hyperbolic equation. Part II will report the use of simulation models to fit the release profiles from the experiments of Kuhn *et al* (2017), Mateus *et al* (2007b) and Mestdagh *et al* (2014).

5.4 Outlook

To improve the connection between the physical parameters of brewing and the composition of brew, there is a great need for further experiment. Further experiments include: dilute batch conditions, maximal yields, questions such as swelling and water quality (including experiments with de-ionised water), how yield varies in detail with applied temperature, the role of air and gas in the beds and some very challenging questions on the packing structure and its evolution under fluid flow of coffee beds, in particular on the relevant path lengths of diffusion, the role of fines and particle shape and how this varies with grinding. Often such investigative experiments are neglected in commercial laboratories due to the time pressures requiring teams to focus on directly commercially relevant conditions. Whilst the motivation to study and compare across different brewing methods is clear, these studies often muddle grind size, flow rate, pump characteristics and temperature effects. The recent work of Lopes *et al* (2019) is an example of a more controlled and systematic approach to correlating brewing parameters and brew composition data. The development of custom experimental rigs such as that of Corrochano *et al* (2015, 2017), Mateus *et al* (2007b) and Kuhn *et al* (2017) combined with systematic studies of grind size, flow field and temperature effects should provide reference experiments and maps (cf. Melrose *et al* 2018a) on these basic fields from which systematic work on the impact of roast, varieties and eventual correlation with sensory can be made. These experiments will allow a better quantitative assessment of release profiles and their variation. Such, controlled experiments combined with the further development of techniques for real time analysis of headspace and brew components (Sánchez-López *et al* (2014, 2016), Mestdgah *et al* (2014)) would greatly improve knowledge on one of the world's most consumed beverages.

5.5 Additional resources

An extensive resource (27 conferences by 2019) of coffee science are the conference proceedings of ASIC conference, the organisation costs little to join (€ 200 bi-annual) and then downloads are free. https://www.asic-cafe.org/. An introduction to the broader science of espresso coffee brewing is given by Illy *et al* (2005). It is hoped that these and the references in the outlook section should motivate the reader to join the next stage of the science of coffee brewing.

References

Albanese D, Di Matteo M, Poiana M S and Spagnamusso S 2009 Espresso coffee (EC) by POD: Study of thermal profile during extraction process and influence of water temperature on chemical–physical and sensorial properties *Food Res. Inter.* **42** 727–32

Anderson B A, Shimoni E, Liardon R, Theodore P and Labuza T P 2003 The diffusion kinetics of carbon dioxide in fresh roasted and ground coffee *J. Food Eng.* **59** 71–8

Bartel C, Mesias M and Morales F J 2015 Investigation on the extractability of melanoidins in portioned espresso coffee *Food Res. Inter.* **67** 356–65

Bekedam E K 2008 Coffee brew melanoidins Structural and Functional Properties of Brown-Colored Coffee Compounds *PhD Thesis* Wageningen University

Beverly D, Lopez-Quiroga E, Farr R, Melrose J, Henson S, Bakalis S and Fryer P J 2020 Mathematical modelling of mass and heat transfer in multiphase coffee aroma extraction systems *Chem. Eng. Sci.* submitted

Cammenga H K, Eggers R, Hinz T, Steer A and Waldmann C 1997 Extraction in coffee-processing and brewing *ASIC(ed.) Proc. of the 17th Int. conference on Coffee Science (Nairobi, Kenya)*

Caprioli G, Cortese M, Cristalli G, Maggi F, Odello L, Ricciutelli M and Vittori S 2012 Optimization of espresso machine parameters through the analysis of coffee odorants by HS/SPME/GC/MS *Food Chem.* **135** 1127–33

Caprioli G, Cortese M, Maggi F, Minnetti C, Odello L, Sagratini G and Vittori S 2014 Quantification of caffeine, trigonelline and nicotinic acid in espresso coffee: the influence of espresso machines and coffee cultivars *Int. J. Food Sci. Nutr.* **2** 465–9

Caporaso N, Genovese A, Canela M D, Civitella A and Sacchi R 2014 Neapolian coffee brew chemical analysis in comparison to espresso, moka and American brews *Food Res. Intern.* **61** 152–60

Clark R J and Macrae R 1985 *Coffee vol 1 Chemistry* (Amsterdam: Elsevier)

Clark R J and Vitzthum O G 2001 *Coffee: Recent Developments* (Oxford: Blackwell)

Cordoba H, Pataquiva L, Osorio C, Moreno Moreno F L and Ruiz R Y 2019 Effect of grinding, extraction time and type of coffee on the physicochemical and flavour characteristics of cold brew coffee *Nat. Scient. Rep.* **9** 8440

Corrochano B R, Melrose J R, Bentley A C, Fryer P J and Bakalis S 2015 A new methodology to estimate the steady-state permeability of roast and ground coffee in packed beds *J. Food Eng.* **150** 106–16

Corrochano B R 2017 Advancing the engineering understanding of coffee extraction *EngD Thesis* University of Birmingham Available at: http://etheses.bham.ac.uk/7176/ (Accessed: 06 July 2018)

Cussler E L 2009 Fundamentals of mass transfer *Diffusion Mass Transfer in Fluid Systems* ed E L Cussler (New York: Cambridge University Press)

Harrison M and Hills B 1997 Mathematical Model of Flavor Release from Liquids Containing Aroma-Binding Macromolecules *J. Agricul. Food Chem.* **45** 1883–90

Illy A, Viani R and Suggi Liverani F 2005 *Espresso Coffee: The Science of Quality* (Amsterdam: Elsevier)

Jaganyi D and Madlala S P 2000 Kinetics of coffee infusion: a comparative study on the extraction kinetics of mineral ions and caffeine from several types of medium roast coffees *J. Sci. Food Agric.* **80** 85–90

Joffe F M 1970 Flaked Coffee and products produced there from *US Patent* 3615667

Kuhn M, Lang S, Bezold F, Minceva M and Briesen N 2017 Time resolved extraction of Caffiene and trigonelline from finely-ground espresso coffee with varing particle sizes and tamping pressures *J. Food Eng.* **206** 37–47

Lee T A, Kempthorne R and Hardy J K 1992 Compositional changes in brewed coffee as a function of brewing time *J. Food Sci.* **57** 1417–9

Lockhart E E 1957 The soluble solids in beverage coffee as an index to cup quality *Tea and Coffee Trade J.* **113** 12

Lopes G R, Passos C P, Rodrigues J A, Coimbra A T and Coimbra A M 2019 Modulation of infusion processes to obtain coffee-derived food ingredients with distinct composition *Europ. Food Res. Tech.* **245** 2133–46

Mateus M L, Champion D, Liardon R and Voilley A 2007a Characterization of water mobility in dry and wetted roasted coffee using low-field proton nuclear magnetic resonance *J. Food Eng.* **81** 572–9

Mateus M, Lindinger C, Gumy J-C and Liardon R 2007b Release kinetics of volatile organic compounds from roasted and ground coffee: online measurements by PTR-MS and mathematical modeling *J. Agric. Food Chem.* **55** 10117–28

McSwiggin J R and Brandt C H 1969 Method of making flaked roast and ground coffee *US Patent* 3660106

Melrose J R, Corrochano B, Norton M, Silanes-Kenny J, Fryer P and Bakalis S 2012 Optimising Coffee Brewing using a Multiscale approach *ASIC(ed.) Proc. of the 24rd Conf. Int. Conf. on Coffee Science San Jose (Costa-Rica)*

Melrose J R, Corrochano B, Montoya-Guerra M and Bakalis S 2018a Towards a new brewing control chart for the 21st century *J. Agric. Food Chem.* **30** 5301–9

Melrose J R, Andrews R J, Harpley P J and Graaff G K 2018b Capsule, System and use of the system for preparing double beverages like a double espresso, double lungo and a double ristretto *Patent Application* WO 2018/026280 A1

Melrose J R, Corrochano B and Bakalis S 2019a Polydisperse diffusion kinetics of Coffee Brewing - modelling batch data; Pre-print: www.reserachgate.com under John Melrose, Coffee Brewing project. DOI: 10.13140/RG.2.2.23545.31841

Melrose J R, Corrochano B and Bakalis S 2019b Polydisperse diffusion kinetics of Coffee Brewing —modelling packed beds. Pre-print www.reserachgate.com under John Melrose, Coffee Brewing project. DOI: 10.13140/RG.2.2.13478.98880

Mestdagh F, Davidek T, Chaumonteuil M, Folmer B and Blank I 2014 The kinetics of coffee aroma extraction *Food Res. Int.* **63** 271–4

Moroney K M, Lee W T, O'Brien S B G, Suijver F and Marra J 2015 Modelling of coffee extraction during brewing using multiscale methods: An experimentally validated model *Chem. Eng. Sci.* **137** 216–34

Navarini L, Cappuccio R and Suggi Liverani F 2004 The Body of the *Espresso* Coffee: the Elusive Importance *ASIC(ed.) Proc. of the 20th conf. (Bengaluru, India, 2004)*

Navarini L and Rivetti D 2010 Water quality for Espresso coffee *Food Chem.* **122** 424–8

Parenti A, Guerrini L, Masella P, Spinelli S, Calamai L and Spugnoli P 2013 Comparison of Espresso Coffee brewing techniques *J. Food Eng.* **121** 112–7

Ribeiro J S, Augusto F, Salva T J G, Thomaziello R A and Ferreira M M C 2009 Prediction of sensory properties of Brazilian Arabica roasted coffees by headspace solid phase micro-extraction-gas chromatography and partial least squares *Anal. Chim. Acta* **634** 172–9

Ribeiro J S, Ferreira M M C and Salva T J G 2011 Chemometric models for the quantitative descriptive sensory analysis of Arabica coffee beverages using near infrared spectroscopy *Talanta* **83** 1352–8

Rivetti D, Navarini L, Cappuccio R, Abatangelo A, Petracco M and Suggi-Liverani F 2001 Effect of water composition and water treatment on espresso coffee percolation, ASIC *Proc. 19th colloquium (Trieste, Italy)*

Sanz C, Maeztu L, Zapalena M J, Bello J and Cid C 2002 Profiles of volatile compounds and sensory analysis of three blends of coffee: influence of different proportions of Arabica and Robusta and influence of roasting coffee with sugar *J. Sci. Food Agricul.* **82** 840–7

Sánchez-López J A, Wellinger M, Gloess A N, Zimmermann R and Yeretzian C 2016 Extraction kinetics of coffee aroma compounds using a semi-automatic machines: on-line a analysis by PTR-ToF-MS *Int. J. of Mass Spec.* **401** 22–30

Sánchez-López J A, Zimmermann R and Yeretzian C 2014 Insight into the time-resolved extraction of aroma compounds during espresso coffee preparation: online monitoring by PTR-ToF-MS *Anal. Chem.* **86** 11696–704

Schenker S, Handschin S, Frey B, Perren R and Escher F 2000 Pore structure of coffee beans affected by roasting conditions *J. Food Sci.* **65** 452–7

Schenker S 2000 Investigations on the hot air roasting of coffee beans *Doctorate of Technical Sciences Thesis* Swiss Federal Institute of Technology, Zurich, Switzerland

Severini C, Derossi A, Fiore A G, De Pilli T, Alessandrino O and Del Mastro A 2015 How the variance of some extraction variables may affect the quality of espresso coffees served in coffee shops *J Sci Food Agric. 2016* **96** 3023–31

Sobolík V, Žitný R, Tovcigrecko V, Delgado M and Allaf K 2002 Viscosity and electrical conductivity of concentrated solutions of soluble coffee *J. Food Eng.* **51** 93–8

Spiro M and Page C M 1984a The Kinetics and mechanism of caffeine infusion from coffee: Hydrodynamic aspects *J. Sci. Food Agric.* **35** 925–30

Spiro M and Selwood R M 1984b The kinetics and mechanism of caffeine infusion from coffee: the effect of particle size *J. Sci. Food Agric.* **35** 915–24

Spiro M 1993 Modelling the aqueous extraction of soluble substances from ground and roast coffee *J. Sci. Food Agric.* **61** 371–2

Spiro M, Toumi R and Kandiah M 1989 The kinetics and mechanism of caffine infusion from coffee: the hindrance factor in intra-bean diffusion *J. Sci. Food Agric.* **46** 349–56

Stapley G F 2002 Modelling the kinetics of tea and coffee infusion *J. Sci. Food Agric.* **82** 1661–71

Voilley A and Simatos D 1979 Modeling the solubilization process during coffee brewing *J. Food Process Eng.* **3** 185–98

Wang X, William J, Fu Y and Loong-Tak L 2016 Effects of caspule parameters on coffee extraction in single serve brewer *Food Res. Int.* **89** 797–805

Wilson A J, Petracco M and Illy E 1997 Some preliminary investigations of oil biosynthesis in the coffee fruit and its subsequent re-distribution within green and roasted beans *17th Int. Coll. on the Chemistry of Coffee* 92–9 (Paris: ASIC)

Zanoni B, Pagliarini E and Peri C 1992 Modelling the aqueous extraction of soluble substances from ground roasted coffee *J. Sci. Food Agric.* **58** 273–9

IOP Publishing

Physics in Food Manufacturing
Case studies in fundamental and applied research
Megan J Povey

Chapter 6

Part II: Modelling coffee brewing release profiles

John R Melrose

The release of flavour and aroma molecules from food matrices is a complex problem. This chapter uses multi-scale modelling to give insights into the underlying physical chemistry that determine release profiles of molecules during coffee brewing. It gives a quantitative basis to assess the roles of hindered diffusion due to structure and co-solutes and interactions such as adsorption and partitioning between phases internal to the coffee particles. Modelling may also be used to assess the impact of the imposed engineering conditions, the bed geometry, and the gas and fluid flows through the beds—although this is not the focus of this review. The main parameters used to fit the data are: the choice of length-scales to approximate the particle size distribution, the hindrance of diffusion internal to coffee particles and the maximal yields. Others parameters such as bulk diffusion constants, volatility constants, oil content and partition coefficients are taken from literature data. The review shows where to find and how to use such data.

6.1 Introduction

Coffee particles have a heterogeneous internal structure of cell-pockets, nano-porous cell walls and oil phases—the current knowledge of this is reviewed in part I (chapter 5). The molecules released from particles during brewing, transport through the internal structure to the outer particle surface; for some molecules the structure and other co-solutes simply hinder their diffusion, but others also interact with macro-molecules and oil phases within the particles by adsorption and partitioning Chemical reactions between diffusing species may also occur. Although there have been a few reported measurements of release profiles in the past, in recent years new techniques are giving more detailed and extensive data sets (see part I). It is important to be able to model this data to give insight into the underlying physics and physical chemistry influencing release profiles. Multi-scale modelling approaches will be reviewed and, for a few selected cases, fits of data by the authors modelling approach methods will be shown.

doi:10.1088/978-0-7503-2596-7ch6

6.2 Background

6.2.1 Modelling approaches

Many authors have fit release profiles with analytical expressions: solutions of the diffusion equation, Zanoni *et al* (1992), Spiro and co-workers (1984a, 1984b, 1989, 1993), Stapley (2002), and Mateus *et al* (2007); stretched exponentials, Mateus *et al* (2007); and hyperbolic expressions, Bartel *et al* (2015) and Mestdagh *et al* (2014). Numerical solutions of more complete models, however, are more powerful and potentially give an *ab initio*, approach. Recently, several multi-scale models of brewing from packed beds of particles have been developed. Melrose *et al* (2012, 2018a, 2018b), Moroney *et al* (2015) model release from particles into a fluid continuum with advection through and out of the bed solved by a simple finite volume scheme for flow through the bed; Moroney *et al* (2019) extend this to full computational fluid dynamics for the flow. Ellero and Navarini (2019) use smoothed particle hydrodynamics for the fluid phase; this also allows motion of the coffee particles, although in dilute conditions in a 2D bed. Cameron *et al* (2020) report a multi-scale model similar to that used here, and fit it to espresso data. They use the model to identify anomalous (w.r.t. to model predictions) low yields in beds of fine grinds—they attribute this to the onset of in-homogeneous flow. They introduce a saturation limit to the total concentration in the boundary conditions at the surface of the coffee particles—but it is unclear what a such an overall saturated value for a mixture of molecules in solution means. Sano *et al* (2019) report a volume averaging of the model of Moroney *et al* (2015, 2019) and discuss the importance of dimensionless variables, similar to those introduced earlier by Melrose *et al* (2018a).

In contrast to the model of Moroney *et al* (2015, 2019) and Ellero and Navarini (2019), the multi-scale model used in this review solves the diffusion equation within the model particles to give the time evolution of concentration profiles inside the particles as well as in the bed; it is therefore straightforward to add source and sink terms to model interaction effects inside particles. Previous publications used the model to propose a new version of the brewing control chart Melrose *et al* (2018a), to model capsule shape effects in on-demand coffee brewing systems, Melrose *et al* (2018b), Corrochano *et al* (2014) and to fit profiles of the yield of total mass in batch and bed brewing Melrose *et al* (2019a, 2019b). Recently, particle-based models have been developed to include interactions within particles and applied to the case of steam stripping in industrial process of instant coffee manufacture Beverly *et al* (2020). Here this is applied to fit published data for the release profiles of individual molecules in both gas stripping and espresso style brewing. The model includes hindered diffusion due to both structure and co-solutes and an internal oil and matrix phases to which molecules may partition or adsorb/desorb from the water phases.

These models have been used to simulate the effects of driving forces in brewing such as pumps and capsule shape, Melrose *et al* (2018b). Moroney *et al* (2019) report on the influence of converging walls in capsule systems.

6.2.2 Brew yield and maximal yield

The strength $S(t)$ and yield $Y(t)$ profiles of the brew at time t are normally expressed in the coffee literature as the following percentage values:

$$S(t) = \frac{100 * M_e(t)}{M_{\text{brew}}(t)} \tag{6.1}$$

$$Y(t) = \frac{100 * M_e(t)}{W} \tag{6.2}$$

where $M_e(t)$, is the total mass of soluble molecules and colloids released by the coffee particles into the brew, $M_{\text{brew}}(t)$ is the mass of the brew and W is the weight of coffee grind used in the brewing. These equations can be adapted for the profile of a single species with $M_e(t)$ replaced by the release mass of just that species. On occasion in this chapter yield will be quoted as a yield fraction or a ratio of masses rather than a percentage and concentrations rather than strength will be plotted in absolute units, mass per volume.

A key input parameter for modelling is the *maximal yield,* Y_{max}, of either the total mass or mass of an individual species. This is the yield in the long-time limit of the mass released from coffee particles in a very dilute bulk solution. *Maximal yields are not commonly measured and reported in the literature.* For a range of grind sizes, Corrochano (2017) measured it for the total mass released in both dilute batch conditions and bed brewing., The maximal yield *decreased* with increasing grind size; data is shown in Melrose *et al* (2019a)—it decreases from 32% for the finest grind studied, $d_{4,3}$ = 100 μm, to values between 30% and 27% for espresso style grinds $d_{4,3}$ = 200–400 μm, to 26%–24% for large grinds, $d_{4,3}$ = 600–1350 μm. For individual molecules, maximal yield may equate to the content of the molecules in dry coffee particles, e.g. as measured directly by analytical methods. However, in brewing conditions irreversible processes inside wetted grains may reduce the amount actually released into the brew; these may be adsorption, reaction or physical entrapment (especially for larger species); clearly, the impact of these mechanisms on what can reach the surface of particles before releasing into the brew will scale (increase) with particle size. *Indeed, the decrease of maximal yield with grind size is suggestive of such interactions affecting a significant portion of the molecules and colloids.* Practical complications in measuring yield can arise for volatile molecules which may further transport from the brew into any headspace.

6.2.3 Modelling transport and interactions inside particles

All the experiments modelled later in this chapter involve either brewing with water flow or gas stripping from packed beds of particles. The multi-scale model divides a bed into layers, typically N = 10–50 layers, in the simplest case of axial symmetry, a set of cylindrical layers is used. Convection between the bed pore spaces of the layers is solved and the exit from the last layer is integrated over time to give concentrations and yield. Equations for the finite volume scheme are given in Melrose *et al* (2018a, 2018b). Within each layer there are a set of particles which are source terms for the release of species into the bed pore space model; here just one fine and one coarse particle with fluxes scaled by the volume fraction of grind in the pore space. Each particle is modelled as a sphere and a radially symmetric diffusion equation with sink and source terms is solved for the internal concentrations and fluxes out of the particle surface. Finite volume techniques are used with meshes from 50 to 500 radial elements per particle depending on particle size. Simpler models that do not model in detail the concentration variation within particles such as first-order rate equations or shrinking core models are computational less intense and therefore allow a more detailed resolution at the bed scale, Moroney *et al* (2019). In this review, however, spatially modelling concentrations within the particles is necessary to model coupled physical chemistry effects in their interior.

The following equations are applied in the interior of the particles. They model partitioning between the internal oil and water phases and the dilute regime of adsorption/desorption between the internal water phase and macromolecular phases, referred to here as a matrix. A version of this model was developed for gas stripping by Beverly *et al* (2020). The notation adopted here, is that of Bruna and Chapman (2015). The model assumes spherically particles and is written in terms of the *intrinsic concentrations* $\bar{c}_i$ in a phase of volume fraction ε_i within the particles. It is useful, however, to define the concentrations at $t = 0$ via the particle volumetric concentration of the species, where y_i is that fraction the species is of the total particle mass

$$C_i = y_i \rho_g. \tag{6.3}$$

(Note y_i may not be equal to the maximum extractable of the species—for example, on wetting and transport through the particles some proportion of the species may irreversibly adsorb into one of the phases internal to the particles.) At $t = 0$ the species may be distributed across the phases in fractions μ_i, so

$$\bar{c}_i(0) = \mu_i \frac{y\rho_g}{\varepsilon_i}. \tag{6.4}$$

For clarity, dropping the species index i, the model is

$$\begin{aligned}
\frac{\varepsilon_w \partial \bar{c}_w}{\partial t} &= -D_p \varepsilon_w \left(\frac{\partial^2 \bar{c}_w}{\partial^2 r} + \frac{2}{r} \frac{\partial \bar{c}_w}{\partial r} \right) \\
&\quad + a_{wm}(v_{mw}\bar{c}_m - v_{wm}\bar{c}_w) + a_{wo}(v_{ow}\bar{c}_o - v_{wo}\bar{c}_w) \\
\frac{\partial \bar{c}_m}{\partial t} &= -\frac{a_{wm}}{\varepsilon_m}(v_{mw}\bar{c}_m - v_{wm}\bar{c}_w) \\
\frac{\partial \bar{c}_o}{\partial t} &= -\frac{a_{wo}}{\varepsilon_o}(v_{ow}\bar{c}_o - v_{wo}\bar{c}_w)
\end{aligned} \tag{6.5}$$

where $\bar{c}_w$, $\bar{c}_m$ and $\bar{c}_o$ are the intrinsic concentrations of the water (wet pore space), matrix and oil phases, respectively; ε_w, ε_m and ε_o are the phase volume fractions of these phases—these are assumed constant across the particle, so ε_w cancels in the first term of equation (6.5); a_{wo} and a_{wm} are the interfacial areas per unit volume of the particle between the water and oil phases and the water and matrix phase; v_{ow} is the transfer 'velocity' from the oil to water phase, v_{wo} from the water to the oil phase and similarly for v_{mw} and v_{wm}. Chemical reactions between species are ignored.

The diffusion constant inside the particle, D_p, is likely lower (hindered) than that in bulk solution; as discussed in part I this is modelled as a hindrance factor between the bulk and particle diffusion constants.

The equations are discretized by a finite volume scheme with each numerical element being a shell. The diffusion constant D_p is assumed constant in equation (6.5), however, a code allowing variable D was developed. The case $D_p(c(r))$ is handled in the finite volume scheme by evaluating the solute hindrance factors at each element boundary from a concentration averaged over the neighbouring elements. The solute hindrance is assumed determined by the total solute concentration, so to estimate this, a model for the total mass (see Melrose *et al* 2019a, 2019b) was simulated simultaneously with one for any specific species. At the centre of the particle the following boundary condition holds

$$t = t; D_p \frac{\partial \bar{c}_w}{\partial r} = 0; (r = 0). \tag{6.6}$$

Whilst there may be transport boundary layer effects in the beds, there is quite some ambiguity as to how to define boundary layer thickness in a packed bed. Therefore, rather than introducing more parameters, for the external boundary condition it is assumed that any boundary layer is small relative to the particle size and concentration boundary conditions are applied at the outer surface of the particles: the concentration at the particle surface is set to that of the external bed pore space.

$$t = t; \; \bar{c} = \frac{c_b}{K}(r = R_p) \tag{6.7a}$$

where c_b is the external concentration—that of the pore space of the bed and K allows for a possible partition coefficient between the bulk and particle pore space.

On the scale of the particle pores, there is a jump in porosity at the particle surface, the flux *into* the bed pore space is

$$j = -\epsilon_w D_p \frac{\partial \overline{C}_w}{\partial r}(R). \tag{6.7b}$$

Estimates can be made of the parameters in equation (6.5). For the interaction terms, consider the case of a molecule which does not interact with the matrix but does partition between the oil and water phases with a partition coefficient P_{ow}:

$$P_{ow} = \frac{\overline{C}_o^{eq}}{\overline{C}_w^{eq}} \tag{6.8}$$

where in the absence of flux at the surface of the particles, $\overline{C}_o^{eq}$ $\overline{C}_w^{eq}$ are the intrinsic concentrations at equilibrium. In practice, P_{ow} will be taken as the measured octanol–water partition coefficient. The porosities inside the particle are assumed constant. For this case, equation (6.5) reduces to

$$\frac{\partial \overline{C}_w}{\partial t} = -D_p\left(\frac{\partial^2 \overline{c}_w}{\partial^2 r} + \frac{2}{r}\frac{\partial \overline{c}_w}{\partial r}\right) + \frac{\varepsilon_o}{\varepsilon_w} T_{wo}(\overline{c}_o / P_{ow} - \overline{c}_w)$$
$$\frac{\partial \overline{C}_o}{\partial t} = -T_{wo}(\overline{c}_o / P_{ow} - \overline{c}_w) \tag{6.9}$$

where the transfer rate is

$$T_{wo} = \frac{v_{wo} a_{ow}}{\varepsilon_o}. \tag{6.10}$$

Estimates of the oil phase volume, ε_o, can be made from the ratio of oil weight to that of the grain f_o, and the particle density.

$$\varepsilon_o = \frac{f_o \rho_g}{\rho_o} = \frac{f_o(1 - \varepsilon_g)\rho_I}{\rho_o} \tag{6.11}$$

where ε_g is the (dry) particle porosity, ρ_I intrinsic density of the coffee particles and ρ_o is the density of the oil. If the geometry of the oil phase is assumed to be spheres of radius r_o (see the observations of Schenker (2000)) then one can estimate that

$$a_{ow} = \frac{3\varepsilon_o}{r_o}. \tag{6.12}$$

The 'transfer velocity' is estimated as diffusive transport across some boundary layer of thickness λ:

$$v_{wo} = \frac{D}{\lambda}. \tag{6.13}$$

Combining the above

$$T_{wo} = \frac{D}{r_o \lambda}. \tag{6.14}$$

It is unclear how to set D and λ. Assuming D is a bulk diffusion constant of $O(10^{-9})$ m^2 s^{-1} or possibly a hindered diffusion constant $O(10^{-10})$ m^2 s^{-1} and given that the boundary layer thickness λ must at most be $O(10)$ μm (cf. a cell-pocket size) suggests T_{wo} to be in the range 10–103 s^{-1}. However, the surface of the oil drops are most likely covered with a molecular coat which may introduce additional effects on the transfer, although it is un-clear how the oleosin coats of coffee bean oil bodies are affected by the roast process. The sensitivity of model predictions to T_{wo} will be investigated in section 6.3.5.

6.3 Modelling selected data sets

In part I, measurements by many authors on release profiles were reviewed. Here the simulation model described in section 6.2.3 will be used to fit the release profiles from the experiments of Kuhn *et al* (2017), Mateus *et al* (2007) and Mestdagh *et al* (2014).

6.3.1 Fitting models

As discussed in Melrose *et al* (2019a, 2019b), particle based models fit relaxation times for release from the particles to the overall brew profile data. For diffusion problems relaxation times scale as L^2/D. So if the data is fit for some diffusion constant D based on a model with a single particle length scale L appropriate for the system (e.g. a moment of the grind particle size distribution, PSD), then rescaling the length by λ and diffusion constants by λ^2 will be give an equivalent fit.

Consequently, inferred hindrance factors for a fitted D_p relative to a known bulk D_{bulk} also vary with the choice of length scale. However, choosing a size appropriate for the measured PSD should give estimates of diffusion constants and hindrance factors comparable to reality; the length scale of fines and surface features is one length scale dominating the early time (<5 s) behaviour, whilst the volume averaged dimension, $d_{4,3}$ (Co.) and the area averaged dimension $d_{3,2}$ (Co.) of the coarse part of the PSD define longer length scales dominating profiles at later times; in coffee grinds they scale together, Melrose *et al* (2019a). Therefore, ignoring the fines, if fits are done by assuming one of these length scales they can be re-scaled to an alternative choice. In the context of fitting total mass yield, comparisons and examples are given in Melrose *et al* (2019a, 2019b). As discussed in part I, it is assumed that the complexity of bed effects will scale with the underlying length scales of the PSD. This may not be true in practice, e.g. the packing of the faceted coffee particles may vary with their size.

The quality of a fits/predictions will be reported by giving a mean percentage error (MPE). Given a set of n experimental points and fit values for yields:

$$MPE = 100 * \frac{1}{n}\sum_{i=1}^{n}\left|\frac{y_{exp}(t_i) - y_{fit}(t_i)}{y_{exp}(t_i)}\right|. \tag{6.15}$$

6.3.2 Hindered diffusion—caffeine in espresso brewing

The release of a caffeine from coffee particles in dilute batch conditions was measured by Spiro and co-workers (1984a, 1984b, 1989, 1993). They modelled transport of caffeine internal to particles as simple diffusion, but with diffusion constants, D_p hindered from that measured in bulk, $D_p = D_{\text{bulk}}/H$; for particles in the size range 850–1200 μm, they reported hindrance factors of $H = 11$ at 84 °C. Jaganyi and Madlala (2000) reported dilute batch measurements over 90 min for Arabica beans of various origins milled and sieved approx. 1700–2000 μm in dimension. Using diffusion models, they reported for caffeine D_p= 1.6–2.0 $m^2 s^{-1}$ at 80 °C and inferred hindrance factors for caffeine in the range 11–14 for a $D_{\text{bulk}} = 2.2 \times 10^{-9}$ m^2 Price *et al* (1989). At 80 °C in water, Poling *et al* (2008) reported $D_{\text{bulk}} = 1.89 \times 10^{-9}\ m^2\ s^{-1}$.

Kuhn *et al* (2017) report data for the release of caffeine and trigonelline from a commercial coffee brewer (Boxer, Rocket Espresso) with 14 g of R&G in its basket. Their caffeine data set will be modelled by assuming caffeine only has hindered diffusion within the particles. They collected aliquots of brew over time and fit data with an approximate solution of a model similar to that simulated here. Notably, they report data at very early times, with three data points in the first 5 s. The grind used was typical of an espresso coffee with median particle size 240 μm and coarse peak around 300 μm and around 20% fines measured dry. The beans were Nicaragua Arabic catura. The basic grind was sieved to give additional distributions. which they refer to as basic, sieved large and sieved small with median particle sizes 240 μm and 107 μm and lower and higher fines levels, respectively; the sieved small case had a coarse peak around 200 μm. Some of the caffeine profile data is shown in figure 6.1—it was extracted from the solid lines of figure 6.11 of Kuhn *et al* (2017) using the s/w PlotDigitizer™. The original data sets show different runs, variability is highest in the sieved small data set. As part of their approximations, Kuhn *et al* (2017) assumed a linear profile of concentration across the pore space of the bed. In the simulations here, bed profiles are predicted.

A concentration profile across the bed prior to the emergence of brew results from filling the bed with fluid, modelling this is crucial to matching the concentrations of

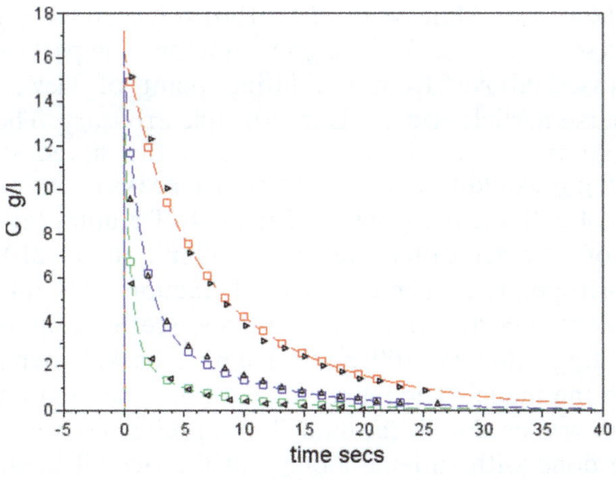

Figure 6.1. Caffeine brew data extracted from Kuhn *et al* co-plotted with some simulated profiles; $t = 0$ is the time at which the brew first exits the bed. Aliquot concentrations of Kuhn *et al* are shown by triangles: right-pointing, up-pointing and left pointing, respectively, the sieved small, basic and sieved large cases. The time points are the mid-times of the aliquot intervals. The model fits are correspondingly red, blue and green lines, the squares are 'aliquot' values derived from the simulations. For *all distributions* these simulations were run with the coarse particle length scale d_{co}= 200 εm and $D_{\text{bulk}} = 1.89 \times 10^{-9}\ m^2\ s^{-1}$, hindrance was varied to find an optimal fit with $H = 16$.

brew at early times There are several complications in modelling the fluid filling of the bed. Firstly, if a bed of inter-particle porosity ε_b of dry porous particles is filled by pumping in fluid at volumetric flow rate Q the filling front velocity may be lower than the case of non-porous particles $Q/(\varepsilon_b A)$ where A is the cross-sectional of the bed—in the porous case, some of the volumetric flow wets the particles rather than the interparticle pore space. Secondly the hydrodynamic resistance of the system changes: one factor being simply the depth of the bed filled; more complex changes in the bed itself such as fines migration and plugging will also lower the bed permeability. However, in the case of the Kuhn *et al* (2017) experiments, the brew system pre-wets the particles for 5 s, so in the modelling below, the particles are assumed wetted at the start of filling and subsequent brewing. The commercial brewer is designed to pump at a constant pressure (approx. 11 bar), the flow rate for the empty chamber was measured at approx. 10 ml s^{-1}; once the bed is filled and brew exits the brew chamber Kuhn *et al* (2017) find a relatively constant flow rate, which is estimated from their figure 10 to be 1.6, 3.8, 9.2 ml s^{-1}, respectively, for the sieved small, basic and sieved large grind distributions. It will turn out that the flow rate variation between the basic and sieved cases is the dominant effect on the brew profiles.

The following model parameters were used. To model the filling, the flow rate was assumed to drop from 10 ml s^{-1} to the brewing flow rate in linear proportion to the height of the bed filled (the results are not that sensitive to this assumption). Once filled, the flow rate is kept constant in the brewing phase at the values given above. The model PSD was restricted to one coarse particle of size d_{co} and a fraction of fines of scale $d_{fi} = 40$ μm. As discussed in part I the fines give a fast release, however, this fast-release in the packed bed is suppressed relative to dilute batch conditions. Following the findings of Melrose *et al* (2019b) for espresso grinds in beds, the fraction of fines is set at $\theta_{fs} = 0.15$ of the grind volume. The particle porosity was set at 0.5. As discussed above, from the fitting point of view, the choice of a representative coarse particle size, d_{co} is in principle arbitrary. The choice of length scale d_{co}= 200 μm is around the coarse peak in the sieved small distribution; although d_{co}= 300 μm would be closer to the coarse peaks of the other distributions. A maximal yield of caffeine needs be set. Figure 4 of Kuhn *et al* (2017) shows the cumulative mass of extracted caffeine tending to a limit of 200–210 mg from the 14 g basket, which corresponds to a maximal yield fraction of 0.014 to 0.015. This is consistent with other reported values for caffeine, see for example Navarini *et al* (2008), where limiting values 90–100 mg from a 6.5 g basket are reported; a value of 0.015 was used in the model. The remaining variable, the hindrance of the particle diffusion constant, was used to fit the data. Two approaches were taken: in the first, simulations were done with caffeine alone and the overall hindrance factor kept constant over time but varied to fit the data. In the second approach, the geometric hindrance was fixed but solute hindrance computed, this requires a simulation with both caffeine species and a representative 'total-solute' species both releasing into the brew. The total-solute concentration profile inside the particles was used along with the plot shown in figure 6.3 of part I to compute $H_s(C_{sol.}(r, t))$ from the internal particle concentration profiles $C_{sol.}(r, t)$ at each time point. In the runs, the total solute was simulated to have a $D_{sol.} = 10^{-10}\ \mathrm{m^2\ s^{-1}}$ and a maximal yield fraction of 0.3 (30%) appropriate for espresso grinds (see data in Melrose *et al* (2019a)); although more complex models for the total mass yield with bi-modal distributions of diffusion constants were reported in Melrose *et al* (2019a, 2019b), the simpler model here gives a good single diffusion constant model, see Corrochano (2017).

Figure 6.1 shows the experimental results co-plotted with some fits to be discussed later. (Note the data are the concentrations of aliquots, not the concentration of the cumulative brew up to that point of time.)

Given the restriction of the model to a single length scale, re-scaling this length across the range 300–200 μm appropriate for the grinds, the hindrance factor fits are in the range $H = 7$–16; these are consistent with values reported by Spiro and co-workers (1984a, 1984b, 1989, 1993) and Jaganyi and Madlala (2000) for larger, around 1000 μm, grinds in batch conditions. However, the measurements and model for tortuosity used by Corrochano (2017) suggest that geometric hindrance reduces with grind size and an H approx. 6 was found for fine espresso grinds such as those of Kuhn *et al.* To estimate the effect of solute hindrance, simulations were done keeping d_{co}= 200 μm, setting $H = 6$, but now including the solute hindrance model; figures 6.2 and 6.3 show the results.

The model with solute hindrance included makes an estimate of the relative effects of geometric and solute hindrance. Good fits were found with this more complete model. Based on the length scale d_{co}= 200 μm, fits were found with $H_g = 6$ and the total solute release modelled with $Y_{\max} = 0.3$ (30%), $D_{sol.} = 10^{-10}\ \mathrm{m^2\ s^{-1}}$. The solute hindrance varies with time as solutes are released. There is a profile of hindrance inside the particles, and the concentration profile internal to the particles

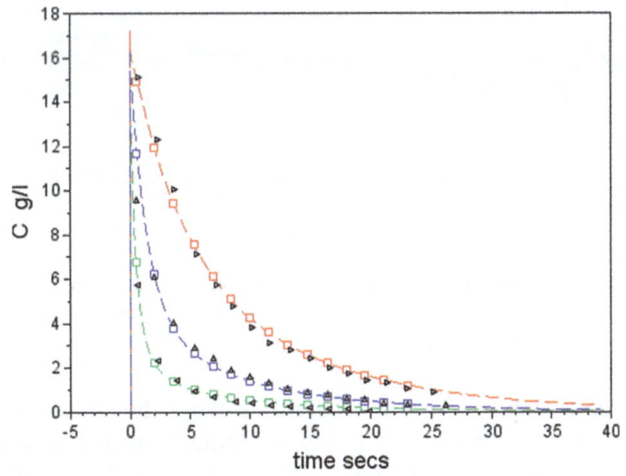

Figure 6.2. Data from simulations runs with solute hindrance included. Caffeine concentration in the brew in $\mathrm{g\ L^{-1}}$, sieved small, basic and sieved large data sets red, blue and green, respectively. The caffeine $H_g = 6$, and H_s (C(r)) species was modelled with $Y_{\max} = 0.014$ and $D_{\text{bulk}} = 1.89 \times 10^{-9}\ \mathrm{m^2\ s^{-1}}$ and simulated with a particle $D_p(r, t) = D_{\text{bulk}}/(H_g H_s(r(t)))$ computed at each time point (see text for details). The MPE for these fits were 3%, 10% and 10%, respectively.

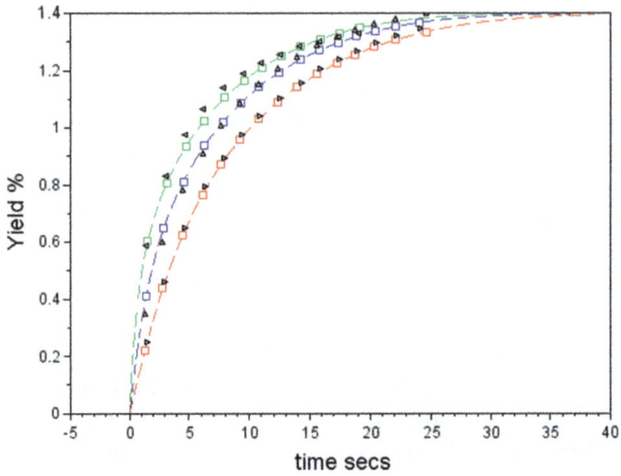

Figure 6.3. Simulation and notation as in figure 6.1. The cumulative plot of caffeine yield. The symbols now show the cumulative mass in the experimental and simulated aliquots of caffeine extracted as a % of the weight of the coffee in the basket.

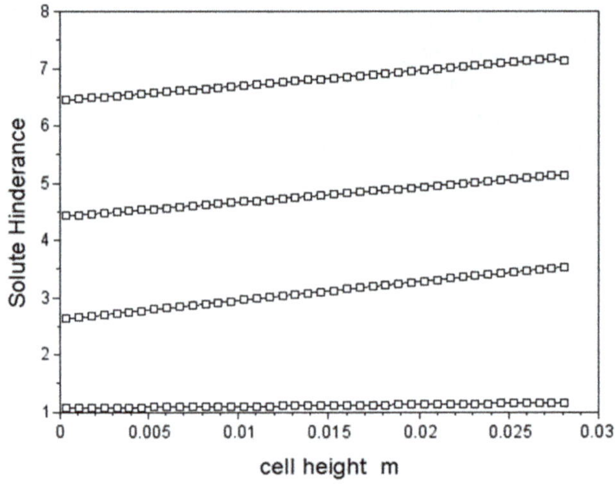

Figure 6.4. Hindrance factors averaged across the coarse particles, plotted against height from flow entrance towards the flow exit of the bed); snapshots at various time points, at the filling of the bed, at 9.5 s, 19 s and 40 s (end of the simulation), top to bottom.

varies along the bed. Figure 6.4 plots some data from the simulations of the sieved large grind. The solute hindrance factors were radially averaged inside each coarse particle and the variation along the bed of these averaged factors plotted at various time instants. The values drop from approx. 6.5–7.5 to 2.5 to 3.5 over the first 19 s of brewing; combined with the geometric factor which in this case was 6, an overall hindrance around 40 occurs at the start of brewing.

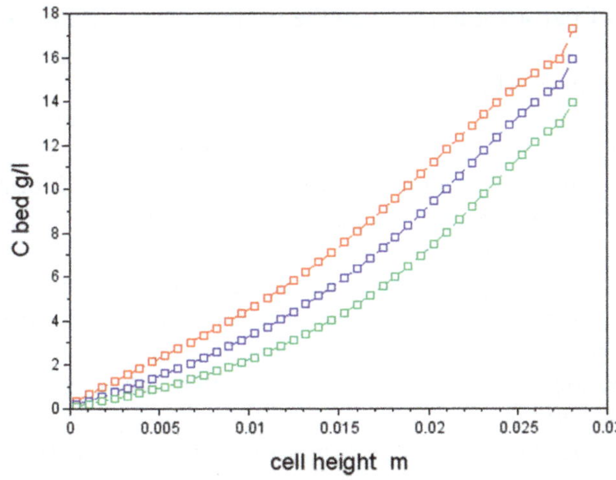

Figure 6.5. Concentration profiles across the bed of height 2.8 cm at the end of filling, for the sieved small, basic and sieved large cases, red, blue and green, respectively. The simulation and all those above were done with the bed divided into $N = 40$ elements.

Figure 6.5 shows profiles of caffeine concentration in the bed pore space just at the end of filling the bed, just before the flow exits the bed, it deviates from linear.

(There is an algorithmic ambiguity in the filling simulation as to whether a bed element has the transport in and out of the particles turned on or not over the time interval during which it is being crossed by the filling front. In the case shown of figure 6.5, it is turned on—for a large enough number of elements the difference is insignificant.)

The concentration of caffeine in the pore space of the wetted coffee particles of porosity around 0.5 is in the range. 8–9 g L^{-1}. If wetted and left to equilibrate in a fluid filled bed of porosity ϵ_{bed}= 0.2 the equilibrium concentration will be 10.5–12 g L^{-1}, this is comparable to the initial aliquot concentrations only in the case of the basic and coarse grind data set (see figure 6.1); the fine bed, which filled at the slowest rate, shows a value higher than this equilibrium; this is due to convection in the bed during filling. Note if the bed simulation is done without filling, e.g. by assuming flow and release start together as the brew first leaves the bed, then brew concentration profiles rise rapidly to a peak at early times before decaying, examples are shown in Melrose *et al* (2018a).

6.3.3 Acetaldehyde in gas stripping

Mateus *et al* (2007) report experiments on gas-stripping from a bed of coffee particles; their data was used to validate the steam stripping model developed by Beverly *et al* (2020). The beans were Columbian Arabica, roasted to light, medium and dark grades; unfortunately, for the data sets reported the roast degree used was not stated. The particle dimensions are stated to be 400, 750 and 1250 μm, it is unclear how this was measured; fines levels will be low in the larger grinds. The experiments were conducted by flowing nitrogen at a rate Q_g = 12.4 ml s^{-1} through a loose bed of 5 g of particles; in one case the particles were dry, in another they were pre-wet.

In the pre-wet case, Mateus *et al* (2007) added 5 ml of water to 5 g of coffee particles. For large 600 μm arabica coffee particles Corrochano (2017) measured the open and total porosity to be 0.4 and 0.6, respectively, see data in figure 6.2 of part I. It is unclear if the wetting penetrates closed pores, so, adopting $\varepsilon_g = 0.5$, the 5 ml of added water will fill the particle pore space and leave an excess 1.3 ml, which is assumed to form a thin layer around the particles. Assuming a loose bed of porosity 0.3–0.5, this layer will occupy a fraction 0.4 to 0.2 of the bed pore space, with the rest occupied by the flowing gas. The releasing volatile molecules diffuse through the water phase internal to the particles, then into the thin external water layer from which they evaporate into the gas phase and are then advected through and out of the bed. *Crucially, and in contrast to brewing, in these experiments, the total solute concentration internal to the particles remains high throughout, hence there is a high solute hindrance throughout the stripping.* The molecules may also be affected by partitioning. Molecules with octanol–water partition coefficient log10 (P_{ow}) approx. 0 and above, will partition significantly between the oil and water phases within the particles, for Arabica coffee oil contents are in the range 0.1 to 0.15 of the particle mass. The model is relatively insensitive to the precise choice of these parameters.

Acetaldehyde (ethanal) is a volatile organic compound, but with a low octanal–water partition coefficient. Mateus *et al* (2007) fitted their data to a stretched exponential form and a series solution: equation (6.43) in Crank (1975) for diffusive release from a set of particles with evaporation at their surface. Using the same expressions and the reported parameters, the author was only able to reproduce the fit shown in figure 6.9 of Mateus *et al* (2007) for the stretched exponential case. The data was fit by the Crank expression with $D_p = 1.0 \times 10^{-11}$ m^2 s^{-1}, in contrast to the reported value of $D_p = 1.86 \times 10^{-11}$ m^2 s^{-1}.

Two simulation models will be used 1) *Model DP: hindered diffusion with boundary conditions as above plus partitioning between the water and oil phase internal to the particles.* The flux out of the particles is that set by zero concentration at the particle surface; this flux is put into the gas stream of the bed pore space and then advected. For acetaldehyde, the literature gives various values of log10(P_{ow}): PubChem (https://pubchem.ncbi.nlm.nih.gov), and ChemSpider gives –0.34 and Sangster (1989) gives 0.45. The oil partitioning model assumes a fraction of oil phase 0.15 the weight of the grinds, appropriate for Arabica. 2) *Model WGT: hindered diffusion with oil partitioning, plus mass transfer by evaporation from the external water layer to the gas phase.* As in Beverly *et al* (2020) the flux between the external water and gas phase is modelled by a term:

$$j = k_{wg}\left(C_w K_H^{wg} - C_g\right) \tag{6.16}$$

where C_g and C_w are the concentrations in the pore space, the gas and water phases, respectively, K_H^{wg} is the dimensionless (conc.-conc.) Henry's law volatility constant, and k_{wg} is the interfacial transfer coefficient. The latter depends on the gas flow rate. The value for K_H^{wg} for acetaldehyde at the experimental temperature is found from averaging the parameters given in the review by Sanders (2015): the simulation used the parameters on p. 4529 and equations (22), (19) and (16) of that reference to set K_H^{wg}: with $H^{cp} = 0.14$ mol m^{-1} Pa^{-1} and d$\ln H^{cp}$/d$(1/T) = 5500$ K giving $K_H^{wg} = 0.057$ at 90 °C; k_{wg} is given by Carberry (1960), see Beverly *et al* (2020). The Biot number, Bi, is the ratio of the rates (s^{-1}) of the evaporation into the gas phase to that of mass transfer internal to the particles, in this case

$$Bi = \left(k_{gw} K_H^{gw} R\right)/D. \tag{6.17}$$

The acetaldehyde parameters above give $Bi > 1000$, and hence imply that diffusion r internal to the particle is rate limiting.

The data sets of 750 μm will be simulated below with the model length scale set at $d_{co} = 750$ μm and fines will be ignored. Fixing the particle size, log10(P_{ow}) and K_H^{wg} for the model WGT, leaves two parameters, the particle diffusion constant, $D_p = D_{\text{bulk}}/H$, and, to match the data in absolute units, the content by weight of acetaldehyde in the coffee particles, C_{Act}^{w}; this is assumed to set to the maximal yield of the molecule; it is not explicitly reported by Mateus *et al* (2007). However, their stretched exponential form with the parameters quoted, fits the data if $C_{Act}^{w} = 32.7$ μgg^{-1} consistent with some reported measurements Ueblacker and Lachenmeier (2011) give 30–40 μgg^{-1} (although Grosch (2001) reported a much higher value $C_{Act}^{w} = 120$–140 μgg^{-1}).

The values for D_p and C_{Act}^{w} in the simulation were varied to find optimal fits, the MPE given in equation (6.15) was used to characterise the fit. Figure 6.6 shows the MPE landscape, in the case log10(P_{ow}) = −0.34 as this marginally gave the best fit.

Over the values scanned, the best fit for both *Model DP* and *Model WGT* to the data set for the wet case was at $D_p = 1.0 \times 10^{-11}$ m^2 s^{-1} and $C_{Act}^{w} = 30$ μgg^{-1} with MPE = 6.9%. (For log10(P_{ow}) = 0.45 a minimum with MPE = 7.3% at $D_p = 1.0 \times$

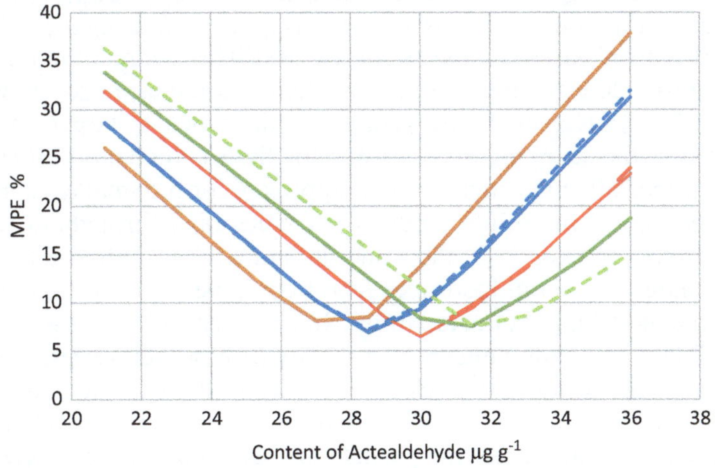

Figure 6.6. For the wet data set and $D_p = 1.4, 1.2, 1.0, 0.8 \times 10^{-11}$ m^2 s^{-1} orange, blue, red green, Model DP, *solid curves; Model WGT* water/gas *dashed curves*. The bed porosity was set at 0.5.

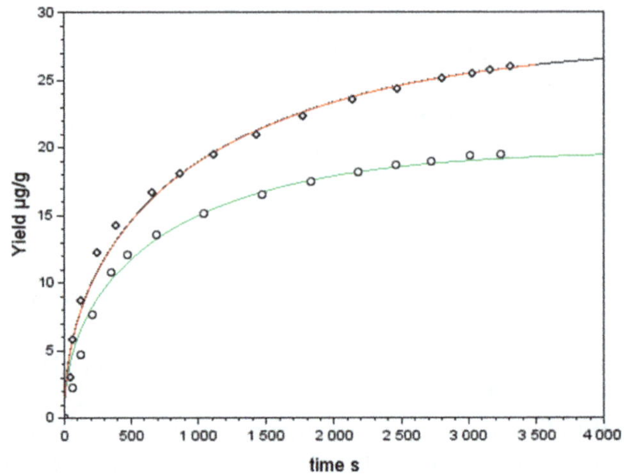

Figure 6.7. Simulations of stripping versus the results for Mateus *et al* (2007) for the case $d_{4,3} = 750\ \mu m$, dry/wet stripping circles/diamonds. For *Model DP* (see text): red curve: using $D_p = 1.0 \times 10^{-11}$ m^2 s^{-1} and $Y_x = 3.0$ μgg^{-1} predicts close to the wet case; green curve, using $D_p = 1.35 \times 10^{-11}$ m^2 s^{-1} and $Y_x = 2.0$ μgg^{-1} predicts the dry case. Also shown by the black curve is a wet case with Model WGT (see text) with $D_p = 1.0 \times 10^{-11}$ m^2 s^{-1} and $Y_x = 2.9$ μgg^{-1} including water gas phase partitioning and mass transfer with parameters given in the text—it is almost indistinguishable from the model DP.

10^{-11} m^2 s^{-1} and $C^w_{Act} = 40.5$ μgg^{-1} was found.) As expected for a high Biot number, the *Model WGT* including evaporation gives very close results to the *Model DP*. To simulate the dry case, the *Model DP* was used with oil partitioning turned off (this assumes that post roast, acetaldehyde is not left partitioned into the oil phase), here a good fit was found with using $D_p = 1.35 \times 10^{-11}$ m^2 s^{-1} (as also reported by Mateus *et al*) and $C^w_{Act} = 20.0$ μgg^{-1}. Figure 6.7 shows these fits and some points sampled from the published data using the software plotdigitizer™.

Given the bulk diffusion constant of acetaldehyde at 90 °C is approx. 10^{-9} m^2 s^{-1}, the hindrance factor from modelling the gas stripping data using $d_{co} = 750$ μm with $D_p = 1.0 \times 10^{-11}$ m^2 s^{-1} is estimated as approx. 100. This is larger than the hindrance factors of approx. 20 estimated above for caffeine in the espresso brewing data of Kuhn *et al* (2017). However, the geometric hindrance is larger for the larger particles, used in the stripping experiments; for the Corrochano (2017) grinds it is approx. 10 for the 750 μm particles, versus approx. 6 for the smaller espresso grinds. Also, as noted above, the solute hindrance remains high throughout the stripping experiments, likely approx. 7, cf. the values in figure 6.7 just after filling of the bed. By these estimates an overall hindrance factor could be 70, lower than the estimate found from the modelling, but comparable. In both dry and wet cases, however, the experimental data also shows unexpected scaling with particles sizes, in the dry case the fitted diffusion constants reported by Mateus *et al* (2007) actually *reduce* with particle size. Mateus *et al* (2007) suggest that the relevant length scale for diffusion to the water–gas interface is not the particle size, but some size of aggregates in the bed. It is unclear how the particles, excess water and gas flow in the bed packing can give rise to this; further experiment is required, e.g. varying the water content. If such a bed effect is verified, this is clearly of commercial interest for steam stripping to separate aromas in instant coffee production.

The maximal yield required to fit the data is notably lower in the case of dry stripping, by eye this is also evident in the data sets. This could be because in the dry state some of the acetaldehyde is bound; but this could also be due to physical entrapment. The cell-pockets are not percolating, therefore vapour in the dry state must diffuse through the cell walls. As discussed in part 1 it could be that in the dry state there are 'closed' pores from which vapour does not diffuse on the timescales of the experiment; Mateus *et al* (2007) comment that in dry conditions the total amount recovered increases with decreasing grind size, this is consistent with a drop in closed porosity with grind size, Corrochano *et al* (2015).

6.3.4 Acetic-acid—gas stripping from a wet bed

For dry conditions, Mateus *et al* (2007) found acetic acid to be fit well by a diffusion model with the release rate limited by hindered diffusion of vapour though the particles. In the wet case, however, a diffusion model did not fit well. The value for K_H^{wg} for acetic acid given by Sanders (2015) p. 4550 is $H^{cp} = 40.0$ mol m^{-3} Pa^{-1} and $\mathrm{dln}H^{\mathrm{cp}}/\mathrm{d}(1/T) = 6200$ K giving $K_H^{wg} = 0.000\,27$ at 90 °C; it is thus far less volatile than acetaldehyde, volatility now effects the profile (see below), and the model WGT needs be used. Partitioning into the oil phase can, however, be ignored.

The complication is that only the *undissociated acid species in solution is volatile*; because the acid sits in solution with many other acids and bases, the concentration of undissociated acid [AH] needs be estimated. For acetic acid in particular, this is very sensitive to mineral content of the coffee which sets the pH as will be now discussed.

The author has developed a numerical tool for finding the equilibria of solutions of weak poly-acids and strong bases for coffee applications, this will be reported in detail elsewhere, but used here. Maier (1987) reported data for the composition of acids in R&G coffee particles, if this composition is solved for equilibrium in water then a pH of approx. 2.5 is found, in contrast to the measured values, approx. 5–5.5, typical at the start of brews (temperature is not always quoted, but Kuhn *et al* (2017), measure at 80 °C an initial brew pH of 5; for a packed bed, this will be close to that of the water phase internal to the coffee particles. This discrepancy can be understood: by using ion-exchange to replace all cations by H^+ in a brew, Maier (1987) measured the lower pH; it is thus assumed that the mineral cation content of coffee (mainly potassium) is proportional to a strong-base content, which neutralises some of the acids, resulting in the observed much higher brew pH. (Asides: neutralisation by bases provided by the input water are insufficient for the discrepancy observed, these are, however, neglected in the estimates here; it is somewhat surprising that the importance of R&G mineral content to brew pH is not generally recognised in the current literature. It should be noted that the correlation between acid content and brew sensory is via the titratable acidity to mouth pH rather than the brew pH, see Maier (1987)).

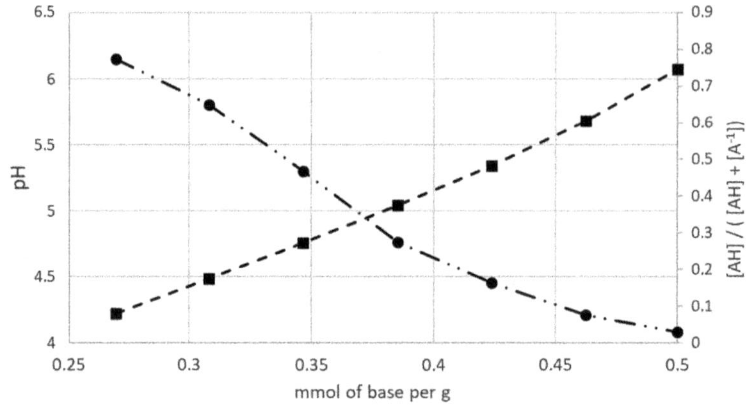

Figure 6.8. Results from using a model for multi-acid string base equilibria and a compositions model based on the data of Maier (1987).

From the data of Jaganyi and Madlala (2000), the mineral content of arabica coffee particles is estimated to be in the range 0.3–0.5 mmol g^{-1}. Estimates were made using the acid composition data of Maier (1987), the pKa and valence states for each acid (there is some ambiguity in assigning the acid sites on high MW components) and assuming that the mmol of mineral cations corresponds 1:1 to mmol of a strong uni-valent base. Figure 6.8 shows, for the expected range of mineral content, the estimate of the equilibrium pH and for acetic acid the ratio of [AH] to total acid $f_{cc} = [AH] / ([AH] + [A^{-1}])$.

The mineral range 0.3–0.5 mmol g^{-1} does correspond to the range observed for initial brew pH 4.5–5.5 (usually measured at room temperature); the fraction of undissociated acetic acid is very sensitive to the mineral content; this is because its pKa, approx. 4.7 at room temperature, is close to the pH of the brew equilibrium. The water–gas volatility constant is reduced to $K_H^{wg} = f_{cc} K_H^{wg}(Ac.\ Ac)$ according to the equilibrium partition shown in figure 6.8. (For a single species of pKa in solution at pH, this factor would be $f_{cc} = 1/(1 + 10^{(\mathrm{pH}-\mathrm{pKa})})$, due to dissociation alone, however, neutralisation will lower this factor further. In principle f_{cc} will vary with time because the equilibrium pH inside the particles will shift as volatile acid components are lost, this is ignored in the modelling.)

The total acetic acid content of R&G particles varies with the degree of roast with a maximum at medium roast. Maier (1987) shows a maxima value of approx. 6 g kg^{-1} at 10% organic roast lost; Balzer (2001) report 5–6 g kg^{-1}; in espresso brews Navarini *et al* (2008) reported yield values approx. 3 g kg^{-1}. To model the Mateus *et al* data, simulations were done with a starting content of C_{AAc}^{w} of acetic acid somewhat lower than the maxima reported by Maier (1987).

Figure 6.9 shows simulations done with $C_{AAc}^{w} = 2.5$ g kg^{-1}, using the Henry's law parameters given above, but multiplied by a factor f_{cc} which is varied to fit the data (for the moment ignore the values $f_{cc} > 1$).

The value $f_{cc} = 1$ using the Henry's law parameters as given above, overpredicts the data. The value $f_{cc} = 0.5$ models the data well, and this is consistent with

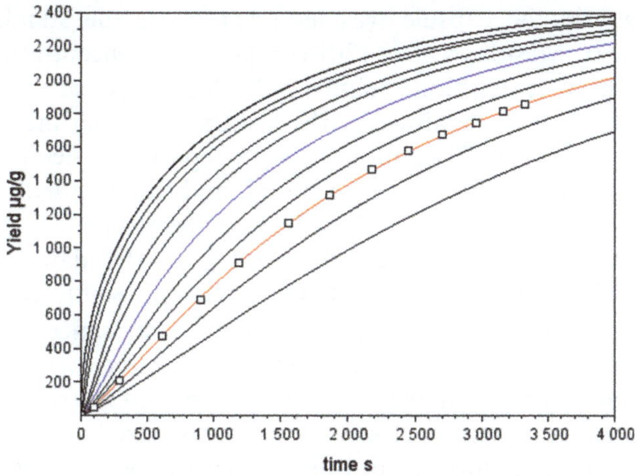

Figure 6.9. Simulations for various K_H^{wg} based on scaling the acetic acid $K_H^{wg} = f_{cc}K_H^{wg}(Ac.\ Ac) = f_{cc}$x0.00027, with $f_{cc} = 200$, 100, 10, 5, 2, 1.5, 1, 0.75, 0.6, 0.5, 0.4 and 0.3 top to bottom (the 100 and 200 profiles lie on top of each other). The blue and red curves are $f_{cc} = 1.\ 0$ and 0.5, respectively, $D_p = 1.0 \times 10^{-11}$ m^2 s^{-1} and 750 µm particles. The squares show data points extracted from figure 6.9 of Mateus *et al* (2007).

figure 6.8 at a mineral content 3.4 mmol g^{-1} pH approx. 4.6, towards the lower end of the expected range. Figure 6.9 also shows how increasing $K_H^{wg} = f_{cc}K_H^{wg}(Ac.\ Ac)$ by increasing the simulated volatility relative that of acetic acid $f_{cc} = 1$, 1.5, 2, 10, 100 and 200 brings the profiles together with increased volatility, the high volatility case is where diffusion internal to the particles is limiting the release rate—it is approached at approx. $fcc = 100$, $K_H^{wg} > 0.03$ (cf. the case of acetaldehyde).

Note, the acetic acid data is in a region where both volatility and diffusion influence the profile. The mechanism is that the low volatility results in enough concentration in the water phase at the surface of the particles to lower the diffusive flux out of the particles. At early times, there is a close to linear increase in yield, the rate of which depends on the volatility, this lasts for longer the lower the volatility.

6.3.5 Oil partitioning interactions

Mestdagh *et al* (2014) report on brewing from a home on-demand capsule system. They measured the release profiles of a range of molecules of varying polarity. This provides a data set against which to compare the model. The full set of parameters in the experiments are not given, so estimates need be made. The brewer was a Nespresso™ Citiz operating at 20 bar. The pump was modified from the commercial brewer to operate at constant pressure, but assuming an ULKA™ -4 pump characteristic typically used in these brewers, this would correspond to a flow of 1–1.5 ml s^{-1}–1.5 ml s^{-1} assumed in the simulations. The capsule was reported to be a Nespresso™ lungo (80 ml) preparation; the grinds in these typically have a volume averaged dimension, $d_{4,3}$, in the range 300–400 µm, the simulations below were done with $d_{co} = 300$ µm and the level of fines was set at $\theta_{fs} = 0.15$ of the grind volume. The oil fraction was set as 10% of the weight, towards the lower end for Arabica beans.

The coffee weight in the capsule was assumed to be the standard 5 g of the Nespresso™ system. Mestdagh *et al* (2014) report the concentration of individual molecules versus brew (extraction) volume. Absolute values are not reported, they just show a few data sets normalised by the last data point at the target volume of 150 ml (144 ml in practice). Three molecules are reported in detail: the hydrophilic 2,3-butanedione, hydrophobic 2-methylbutanal, and the very hydrophobic β-damasceone with log10(P_{ow}) of −1.34, 1.27 and 3.0, respectively.

Figure 6.10(a) shows qualitative agreement with the data as plotted, with $D_p = 3x10^{-10}$ m^2 s^{-1}—the bulk diffusion constants of these molecules at 80 °C are $O(10^{-9})$ m^2 s^{-1}, so using d_{co} = 300 μm, the estimated hindrance is O(10) consistent with the earlier cases. Note, however, that the normalisation procedure of Mestdagh *et al* (2014) obscures trends in the data. This is most clearly seen for the most hydrophobic molecule in figure 6.10(a). In their case, over the experimental time scales, its release profile is linear, but normalising by the yield at the last volume point V_n will collapse all linear profiles onto a slope 1 / V_n! Figure 6.10(b) shows the simulations of figure 6.10(a) plotted against time, but normalised by the maximal yield (unknown) at very long times; for the hydrophobic case log10(P_{ow}) = 3.0, the dependence on D_p, although weak, is as expected: lower yield for smaller D_p. The yields of the very hydrophobic molecules evolve slowly, their concentration in the water phase internal to the particles is maintained low by the high partitioning into the oil phase internal to the particles; although in the absence of other interactions all the molecules will eventually release. To convert the model predictions into absolute units, the maximal yield needs be known either through explicit experiment, e.g. by brewing over a long time, or by direct chemical analysis of the R&G particles. The slow release of the hydrophobic molecules suggests they form part of the slow diffusion mode found in the fitting of total mass by Melrose *et al* (2019b)—if fitted by a pure diffusion model the diffusion constant would be very small.

As discussed in section 6.2.3 the mass transfer rate, T_{ow}, at the oil–water interface is unknown, although estimates were made.

Figure 6.11 shows that the predictions are not sensitive to the transfer rate if $T_{ow} > 101/s$, towards the lower end of the estimates made in section 6.2.3.

Mestdagh *et al* (2014) characterise the dependence on log10(P_{ow}) by plotting the initial rate of increase of yield (normalised by the maximal yield) as $t \to 0$. Due to their normalisation procedure their results are spurious: the slope as $t \to 0$ depends on their normalisation volume point.

6.4 Summary and discussion

The results of section 6.3 give some insight on how the release profiles of molecules in coffee brewing are affected by the internal physics and physical chemistry of the coffee particles; they show that these effects can be estimated and modelled. From the engineering point of view, the multi-scale modelling allows estimates on how flow fields, bed geometry, grind size and contact times can also be used to manipulate the different profiles—Beverly *et al* (2020) do this for the case of steam stripping of aromas in instant coffee production. The model used for water–air

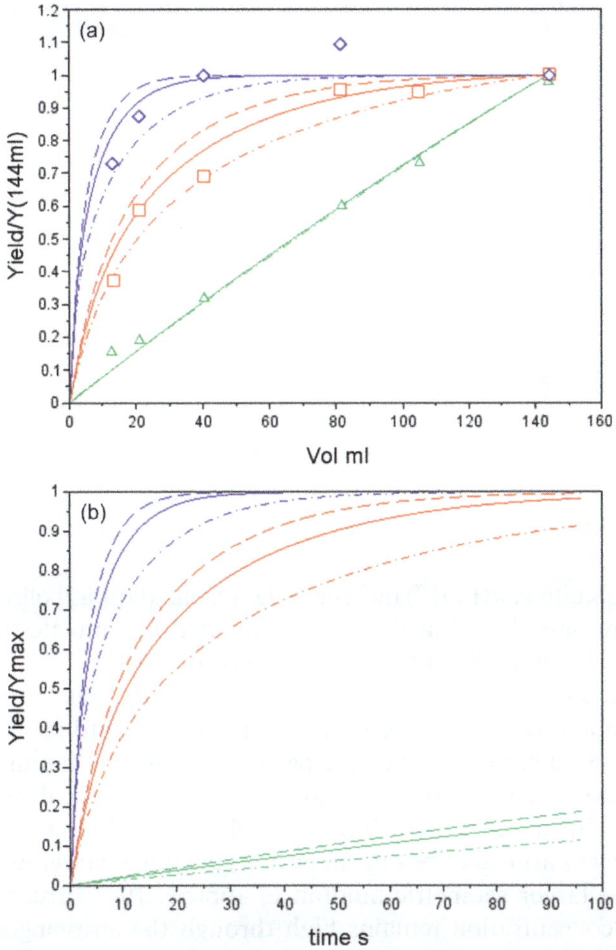

Figure 6.10. Simulations for log10(P_{ow}) = −1.34, 1.27, 3.0 shown in green, red and blue, respectively; for $D_p = 4$, 3 and 210^{-10} m^2 s^{-1} shown by dashed, solid and dash dot curves, respectively. The transfer velocity was fixed at $T_{ow} = 10^2$ s^{-1}. (a) versus vol and normalised as Mestdagh *et al* (2014), the symbols shown are the respective data points for Mestdagh *et al* (2014); (b) against time.

transfer in stripping cases may be directly adapted to model volatiles in head space above coffee brews.

Two cases where hindered diffusion internal to the particles was expected to be the limiting rate determining profiles were modelled above. Hindrance was estimated as a product of geometric and solute hindrance factors. To compute the solute hindrance, which varies over time, a model for the total mass yield was simulated to give a profile of solute concentration inside the particles. In the first case, release profiles of caffeine in espresso brewing, reported by Kuhn *et al* (2017) were modelled well by diffusive release with expected parameters: geometric hindrance approx. 6 for the espresso grinds (see Corrochano 2017) and solute hindrance varying from around 6–8 at the start to 1–2 over 40 s of brewing. The data, however, was also fit well with a constant

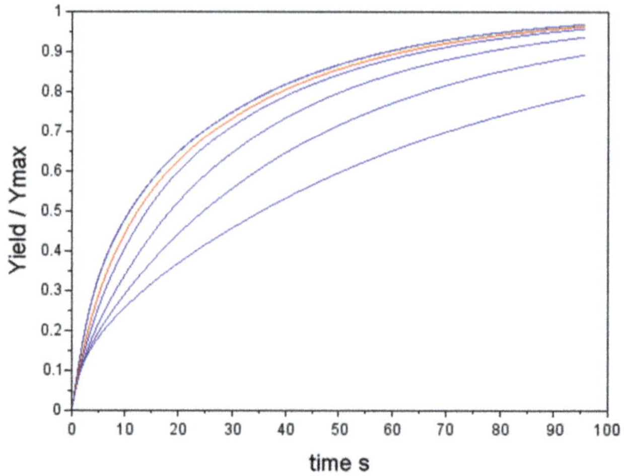

Figure 6.11. Simulations as above; for log10(P_{ow}) = 1.27 but in this case $D_H = 3x10^{-10}$ m^2 s^{-1} and T_{ow}= 0.5, 1, 2, 5, 10, 100 and 200 s^{-1} and bottom to top; the latter two are indistinguishable in the plot.

hindrance of 16. Kuhn *et al* (2017) measured data from aliquots collected at very early times; these were only fit well if the model was used to simulate the concentration profile in the bed resulting from the fluid filling of the bed.

The second case was that of Mateus *et al* (2007) the gas stripping of volatile acetaldehyde from a wet and dry bed. This was also shown to be fit well by a model based on diffusion internal to the coffee particles being rate limiting; it was shown that including the water–gas transfer gave an almost identical result to the pure diffusion model. In this case a significantly higher overall hindrance, approx. 100, was found. This was argued to be due, in part, to the much larger particles which are expected to have larger geometric hindrance, approx. 10, Corrochano (2017), and that the solute concentration remains high through the stripping, give a constant solute hindrance of approx. 7–8. As noted by Mateus *et al* (2007), however, there is some ambiguity as to the relevant diffusion length scale in their beds. This requires further investigation.

A further two cases involving the effect of interactions beyond hindrance were modelled. The first case, acetic acid in the stripping experiments of Mateus *et al* (2007), gave an example where volatility was influencing the profile, but a good fit was only found if estimates were made of the content of the undissociated acetic acid in solution within the particles. This was estimated on the basis that minerals in coffee particles could be equated with a base content neutralising the acids, and the use of an overall acid content model. More details of this work will be published elsewhere. This case was in a mixed regime where both volatility and diffusion were relevant, the hindrance was comparable to that estimated for acetaldehyde. In the second case, the data of Mestdagh *et al* (2014) from espresso brewing was modelled for three molecules ranging from high to low octanol–water partition coefficients, P_{ow}. Due to the limited nature of the original data reporting, only qualitative agreement can be claimed. For the molecules with high P_{ow} the partitioning affects release by forcing a low concentration in the water phase internal to the particle. It was shown that estimates of the oil–water transfer rate were in the regime where this parameter was not rate limiting.

6.5 Outlook

The combination of release profile measurement and multi-scale modelling is posed to give a new level of understanding of the physical chemistry underlying release profiles in coffee brewing and ultimately the time evolution of the composition of coffee brews. This should improve the technology for manipulation of flavour evolution (grinding, flow rates, capsule shape, particle structure…) and quantify insights into the progression of coffee brews from under, balanced through to over extracted flavour. This approach, now building pace, offers a much needed bridge between brewing and the extensive data sets available of flavour molecule development in roasting. There is much work to be done to classify the different release effects. The multi-scale models, such as those of the author and collaborators, Melrose *et al* (2012, 2018a, 2018b, 2019a, 2019b), Beverly *et al* (2020) and that of Moroney *et al* (2015, 2019) which directly connect independently measurable physical-chemical parameters are to be preferred to classification by fitting analytic expressions. The numerical simulation approach is quite feasible with the cheap computing power now available. One key question is to re-examine at a profile level, the relationship between yield and composition. It is clear that the flavour relationship expressed in the control chart of Lockhart (1957), can only be approximate, what degree of variation in brew composition is there between different routes to the same yield? Ultimately, a direct relationship to flavour and sensory panel response will, however, be very challenging due to the need to overlay sensory thresholds, odour activities and sensory interactions—see the review by Sunarharum *et al* (2013).

References

Balzer H H 2001 Ch 1b 2001 *Coffee Recent Developments* ed R J Clark and O G Vitzthum (Oxford: Blackwell)

Bartel C, Mesias M and Morales F J 2015 Investigation on the extractability of melanoidins in portioned espresso coffee *Food Res. Int.* **67** 356–65

Beverly D, Lopez-Quiroga E, Farr R, Melrose J, Henson S, Bakalis S and Fryer P J 2020 Mathematical modelling of mass and heat transfer in multiphase coffee aroma extraction systems *Chem. Eng. Sci.* submitted

Bruna M and Chapman S J 2015 Diffusion in spatially varying porous media *SIAM J. Appl. Math.* **75** 1648–74

Cameron M I, Morisco D, Hofstetter D, Uman E, Wilkinson J N, Kennedy Z C, Fontenot S A, Lee W T, Hendon C H and Foster J M 2020 Espresso: Insights from mathematical modeling and experiment *Matter* **2** 1–18

Carberry J 1960 A boundary-layer model of fluid-particle mass transfer in fixed beds *Am Inst. Chem. Eng. J.* **6** 460–2

Corrochano B, Melrose J R and Bakalis S 2014 Kinetics of coffee extraction and particle microstructure: numerical modelling and experimental validation in slurry experiments; ASIC *Proc. of the 25th International Conf. on Coffee Science (Armenia, Columbia)*

Corrochano B R, Melrose J R, Bentley A C, Fryer P J and Bakalis S 2015 A new methodology to estimate the steady-state permeability of roast and ground coffee in packed beds *J. Food Eng.* **150** 106–16

Corrochano B R 2017 Advancing the engineering understanding of coffee extraction *EngD Thesis* University of Birmingham Available at: http://etheses.bham.ac.uk/7176/ (Accessed: 06 July 2018)

Crank J 1975 *The Mathematics of Diffusion* 2nd edn (Oxford: Clarendon)

Ellero N and Navarini L 2019 Mesoscopic modelling and simulation of espresso coffee extraction *J. Food Eng.* **263** 181–94

Grosch W 2001 Ch 3 table 3.6, 3.11 *Coffee Recent Developments* ed R J Clarke and O G Vitzthum (Oxford: Blackwell)

Jaganyi D and Madlala S P 2000 Kinetics of coffee infusion: a comparative study on the extraction kinetics of mineral ions and caffeine from several types of medium roast coffees *J. Sci. Food Agric.* **80** 85–90

Kuhn M, Lang S, Bezold F, Minceva M and Briesen N 2017 Time resolved extraction of Caffiene and trigonelline from finely-ground espresso coffee with varing particle sizes and tamping pressures *J. Food Eng.* **206** 37–47

Lockhart E E 1957 The soluble solids in beverage coffee as an index to cup quality *Tea and Coffee Trade J.* **113** 12

Lockhart E E 1957 *Coffee and Tea Industries* **80** 16 See discussion in Sivetz M and Foote H E 1963 *Coffee Processing Technology* Westport, Comm.: AVI Pub. Co.

Maier H G 1987 The acids in coffee; ASIC *Proc. of the 12th Int. Conf. on Coffee Science (Montreux, France)* p. 229

Mateus M, Lindinger C, Gumy J-C and Liardon R 2007 Release kinetics of volatile organic compounds from roasted and ground coffee: online measurements by PTR-MS and mathematical modeling *J. Agric. Food Chem.* **55** 10117–28

Melrose J R, Corrochano B, Norton M, Silanes-Kenny J, Fryer P and Bakalis S 2012 Optimising coffee brewing using a multiscale approach; ASIC *Proc. of the 24th In. Conf. on Coffee Science (San Jose, Costa-Rica)*

Melrose J R, Corrochano B, Montoya-Guerra M and Bakalis S 2018a Towards a new brewing control chart for the 21st century *J. Agric Food Chem.* **30** 5301–9

Melrose J R, Andrews R J, Harpley P J and Graaff G K 2018b Capsule, System and use of the system for preparing double beverages like a double espresso *double Lungo and a double ristretto Patent* Application WO 2018/026280 A1

Melrose J R, Corrochano B and Bakalis S 2019a Polydisperse diffusion kinetics of Coffee Brewing – modelling batch data Pre-print www.reserachgate.com *under John Melrose, Coffee Brewing project*

Melrose J R, Corrochano B and Bakalis S 2019b Polydisperse diffusion kinetics of Coffee Brewing—modelling packed beds Pre-print www.reserachgate.com *under John Melrose, Coffee Brewing project* doi: 10.13140/RG.2.2.13478.98880

Mestdagh F, Davidek T, Chaumonteuil M, Folmer B and Blank I 2014 The kinetics of coffee aroma extraction *Food Res. Int.* **63** 271–4

Moroney K M, Lee W T, O'Brien S B G, Suijver F and Marra J 2015 Modelling of coffee extraction during brewing using multiscale methods: An experimentally validated model *Chem. Eng. Sci.* **137** 216–34

Moroney K M, O'Connell K, Meikle-Janney P, O'Brien S B G, Walker G M and Lee W T 2019 Analysing extraction uniformity from porous coffee beds using mathematical modelling and computational fluid dynamics approaches *PLoS ONE* **14** e0219906

Navarini L, Colomban S, Lonzarich V, Rivetti D, Brolloa F and Suggi-Liverani F 2008 Hyper espresso coffee extraction: adding physics to chemistry; ASIC *Proc. of the 22nd Int. Conf. on Coffee Science (Campinas, Brazil)*

Poling B E, Thomson G, Friend D G, Rowley R L and Wilding W 2008 Physical and chemical data *Perry's Chemical Engineers' Handbook* 8th edn ed D Green and R Perry (New York: McGraw-Hill)

Price N E, Trickett K A and Harris K R 1989 Association of caffeine in aqueous solution: effects on caffeine intra-diffusion *J Chem. Soc, Faraday Trans. 1* **85** 3281–88

Sanders R 2015 Compilation of Henry's law constants (version 4.0) for water as solvent *Atmos. Chem. Phys.* **15** 4399–981

Sangster J 1989 Octanol-water partition coefficients of simple organic compounds *J. Phys. Chem. Ref. Dat* **18** 1111–227

Sano Y, Kubota S, Kawarazaki A, Kawamura K, Kashiwai H and Kuwahara F 2019 *J. Food Eng.* **263** 1–12

Schenker S 2000 Investigations on the hot air roasting of coffee beans *Doctorate of Technical Sciences Thesis* (Zurich, Switzerland: Swiss Federal Institute of Technology)

Spiro M and Page C M 1984a The kinetics and mechanism of caffeine infusion from coffee: hydrodynamic aspects *J. Sci. Food Agric.* **35** 925–30

Spiro M and Selwood R M 1984b The kinetics and mechanism of caffeine infusion from coffee: the effect of particle size *J. Sci. Food Agric.* **35** 915–24

Spiro M 1993 Modelling the aqueous extraction of soluble substances from ground and roast coffee *J. Sci. Food Agric.* **61** 371–2

Spiro M, Toumi R and Kandiah M 1989 The kinetics and mechanism of caffine infusion from coffee: the hindrance factor in intra-bean diffusion *J. Sci. Food Agric.* **46** 349–56

Sunarharum W B, Williams D J and Smyth H E 2013 Complexity of coffee flavor: a compositional and sensory perspective *Food Res. Int.* **62** 315–25

Stapley G F 2002 Modelling the kinetics of tea and coffee infusion *J. Sci. Food Agric.* **82** 1661–71

Ueblacker M and Lachenmeier D W 2011 Quantitative determination of acetaldehyde in foods using automated digestion with simulated gastric fluid followed by headspace gas chromatography *J. Autom. Methods Manag. Chem.* 907317

Zanoni B, Pagliarini E and Peri C 1992 Modelling the aqueous extraction of soluble substances from ground roasted coffee *J. Sci. Food Agric.* **58** 273–9

Chapter 7

Crystal engineering approaches for the food industry

Elena Simone

7.1 Introduction

Crystallization is one of the oldest unit operations in the chemical industry and can be carried out in several ways [1], either from melt or from solution (cooling, antisolvent or evaporation). Many food products are crystalline, including sugars (sucrose, lactose, glucose and fructose), ice, lipids and starches. Furthermore, many compounds considered nutraceuticals (e.g. curcumin, quercetin, rutin) present a crystalline nature and a wide landscape of different solid forms [2, 3].

Crystal properties such as shape, size and polymorphism can have a dramatic effect on the properties of food products in terms of stability, taste and texture as well as bioavailability. Therefore, tailoring crystal properties and controlling crystal nucleation and growth during manufacturing is essential to ensure the final product quality. The crystal engineering approach utilises the understanding of the intermolecular interactions within a crystalline structure for the predictive, systematic design of crystals with specifically tailored properties as well as the crystallization processes necessary to achieve such crystals [4].

The first section of this chapter presents a brief overview of the main mechanisms of crystallization that contribute to determine final crystal properties: nucleation, growth and polymorphism. In the second part of the chapter three recent examples of crystal engineering approaches developed by the Food Crystal Engineering Group at the University of Leeds, for the control of food crystallization are presented. Such examples include:

(1) shape control of succinic acid crystals using temperature cycling [5];
(2) optimal design of lactose crystallization processes [6];
(3) synthonic modelling for the prediction of bulk properties of quercetin hydrates [2].

7.2 Thermodynamic and kinetic aspects of crystallization from solution

During crystallization an ensemble of randomly organized molecules, ions or atoms in a fluid come together to create an ordered and regular three-dimensional structure that is called crystal. The crystallization process can be seen as a purification technique, a separation process or a branch of particle technology.

7.2.1 Solutions and solubility

A solution is a homogenous mixture of two or more substances. It can be in a gaseous, liquid or solid state and its constituents are called solvent (the continuous phase) and solute (the dispersed phase). Water is the most used solvent for the industrial crystallization of inorganic substances because it can dissolve many compounds; it is readily available, cheap and innocuous. For organic chemicals in industrial applications, when water cannot be used, solvents are chosen from the following categories: acetic acid and its esters, lower alcohols and ketones, ethers, chlorinated hydrocarbons, benzene homologues and light petroleum fraction [1]. The solubility of a solute represents the maximum concentration that can be dissolved in a specific solvent at any temperature. Because this parameter is usually strongly affected by temperature solubility curves are built, which report the solubility as a function of temperature. Figure 7.1 shows a hypothetical solubility curve: if the concentration is below the solubility curve the solution is under-saturated and existing crystals dissolve [4].

A solution above the solubility curve is called supersaturated, in which crystals can nucleate and grow. However, at low supersaturations, existing crystals grow but the formation of new crystals is difficult. Nucleation usually requires a critical level of supersaturation. The region between equilibrium and the supersaturation level necessary to induce nucleation is called the metastable zone. The width of this zone depends both on kinetics and thermodynamic factors [7]. Supersaturation is,

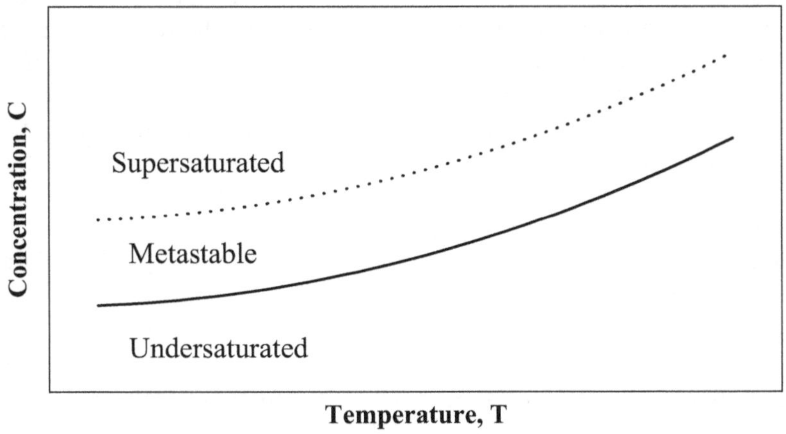

Figure 7.1. Solubility/supersolubility diagram.

therefore, the driving force for nucleation and growth. It is defined as the difference between the chemical potential of the solute molecules in the supersaturated (μ_{ss}) and saturated (μ_{eq}) state, respectively [8].

The expression of supersaturation σ can be written as:

$$\sigma = \frac{\mu_{ss} - \mu_{eq}}{kT}, \tag{7.1}$$

where k is the Boltzmann constant and T is the temperature.

Using the Gibbs–Duhem equation the chemical potentials can be related to the activities, which, for ideal solutions, are independent of concentration:

$$\sigma = ln\left(\frac{a_{ss}}{a_{eq}}\right) \rightarrow ln\left(\frac{x_{ss}}{x_{eq}}\right) \approx \frac{x_{ss} - x_{eq}}{x_{eq}}. \tag{7.2}$$

The letter x refers to the molar fractions of solute in the supersaturated and saturated solution.

In industrial applications the supersaturation is more commonly defined as:

$$\Delta C = C_{ss} - C_{eq}, \tag{7.3}$$

where C is the concentration of solute, ΔC is usually called 'concentration driving force' and is more suitable than the chemical potential for engineering applications [4].

Crystallization from solution can be realized by cooling, solvent evaporation or by antisolvent addition. The aim is the same: creating a supersaturated solution that allows nucleation and/or growth of the crystals. Cooling is the most common technique for generating supersaturation as temperature control is easy and reliable. In antisolvent crystallization a liquid in which the crystallizing compound is poorly soluble is mixed with an undersaturated or saturated solution of the compound itself. This technique can be used for temperature sensitive compounds; however, it is associated with an increase in volume of the total solution and needs to be appropriately controlled. Evaporation is used with substances whose solubility is poorly dependent on temperature. In some applications the solution can be sprayed from the top of a tower in the form of drops that cool below their melting point as they fall and solidify in the form of crystals. Progressive freezing is another possibility: a cooling surface (at a temperature below the melting point of the liquid) is placed in contact with the melt and a layer of solid will form at the surface. This layer will grow while heat is removed at the surface.

Nucleation

Supersaturation is not sufficient for a system to begin to crystallize. Some nuclei or seeds must be present in the solution to act as centres of crystallization. Nucleation can occur spontaneously or it may be induced artificially. It can also be triggered by agitation, mechanical shock, friction and pressure within solutions or melts [1]. Nucleation that happens in systems which do not contain crystals is defined as 'primary'. If the nuclei appear in the vicinity of other crystals in a supersaturated solution the process is then called 'secondary' nucleation. Primary nucleation can be homogeneous if it happens spontaneously in the solution or heterogeneous if it is induced by foreign particles. A schematic summary of the principal types of nucleation is shown in figure 7.2 [4].

There are two theories used to describe homogeneous primary nucleation:

- the classical nucleation theory (CNT);
- the two-step nucleation theory.

According to the CNT, nucleation is a first-order phase transition that requires energy to happen. The thermodynamic explanation of this phenomenon was given by Gibbs in the 19th century. The free energy change for the formation of a stable cluster that can grow is the sum of the free energy required for the phase transformation (ΔG_V), which is proportional to the volume of the nucleus, and the free energy to form a surface (ΔG_S), which is proportional to the surface of the cluster itself. In a supersaturated solution the solid state is more stable than the solution so ΔG_V is a negative term and tends to decrease the barrier energy for nucleation. The ΔG_s instead is associated with the presence of a discontinuity of concentration at the solution–crystal boundary, thus is positive and tends to increase the total ΔG_V as shown in figure 7.3 [9, 10]. In the CNT the activation energy for the nucleation is related to the size of the cluster: a group of molecules becomes stable and can grow when it reaches a critical size. Below that the molecules tend to dissolve again in the solution [7]. Clusters of this critical size are called critical nuclei and the chance of forming nuclei of this size will depend on the height of the free energy barrier. As the supersaturation increases, both the height of the barrier and the value of the critical size decrease. The rate of nucleation for the CNT is defined as the rate at which a cluster grows through this critical size and becomes a crystal. A simple, semi-empirical equation can be derived from an energy balance on a critical nucleus and is reported below [1]:

$$J = A \exp\left[-\frac{16\pi\gamma^3\nu^2}{3k^3T^2(ln\ S)^2}\right]. \tag{7.4}$$

The equation has the typical Arrhenius form, where T is the temperature, S the supersaturation, k the Boltzmann constant, γ is the interfacial tension on the surface of the cluster and ν the molecular volume [4].

The CNT approach was found not to perfectly work for some proteins and a few organic and inorganic molecules, for which the two-step theory seems to work better [11]. According to this theory the formation of a nucleus is preceded by the formation of a stable liquid cluster at higher density than the solution. The high concentration of molecules in this cluster favours the formation of the solid nuclei which then can grow in ordered crystalline structures [12]. Both the theories imply the formation of clusters in the liquid phase where molecules are connected to each other by weak forces (polar, hydrogen bonding, etc). The difference is that for the CNT clusters are metastable, while for the two-step nucleation theory clusters represent a distinct, stable phase. The type of interactions at the cluster level is very important in the determination of the polymorphic form of the nucleated crystals

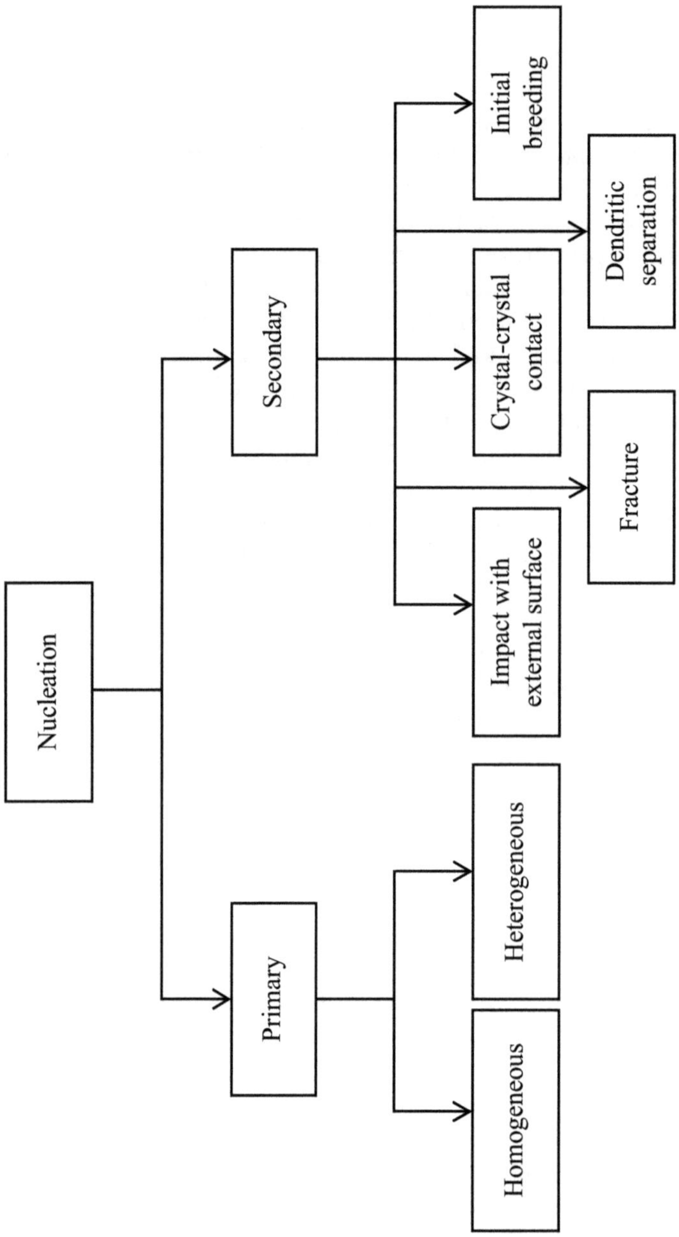

Figure 7.2. Nucleation mechanisms.

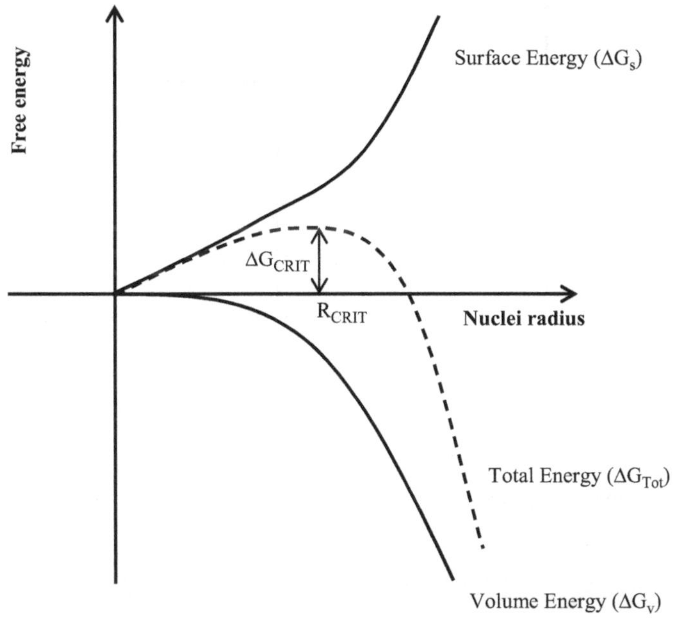

Figure 7.3. Free energy diagram of nucleation.

[11–13]. For this reason, many researchers are currently trying to understand how the nature of the chemical bonding before nucleation affects the polymorphic form nucleated [14–18]. Secondary nucleation happens in the presence of other crystals and it can be affected by many parameters. It can be expressed empirically by a power law function:

$$B = k_B M_T^j N^k \, \Delta C^b; \tag{7.5}$$

where B is the secondary nucleation rate, N is a measure of the fluid mechanics interactions and of the power inputs (a stirrer or a pump impeller), ΔC is the supersaturation and M_T is the concentration of crystals in the solution. k_B, j, k and b can be predicted by semi-empirical models of contact nucleation in crystallizers [4, 7].

7.2.2 Crystal growth

The growth of a crystal is a combination of two stages in series: diffusion and reaction. The molecules must diffuse from the bulk to the surface of the nuclei where they can attach. Therefore, two resistances in series are present: diffusion ($1/k_d$) and particle integration on the surface ($1/k_r$). The expressions of the mass flux to the crystal surface for the two mechanisms are reported below [1]:

$$\frac{dm}{dt} = k_d A(c - c_i), \tag{7.6}$$

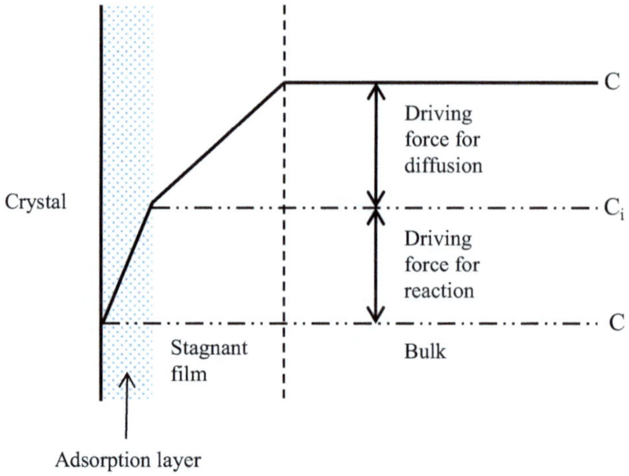

Figure 7.4. Concentration driving forces in crystallization from solution according to the simple diffusion–reaction model [1].

for diffusion,

$$\frac{dm}{dt} = k_r A(c_i - c*), \tag{7.7}$$

for reaction.

Figure 7.4 is a schematic picture of the diffusion–reaction model. It is possible to identify in the figure: (i) the crystal surface, (ii) the absorption layer where the solids in suspension come in contact with the crystal (iii) the stagnant film where the diffusion takes place and (iv) the bulk at constant concentration.

Because of the difficulty in measuring the interface concentration C_i the two equations are normally combined together as follows:

$$\frac{dm}{dt} = K_G A(c - c*)^g, \tag{7.8}$$

where K_G is the overall crystal growth coefficient and g is the order of the overall crystal growth process. If $g = 1$ then:

$$\frac{1}{K_G} = \frac{1}{k_r} + \frac{1}{k_d}. \tag{7.9}$$

A system can have a diffusion controlled growth kinetics if $k_r \gg k_d$ or an integration controlled growth kinetic if $k_d \gg k_r$.

7.2.3 Polymorphism

The capacity of a substance to exist in two or more different crystalline structures is called polymorphism. The first definition of polymorphic form was given by Mitscherlinch in 1821, in relation to arsenates, phosphates and sulphur while, in 1832, Wöhler and Liebig discovered the first example of polymorphism in an organic substance. In 1899, Ostwald concluded that almost every substance could exist in two or more solid phases [7]. Polymorphism can be considered the rule rather than an exception, as also stated by Walter McCrone who, in 1965, wrote: 'Every compound has different polymorphic forms, and, in general, the number of forms known for a given compound is proportional to the time and money spent in research on that compound' [19]. The nucleation of a certain polymorph and the transition between one form to another is an important issue in industry because different polymorphs have different chemical and physical properties. A polymorphic transformation occurs in response to changes in pressure, temperature, composition, pH, etc; therefore, it can happen in several stages of tablet manufacturing, from the crystallization stage, to drying, milling granulation or storage. There are four possible polymorphic transformations: solid state, solution mediated, melt mediated and interface mediated. From a kinetic point of view the solid state transformation occurs much slower than the other three since it requires larger activation energy [4, 20].

The solution mediated transformation is among the most interesting because the solvent can have a strong influence both on the thermodynamic and the kinetic parameters. Besides, crystallization from solutions is widely used in pharmaceutical, food and agrochemical industries and the interest in understanding and controlling this mechanism is very high. For all these reasons, in this section of the literature review, a brief description of solvent-mediated polymorphic transformation is presented.

In a dimorphic system constituted by form I and II, at specific temperature and pressure, the more stable form (II for example) has the lower free energy:

$$G_{II} < G_I, \tag{7.10}$$

which, in terms of chemical potentials can be written as:

$$\mu_{II,\mathrm{solid}} < \mu_{I,\mathrm{solid}}. \tag{7.11}$$

At equilibrium the solution and the solids have the same chemical potential, thus:

$$\mu_{I,\mathrm{solid}} = \mu_{I,\mathrm{solution}} = \mu_0 + RT\ ln\ a_{I,\mathrm{solution}}, \tag{7.12}$$

$$\mu_{II,\mathrm{solid}} = \mu_{II,\mathrm{solution}} = \mu_0 + RT\ ln\ a_{II,\mathrm{solution}}, \tag{7.13}$$

where μ_0 is the standard chemical potential and $a_{I,\mathrm{solution}}$ and $a_{II,\mathrm{solution}}$ the solution activities. Combining the last three equations it can be inferred:

$$a_{II} < a_I. \tag{7.14}$$

Because the activities are proportional to the equilibrium concentration of the polymorph in solution, it can be written [8]:

$$C_{II} < C_I. \tag{7.15}$$

In conclusion, the most stable form will always be the more soluble in every solvent at given temperature and pressure. There are two possible situations for a dimorphic system:

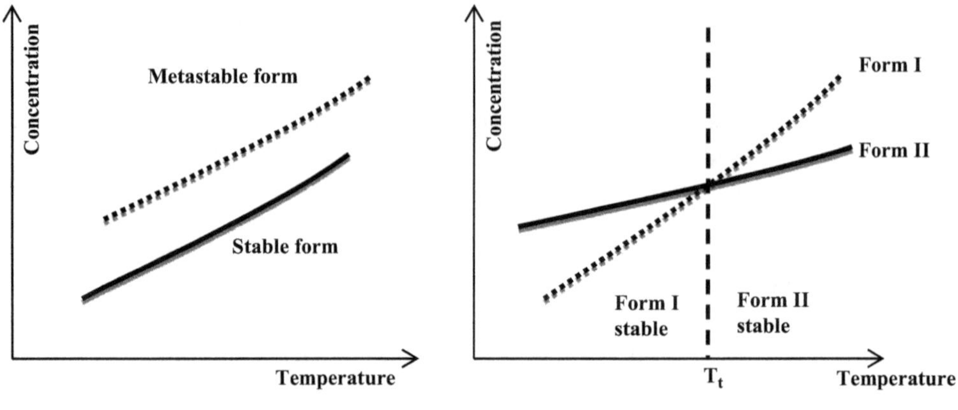

(a) (b)

Figure 7.5. Monotropic system (a) and enantiotropic (b).

- the system is considered enantiotropic if the solubility curves cross each other at a lower temperature (called transition temperature) than the melting points of both form I and II. See figure 7.5(a);
- the system is considered monotropic if the solubility curves do not cross each other in solution. See figure 7.5(b).

The less stable form is called 'metastable' and it can nucleate for kinetic reasons but it will always tend to transform into the thermodynamically stable polymorph. The classical nucleation theory is usually good enough to describe the nucleation of many organic molecules, including their polymorphs, remembering that the competitive nucleation between them can generate a very complex system. Solvent may prove to have a strong effect on the nucleation of one polymorph or another because of the kind of bonds that can form in the clusters [13, 21]. Also, the level of supersaturation affects the polymorph nucleated: generally, high supersaturation favours the nucleation of the metastable form. Additives can also direct the nucleation towards a specific polymorph rather than others [4, 22].

The polymorphic transformation from a metastable to a stable structure is the result of a combined effect of competitive nucleation and crystal growth of the two different polymorphs as shown in the schematic of figure 7.6 [23].

During the polymorphic transformation the metastable form dissolves in the solution while the stable form nucleates. The difference in solubility is the driving force of the process but the rate of transformation depends on many different factors such as temperature, solvent, crystal size, stirring rate, impurities etc [23]. The solute concentration changes in time according to the relative speed of the processes of dissolution of the metastable form and nucleation and growth of the stable one [24]. If the dissolution of the metastable form is considerably slower than growth and nucleation of the stable one the process is defined dissolution-controlled. Conversely, if growth of the stable form is the limiting factor the transformation is called growth-controlled. Figure 7.7 shows the trend of supersaturation for the two limiting and an

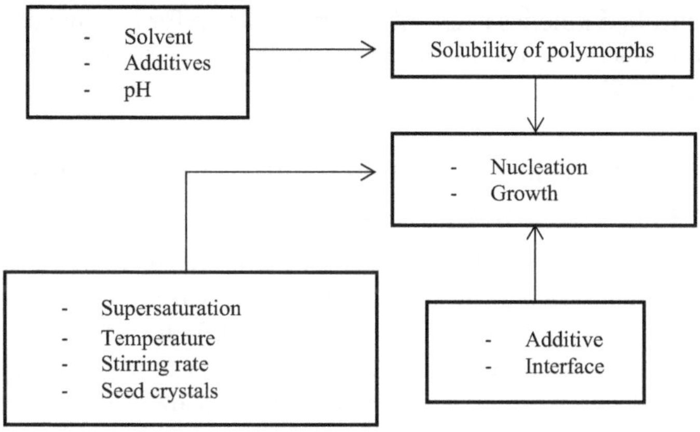

Figure 7.6. Schematic of controlling factors in crystallization of polymorphs.

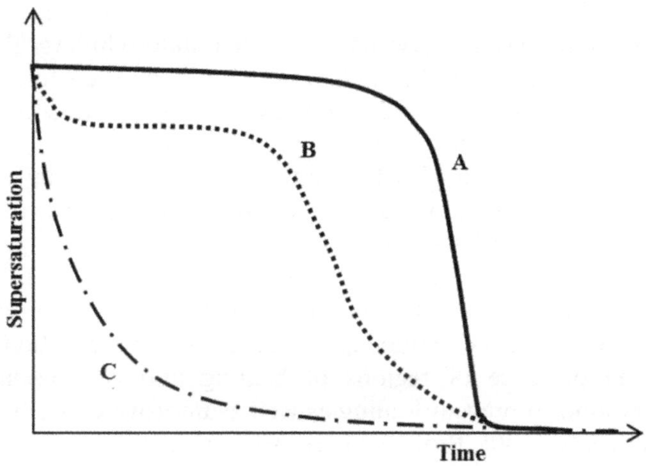

Figure 7.7. Supersaturation profile in the case (A) growth-limited, (B) intermediate case, and (C) dissolution-limited.

intermediate case. If nucleation and growth of the stable form are significantly slower than the dissolution of the metastable form, the supersaturation remains constant for a long time since the dissolution process is able to sustain the growth one. When all the metastable form is dissolved the supersaturation rapidly drops down and the solubility of the solution reaches that of the stable form. In the opposite case, if dissolution is slower, the supersaturation falls rapidly from the initial value to a very low one (solubility close to the equilibrium of the stable polymorph) because of the nucleation and growth of the stable form and then slowly goes to zero as dissolution of the metastable polymorph proceeds. If there is not a controlling factor the supersaturation profile is intermediate between the two limiting cases [24, 25].

Growth controlled kinetics is usually more common for organic compounds than dissolution controlled: piracetam, carbamazepine and glutamic acid present this type of kinetic [26–29].

7.3 Temperature cycling for shape control of succinic acid in batch cooling crystallization processes

Crystal shape has a profound effect on the properties of the final solid-form product as well as on the efficiency of downstream operations. Needles and plate-like crystals are usually considered undesirable since they are difficult to filter and suspend in solution [4]. Including additives during the crystallisation process can be effective in modifying the shape of the final crystals and reducing their aspect ratio, producing improved crystal shapes that enable more effective operation of downstream processes such as filtration [30, 31]. For example, changing the morphology of fat crystals from needle to sphere underlies the industrially important dry fractionation process for the refining of vegetable oils. (*Patent Van Dam, Petrus Henricus J. (Rotterdam, NL), Hogervorst, Wim (Spijkenisse, NL), Kamp, Frans (Spijkenisse, NL), 1999, Fractionation of triglyceride oils United States Unilever Patent Holdings B.V. (Vlaardingen, NL) 5959129). Additives can also be used to enable access to new or elusive polymorphic forms [22, 32]. However, the introduction of additives in crystallizer can compromise the purity of the final product.

Crystal size and shape can also be manipulated through temperature cycling: heating and cooling cycles allow dissolution of fines and growth of the larger crystals; temperature cycling was also found to affect the surface properties of the resulting crystals as well as the degree of solvent inclusion [33–35]. This procedure was applied to both batch and continuous crystallization processes in mixed suspension mixed product removal (MSMPR) configuration [5, 36–38]. Furthermore, the presence of regions of heating and cooling in a plug flow crystallizer was found to prevent fouling as well as improve crystal size distribution of the final product [39, 40]. Prolonged temperature cycling can also have a large impact on the shape of the original crystals, since different faces of the same crystal can have different relative growth and dissolution rates. Consecutive cycles of growth and dissolution can generate shapes that would be difficult or impossible to obtain by growth alone (e.g. linear, natural or programmed cooling) [41, 42]. The effect of temperature cycling on the shape of a single plate-like crystal was modelled and validated experimentally by Lovette *et al* (2012), while Jiang *et al* (2014) and Eisenschmidt *et al* (2015) both estimated the growth and dissolution kinetics using a multi-dimensional population balance model [35, 42–45]. In the first two studies a converging trend of the aspect ratio of crystals was experimentally observed during cycling for a single crystal and a population of crystals in a stirred vessel. However, experiments were not continued for long enough to observe a cycling equilibrium. In particular, Jiang *et al* (2014) used focused beam reflectance measurement (FBRM), infrared spectroscopy (ATR-FTIR) and a particle vision and measurement (PVM) probe to determine the kinetic parameters during a single experiment in which crystals of monosodium glutamate were subjected to a deep but short temperature cycling (only five cycles). A considerable change in shape of the crystal was observed together with changes in the FBRM statistics but the experiment was stopped before any of them could reach a stable value. Eisenschmidt *et al* (2016) also optimized the number and amplitude of cycles during the crystallization process in order to achieve the desired shape of potassium dihydrogen phosphate (KDP) crystals. Despite the simplicity of the model, which includes only dissolution and growth, and the imaging technique used for the estimation of the kinetic parameters the study clearly shows how cycles of growth and dissolution can effectively be used to obtain morphologies that are not attainable by growth only.

More recently, crystals of **succinic acid**, a common ingredient for beverages, have been subjected to different temperature cycles until the oscillating trends of shape and size distributions of the crystals did not change significantly over a relatively long period of time [5]. Such a condition was determined using FBRM and PVM statistics as well as the Raman signal. This piece of work represents a first attempt to describe a complex crystallization process that includes: (1) primary and secondary nucleation, (2) complete dissolution of fine crystals and Ostwald ripening (3) partial, face-specific, dissolution of large crystals (4) face-specific crystal growth. Furthermore, the existence of practical steady-state size and shape of succinic acid crystals reached through temperature cycling is established, at the 350 ml scale. Figures and tables in this section are reprinted with permission from Simone *et al* (2017). Copyright (2019) American Chemical Society [5].

Figure 7.8 shows the trends from PVM, FBRM and Raman spectroscopy during one of the experiments performed. The total counts/s recorded from FBRM shows an oscillating trend indicating the presence of secondary nucleation during the cooling steps of the cycles for the duration of the experiment. However, secondary

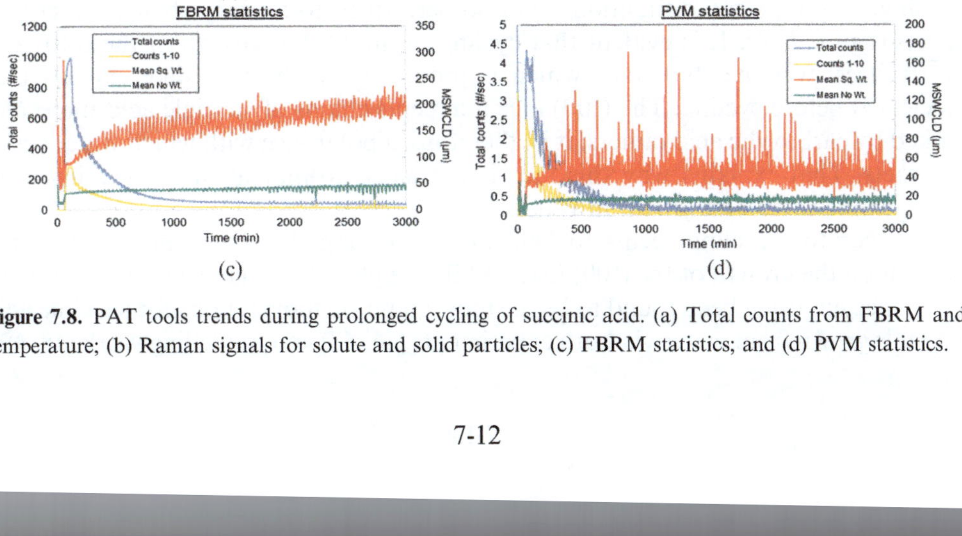

Figure 7.8. PAT tools trends during prolonged cycling of succinic acid. (a) Total counts from FBRM and temperature; (b) Raman signals for solute and solid particles; (c) FBRM statistics; and (d) PVM statistics.

nucleation is counter-balanced by dissolution, growth and Ostwald ripening which contribute to an overall decrease in the total counts to a practical steady-state value.

The smaller particles generated by both primary and secondary nucleation are preferentially dissolved during the heating phases allowing further growth of the larger ones during cooling. Primary nucleation generates a large amount of fines that are dissolved and incorporated into the largest crystals during the first cycles. This generates the rapid decrease in the number of crystals in solution and a quick increase in the mean crystal size [4]. Towards the end of the experiments almost all the fines produced by the initial primary nucleation have been dissolved and incorporated in the larger crystals; the following heating steps primarily dissolve fines produced by secondary nucleation. Dissolution, growth and secondary nucleation are the main phenomena in the presented static cycling experiments while Ostwald ripening is believed to play only a secondary role. These considerations are confirmed by the trends for both PVM and FBRM main statistics, shown in figure 7.8(c) and (d). Total counts/s and fines counts (1–10 μm chord length) are shown together with the mean of the chord length distribution (CLD) and the mean of the square weighted chord length distribution (MSWCLD). While the counts of fines and total particles is decreasing the means of the CLD and the SWCLD are increasing as a result of the growth of the succinic acid crystals.

Figure 7.8(b) shows the trend of two Raman peaks, one for the solute (840 cm^{-1}) and one for the solid particles (937 cm^{-1}). Second derivative and smoothing were applied to all Raman spectra for a better peak identification. While the intensity of the solute peak oscillates between the same minimum and maximum values, the peak corresponding to the solids decreases over time even though it continues to oscillate. During the cycles the number of particles decreases while their size increases as a result of the dissolution of the fine particles. In terms of shape, crystals moved from irregular plates to a tri-dimensional diamond shape. This change indicates that the heating steps contribute to both the complete dissolution of fine crystals produced by primary and secondary nucleation as well as partial dissolution of the larger particles, which allows the observed morphology evolution. Microscopic images of the crystals extracted during the cycling experiments are shown in figure 7.9: the shape is initially plate-like but it gradually converts to diamond-like. The diamond crystals are easier to filter and present a higher uniformity in shape compared to the particles obtained before cycling.

Single crystal x-ray diffraction was also performed to identify changes in crystal facets (figure 7.10). It is evident that cycling promotes the growth of the (110) and ($1\bar{1}0$) faces over the (100) face which is prominent in the plate-like crystals that nucleated before cycling. The (100) face intercepts chains of succinic acid molecules linked by carboxylic acid dimers; it is, therefore, a polar face with which water easily interacts. That explains why this face is normally prominent in the succinic acid crystals nucleated and grown in water.

Temperature cycling seems to help in overcoming the inhibiting effect of the solvent on the growth of the (100) face. While the plate-like shape with predominant (100) face has often been found to be the most common morphology for succinic acid crystals grown from water via both experiments [46, 47] and modelling work [48, 49],

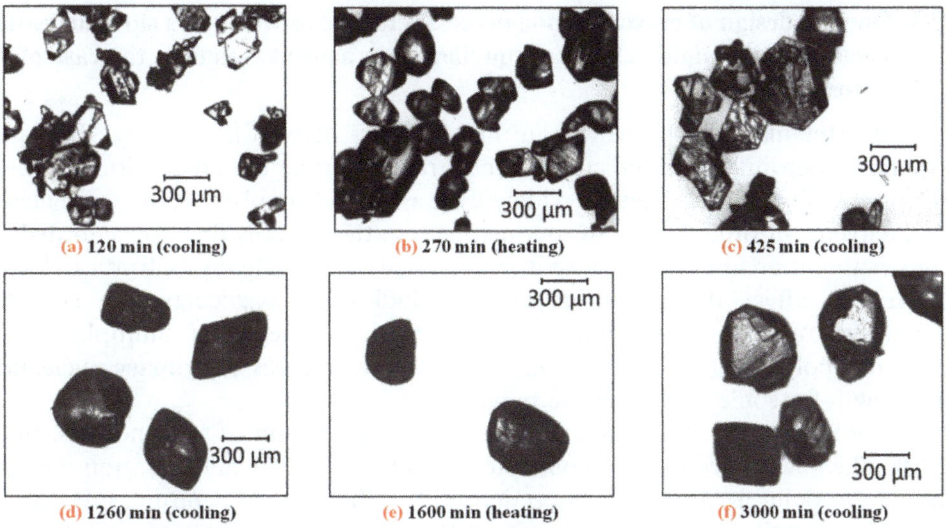

Figure 7.9. Microscopic images of crystals collected at different stages of the cycling experiments. The crystal shape distribution changes from irregular plate-like to uniform diamond shaped.

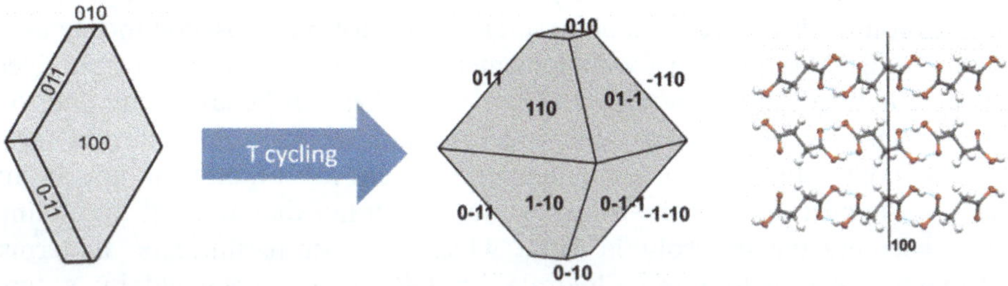

Figure 7.10. Evolution of facets during temperature cycling for succinic acid, from x-ray single crystal indexing.

diamond shaped crystals of succinic acid obtained from water have never been observed before. Three different cycling amplitudes (4.5, 6 and 7.5 °C) were tested in the study, as well as different heating and cooling rates (±0.1, ±0.3 and ±0.5 °C min^{-1}). All crystals have the same diamond shape at the end of the experiments despite the different operating conditions. The time required to obtain a stable oscillating trend of the total counts/s seems to increase with decreasing the amplitude of the cycles; larger cycles of heating and cooling allow a faster dissolution of the small particles and a quicker adjustment to the steady-state shape. Faster heating/cooling rates enable reaching a steady-state oscillating trend in shorter time but the final shape distribution was more irregular, with diamond shaped crystals co-existing with plate-like ones. Most likely, the fast cycling does not allow enough time for the rearrangement of the shape (especially for larger particles) resulting in inhomogeneity within the crystals population.

7.3.1 Optimal design of crystallization processes for the recovery of a slow nucleating sugar with a complex chemical equilibrium in aqueous solution: the case of lactose

Lactose is the major carbohydrate in milk and it is derived from whey, the liquid remaining after milk has been curdled and strained during the production of cheese and yogurt [50–52]. This sugar is widely used as a food supplement or pharmaceutical excipient. In order to obtain purified lactose the whey is first concentrated by water evaporation and then lactose is precipitated via cooling crystallization. Many factors can affect the latter operation including, in particular, the complex equilibrium of lactose molecules in solution, the existence of multiple crystal structures (polymorphs) for this sugar and the slow kinetics of primary nucleation and growth for some of the lactose polymorphs [53].

The molecules of lactose can exist in solution in the form of two anomers, the α and β. The change from one anomer to the other is called mutarotation and the equilibrium point for this conversion depends on the pH, temperature, presence of impurities as well as solvent composition [6]. Both anomers can precipitate from solution in different crystalline structures as shown in figure 7.11. Figures and tables in this section are reprinted with permission from Simone *et al* (2019). Copyright (2019) American Chemical Society [6].

The lactose crystals normally precipitated from water at ambient temperature are α-lactose monohydrate. Because of its stability, this form is most commonly used as a pharmaceutical excipient or food supplement. β-lactose crystals can be obtained in water at a temperature above 93 °C. This crystallization behaviour is due to the different solubility of β-lactose and α-lactose monohydrate in water. In fact, β-lactose is more soluble at ambient temperature but its solubility is not as much affected by temperature as the α-lactose monohydrate; therefore, at high temperature it becomes the less soluble form, which is easier to nucleate. In terms of morphology the α-lactose monohydrate crystals are characterized by a typical tomahawk shape, while β-lactose crystals are normally needle-like with high aspect ratio. Lactose can exist also as anhydrous α-lactose crystals obtained via dehydration of the α-lactose monohydrate form or in amorphous form.

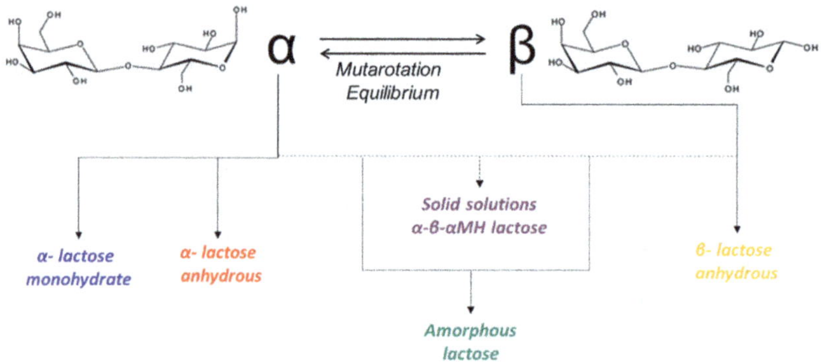

Figure 7.11. Mutarotation equilibrium and possible crystal structures of lactose.

Because of the mutarotation equilibrium both types of anomers are present during the nucleation and growth of a specific crystal structure. This can affect the purity of the precipitated crystals; in fact, even the commercially available α-lactose monohydrate normally contains 1%–5% w/w of β-lactose incorporated in the crystal structure as a solid solution. Despite slow kinetics, the mutarotation equilibrium is not a rate-limiting factor for the growth and nucleation of lactose crystals. However, the β-lactose molecules present in solution can act as both nucleation and growth inhibitors for the α-lactose monohydrate crystals. This determines the very slow kinetics of primary and secondary nucleation as well as growth for this specific crystal structure, although the exact mechanism is still not clear. For all these reasons, it is very difficult and time-consuming for dairy industries to achieve a high yield of recovery and to obtain large crystals with a narrow size distribution and consistent purity and crystal structure.

In recent work by the Crystal Engineering Group at Leeds [6], both seeded cooling and anti-solvent experiments at 400 mL scale were conducted to determine the effect of process parameters on the characteristic properties of lactose crystals, including: morphology, size distribution, level of agglomeration, crystal structure and purity. Furthermore, the effect of such conditions on the overall recovery yield was evaluated. This work is the most comprehensive study conducted on lactose crystallization, looking at all the crystal parameters that can affect product quality and using two different, highly controlled crystallization techniques.

The experimental conditions used to perform the experiments and the final crystal properties and process yield are shown in table 7.1 (seeded experiments in water, with α-lactose monohydrate crystals used as seeds) and table 7.2 (anti-solvent experiments using acetone addition to generate supersaturation).

For seeded experiments only α-lactose monohydrate crystals were obtained at the end of the batch with a volume mean diameter (VMD), estimated via laser diffraction, within 80 and 150 μm. The recovery yield was always below 60%

Table 7.1. Experimental conditions and recovery yield after 23 h for the seeded cooling crystallization experiments.

Run	Initial lactose concentration in solution (g lactose/100 g water)	Stirring rate (rpm)	Seed (% w/w)	Cooling rate (°C min^{-1})	Recovery yield	VMD (μm)
Seed crystals	/	/	/	/	/	71
1	45	400	0.1	−0.25	28%	130
2	45	400	1	−0.25	44%	83
3	45	400	2	−0.25	43%	87
4	45	400	1	−0.15	42%	82
5	55	400	1	−0.25	55%	78
6	45	500	0.1	−0.25	30%	149
7	45	600	0.1	−0.25	24%	110

Table 7.2. Experimental conditions and recovery yield after 2 h for the anti-solvent experiments.

Run	Initial lactose concentration in solution (g lactose/100 g water)	Mass ratio of acetone in the final solvent (w/w)	Flow rate of acetone (ml min^{-1})	Stirring rate (rpm)	Temperature (°C)	Seed (% w/w)	Seeding temperature (°C)	Recovery yield	VMD (μm)	Crystal structure precipitated from WAXS and ATR-FTIR
8	15	60%	/	400	15	/	/	59%	232	β-lactose + α-lactose monohydrate
9	20	60%	/	400	15	/	/	76%	248	β-lactose + α-lactose monohydrate
10	25	60%	/	400	15	/	/	82%	233	β-lactose + α-lactose monohydrate
11	30	60%	/	400	15	/	/	85%	322	β-lactose + α-lactose monohydrate
12	15	50%	/	400	15	/	/	2%	n.a.	α-lactose monohydrate
13	15	70%	/	400	15	/	/	82%	126	β-lactose + α-lactose monohydrate
14	15	80%	/	400	15	/	/	81%	37	β-lactose + α-lactose monohydrate
15	15	60%	/	500	15	/	/	55%	117	β-lactose + α-lactose monohydrate
16	15	60%	/	600	15	/	/	68%	101	β-lactose + α-lactose monohydrate
17	15	60%	/	700	15	/	/	24%	39	α-lactose monohydrate
18	15	60%	/	400	25	/	/	48%	141	α-lactose monohydrate
19	15	60%	/	400	35	/	/	33%	182	α-lactose monohydrate
20	25	60%	5	400	6	/	/	90%	102	β-lactose + α-lactose monohydrate

(Continued)

Table 7.2. (*Continued*)

Run	Initial lactose concentration in solution (g lactose/100 g water)	Mass ratio of acetone in the final solvent (w/w)	Flow rate of acetone (ml min^{-1})	Stirring rate (rpm)	Temperature (°C)	Seed (% w/w)	Seeding temperature (°C)	Recovery yield	VMD (μm)	Crystal structure precipitated from WAXS and ATR-FTIR
21	25	60%	10	400	6	/	/	83%	170	β-lactose + α-lactose monohydrate
22	25	50%	5	400	6	/	/	75%	223	β-lactose + α-lactose monohydrate
23	25	50%	10	400	6	/	/	76%	216	β-lactose + α-lactose monohydrate
24	25	60%	5	400	15	/	/	81%	333	β-lactose + α-lactose monohydrate
25	25	60%	10	400	15	/	/	82%	256	β-lactose + α-lactose monohydrate
26	25	60%	5	400	6	1	27	84%	135	β-lactose + α-lactose monohydrate
27	25	60%	5	400	6	1	6	80%	130	β-lactose + α-lactose monohydrate

despite the length of the experiment (around 23 h). This is due to the slow growth of the α-lactose monohydrate crystals. Higher initial lactose concentrations and amount of seeds led to higher recovery yields, most likely associated with higher rate of secondary nucleation since the VMD value decreases with the increase in seed loading and initial lactose concentration. Stirring rate was found to have negligible effect on process yield and crystal size distribution. Unimodal size distributions could be obtained with low levels of crystal agglomeration using seeded experiments.

Anti-solvent experiments produced higher recovery yields (up to 90%) but with lower crystal purity and multimodal final crystal size distribution. Mixtures of α-lactose monohydrate (tomahawk shape) and β-lactose crystals (needle-like) were obtained in most experiments. This was confirmed by off-line wide angle x-ray scattering (WAXS) and infrared spectroscopy (ATR-FTIR). The rate of nucleation and growth in mixture of water and anti-solvent was considerably higher than in pure water, but the crystals obtained at the end of each batch presented a higher level of agglomeration, as shown in figure 7.12. Similarly to the seeded experiments, increasing the solution concentration generated a higher recovery yield of the lactose crystals both in α-lactose monohydrate and β-lactose forms. The final ratio of solvent to anti-solvent had the strongest effect on the final recovery yield.

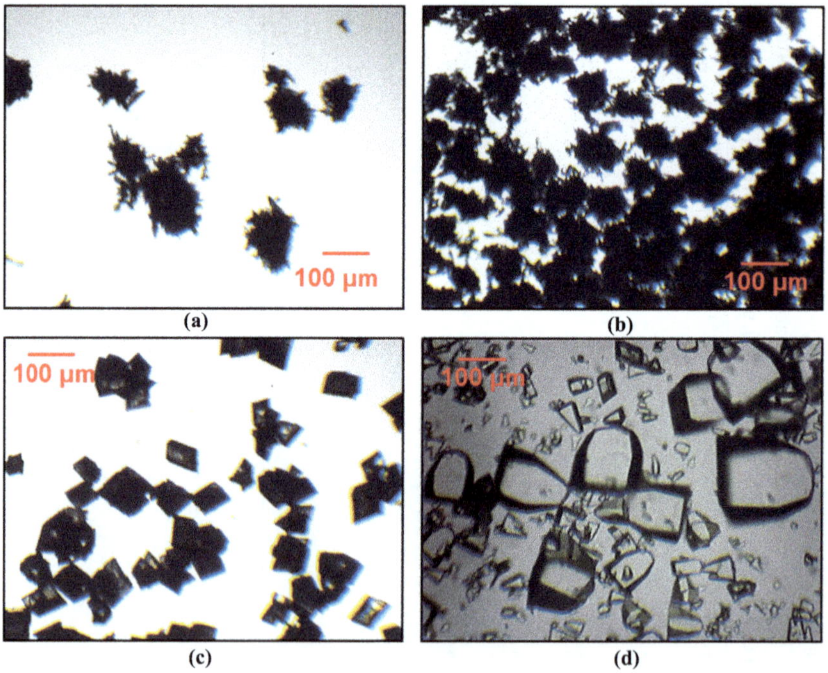

Figure 7.12. (a) Crystals of β-lactose (agglomerates) obtained via anti-solvent addition with final concentration of acetone in the solvent of 60% w/w; (b) crystals of β-lactose (agglomerates) obtained via anti-solvent addition at stirring rate of 600 rpm; (c) crystals of α-lactose obtained via anti-solvent addition with final concentration of acetone in the solvent of 50% w/w; (d) crystals of α-lactose obtained via seeded cooling crystallization of a solution at 45% w/w initial concentration of lactose.

The purity of the crystals obtained in both seeded and anti-solvent experiments was measured with proton NMR [54] and it is shown in table 7.3.

Most samples obtained from anti-solvent crystallization experiments (runs 8–28) contained a high number of β-lactose molecules. This indicates a general difficulty in controlling the outcome of anti-solvent crystallization experiments, probably because lactose crystals are produced via primary nucleation which is notoriously difficult to predict for polymorphic systems in particular. It is worth noticing that β-lactose crystals were always simultaneously nucleated in a physical mixture with α-lactose monohydrate particles, as identified by off-line WAXS and ATR-FTIR. However, these solid samples presenting both β and α-monohydrate crystals were

Table 7.3. Percentages of α and β molecules in the solid samples, quantified with proton NMR.

Run	% α molecules (w/w)	% β molecules (w/w)
Seeds	98%	2%
1	97%	3%
2	98%	2%
3	100%	0%
4	98%	2%
5	100%	0%
6	99%	1%
7	99%	1%
8	67%	33%
9	47%	53%
10	54%	46%
11	57%	43%
12	94%	6%
13	88%	12%
14	44%	56%
15	84%	16%
16	57%	43%
17	96%	4%
18	94%	6%
19	96%	4%
20	62%	38%
21	56%	44%
22	53%	47%
23	62%	38%
24	64%	36%
25	63%	37%
26	58%	42%
27	53%	47%
28	53%	47%

fully dissolved before the NMR analysis. After dissolution it is not possible to distinguish between β molecules contained in the β-lactose solid crystals and the ones incorporated as impurity in the α-lactose monohydrate structure and vice versa. However, since β molecules were found to be incorporated in the pure monohydrate crystals during the seeded experiments, it is reasonable to assume that both solid structures can incorporate molecules of the other anomers as impurity.

7.4 Synthonic modelling of quercetin and its hydrates: explaining crystallization behaviour in terms of molecular conformation and crystal packing

Hydrates are multicomponent crystalline solids that contain both the host molecule and one or more water molecules incorporated in the crystal lattice; approximately a third of organic compounds can form hydrated structures, including many food and nutraceutical molecules. Understanding organic molecules propensity to form hydrates and mapping their thermodynamic stability is of critical importance when formulating particulate products, particularly for the pharmaceutical, food and agrochemical industries [2, 55, 56].

Synthonic modelling and computational methods utilize atomistic force fields drawn from empirical data, to calculate the strength, directionality and dispersive nature of the intermolecular interactions (synthons) within a crystalline structure. This information can help in predicting the physiochemical properties of crystals [57, 58]. Recently the structure and conformation of quercetin molecules in three crystal forms with different levels of hydration have been studied at Leeds. The effect of water on the molecular geometry of quercetin in the hydrated structures has been discussed [2]. Figures in this section are reprinted with permission from Klitou *et al* (2019). Copyright (2019) American Chemical Society [2].

Quercetin, 2-(3,4-Dihydroxyphenyl)-3,5,7-trihydroxy-4H-chromen-4-one, was chosen as the model compound due to the recent interest that the molecule has received from the food and pharmaceutical sectors [59, 60]. Quercetin is a polyphenolic compound found in many fruits and vegetables, with antioxidant, anti-inflammatory, anti-bacterial and anti-hypertensive properties as well as the ability to inhibit the growth of human cancer cell lines. This molecule can exist as anhydrous (QA), monohydrate (QMH) and dihydrate (QDH) forms. In the presented work, the effect of water molecules on the structure, packing energetics and conformation of QA, QMH and QDH was investigated.

The main bulk intrinsic intermolecular synthons in the three structures were computed using HABIT98, a software developed in-house at the University of Leeds (Professor Kevin Roberts' group) [61]. The three strongest intermolecular synthons in each structure, those having the highest energy in kcal mol^{-1}, were calculated and are summarised in table 7.4 and illustrated in figures 7.13, 7.14, and 7.15. In QA, the three strongest interactions in the lattice are found to be mainly hydrogen bonds between quercetin molecules, whereby the QA2 forms an unbroken chain of quercetin molecules running along the a-direction of the lattice. In QA one quercetin molecule was found to form hydrogen bonds with six other quercetin molecules. The

Table 7.4. Summary of intermolecular synthons in QA, QMH and QDH structures.

Quercetin structure	Synthon name	Molecules involved	Main synthon type	Intermolecular distance (Å)	Synthon energy (Kcal mol^{-1})	% contribution to lattice energy
QA	QA1	Quercetin–quercetin	Hydrogen bond	6.93	−4.26	38.4%
	QA2	Quercetin–quercetin	Hydrogen bond	7.57	−2.86	25.8%
	QA3	Quercetin–quercetin	Hydrogen bond	11.24	−1.57	14.1%
QMH	QMH1	Quercetin–quercetin	π–π stacking	4.85	−6.39	24.5%
	QMH2	Quercetin–quercetin	Hydrogen bond	7.99	−5.33	10.2%
	QMH3	Quercetin–water	Hydrogen bond	5.93	−2.55	9.8%
QDH	QDH1	Quercetin–quercetin	π–π stacking	3.67	−7.66	37.8%
	QDH2	Quercetin–water	Hydrogen bond	5.64	−1.61	7.9%
	QDH3	Quercetin–quercetin	Permanent dipole–dipole	9.14	−1.43	3.5%

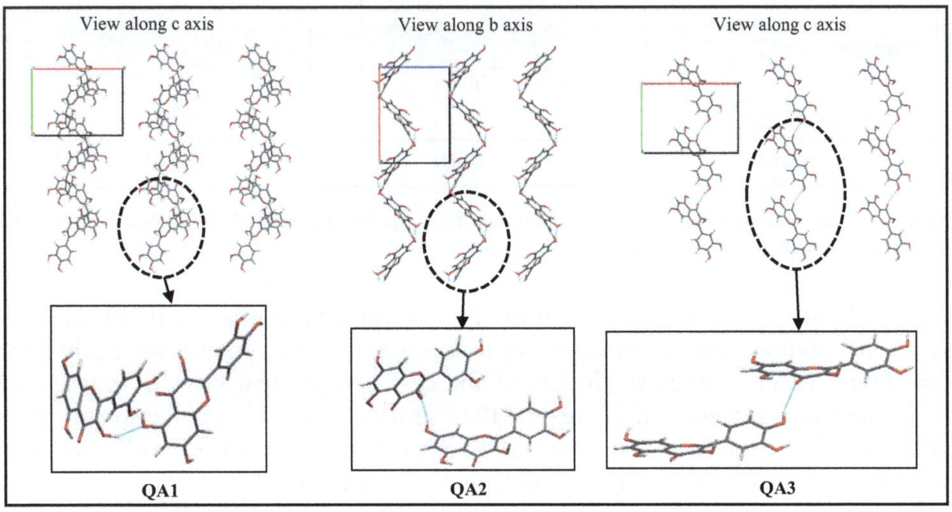

Figure 7.13. Key intermolecular synthons in quercetin anhydrous (QA) ordered by synthon strength. Light blue dotted lines indicate hydrogen bonding.

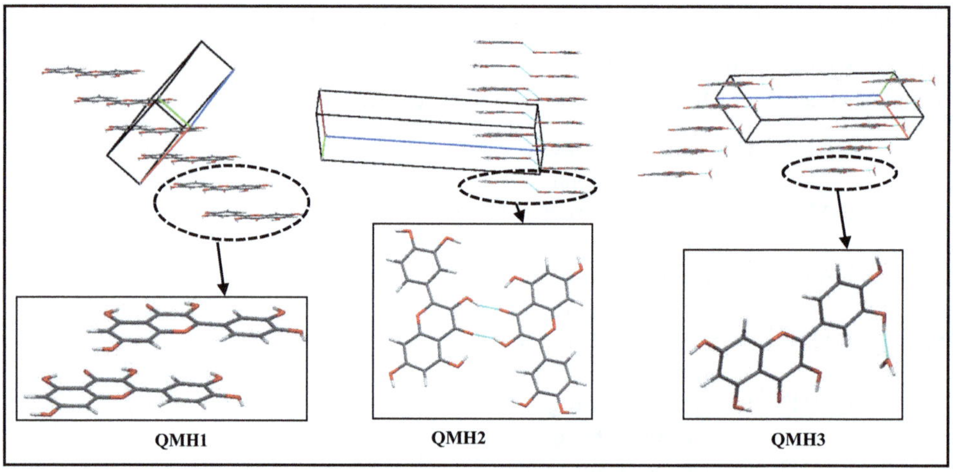

Figure 7.14. Key intermolecular synthons in quercetin monohydrate (QMH) ordered by synthon strength. Light blue dotted lines indicate hydrogen bonding.

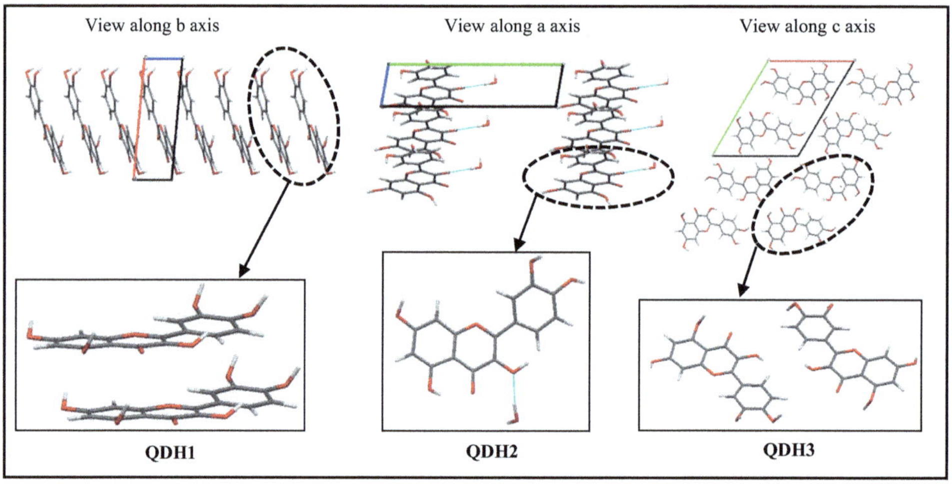

Figure 7.15. Key intermolecular synthons in quercetin dihydrate (QDH) ordered by synthon strength. Light blue dotted lines indicate hydrogen bonding.

carbonyl bond and hydroxyl groups of the quercetin molecule are involved in the hydrogen bonding. The non-planar conformation of the quercetin molecule facilitates close contact between the hydroxyl and carbonyl groups, in order to maximize the number and strength in energy of these interactions. This is demonstrated in figure 7.13, where the twisted conformation of quercetin allows for the close contacts between the hydroxyl and carbonyl groups to form the QA3 synthon.

However, the non-planar conformation of the molecule does not allow the formation of strong π–π stacking interactions which can be observed in the two hydrates structures, as shown in figure 7.14 and figure 7.14. Stacking interactions in the anhydrous form are not found to be among the three strongest interactions in the lattice. This suggests the less closely packed nature of the quercetin molecules in the anhydrous form.

The strongest intermolecular synthons in the QMH and QDH structures, named QMH1 and QDH1, respectively, are π–π stacking interaction between quercetin molecules. The main contribution to this type of interaction comes from the aromatic carbon atoms of the phenyl and pyrone rings of the quercetin molecules, which interact via Van der Waal forces of attraction. In both hydrated structures, these interactions promote the formation of uninterrupted chains of stacked quercetin molecules packed in an offset orientation, thought to maximise the interaction between the negative central aromatic π-system and the positively charged hydrogens on the outer ring. These strong interactions promote the close packing (shorter intermolecular distances) of the quercetin molecules in the two hydrates.

Comparison of the QDH1 to QMH1 shows that the π–π stacking interaction in the dihydrate form is stronger, with shorter intermolecular distances compared to that of the monohydrate. The π–π stacking interaction in the dihydrate is the dominant synthon, contributing to almost 38% of the total lattice energy. Clearly, the addition of the second water molecule in the unit cell of QDH indirectly influences the interactions among quercetin molecules, allowing them to pack closer together by forming stronger bonds. In the two hydrated forms, synthons QMH3 and QDH2 are both hydrogen bond interactions between a hydroxyl group of a quercetin molecule and a water molecule. In both cases the interaction creates a channel of water molecules running parallel to the stacked chain of quercetin molecules.

The contribution of the quercetin and water molecule interaction energies to the total lattice energy of each structure was calculated. As shown in figure 7.16, the contribution of water to the total lattice energy increases with the number of water molecules per unit cell, indicating the tendency of incorporation of water molecules into the lattice, and that the formation of interactions between quercetin and water is favoured. The water molecules are found to contribute to 23% of the total lattice

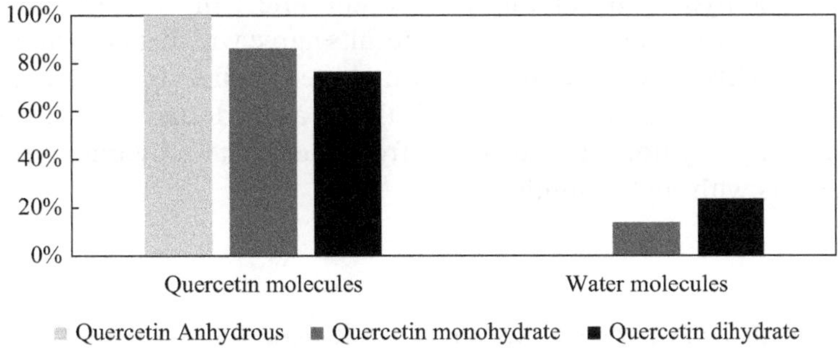

Figure 7.16. % contributions of quercetin and water molecules' interactions to total lattice energy of the three structures.

energy of QDH, highlighting the significance of the quercetin–water interactions in this structure.

In conclusion, the results of the present modelling analysis show that, as the degree of hydration and the number of water molecules in the unit cell increases for the three quercetin structures:

(1) hydrogen bonding in the lattice is more satisfied by interaction with the incorporated water molecules, allowing a more planar conformation for the quercetin molecules in the two hydrate structures;
(2) the contribution of the π–π stacking interactions between quercetin molecules to the stabilization of the crystal lattice increases.

It is obvious from literature that crystallization of quercetin from an aqueous solvent always produces the dihydrate form. This behaviour is explained by the modelling results as follows:

- during crystallization from an aqueous solvent, the water molecules being much smaller in size compared to quercetin molecules, can be positioned close to the polar groups of the quercetin molecule forming hydrogen bonds;
- once hydrogen bonding is satisfied, the quercetin molecules, having a more planar configuration, can pack more closely and efficiently via strong π–π stacking interactions;
- the smaller amount of de-solvation and conformational rearrangement in the dihydrate structure probably results in the easier crystallisation of this form from aqueous solution. This agrees with the higher calculated unit cell density which predicts the greatest stability;
- quercetin in QA must take an energetically unfavourable conformation to satisfy its hydrogen bonding groups, thus ends up having a lower thermodynamic stability, and is not preferentially nucleated from an aqueous solvent.

From the points above, we summarise that the favourable packing of the quercetin–water H-bonding and quercetin–quercetin π–π stacking results in the dominant crystallisation of the dihydrate form.

The modelling results shown in this study highlight the importance of the water molecules in the stabilization of the crystal structures of QMH and QDH, as they can influence the hydrogen bonding pattern and affect the strength and nature of intermolecular interactions formed. These results agree with experimental studies on the relative stability of quercetin and its hydrated forms. In fact, this work can explain why quercetin preferentially crystallizes as hydrated form from aqueous solvents and why polymorphic transitions from the QA to a hydrate are favourable in environments with high humidity.

References

[1] Mullin J W 2001 *Crystallization* vol 4 (Oxford: Butterworth-Heinemann)
[2] Klitou P, Rosbottom I and Simone E 2019 *Cryst. Growth Des.* **19** 4774–83
[3] Ukrainczyk M, Hodnett B K and Rasmuson Å C 2016 *Org. Process Res. Dev.* **20** 1593–602

[4] Simone E 2015 Application of process analytical technology (PAT) tools for the better understanding and control of the crystallisation of polymorphic and impure systems *PhD Thesis* Loughborough University
[5] Simone E, Klapwijk A R, Wilson C C and Nagy Z K 2017 *Cryst. Growth Des.* **17** 1695–706
[6] Simone E, Tyler A I I, Kuah D, Bao X, Ries M E and Baker D 2019 *Org. Process Res. Dev.* **23** 220–33
[7] Davey R and Garside J 2000 *From Molecules to Crystallizers* (Oxford: Oxford University Press) pp 1–52
[8] Mangin D, Puel F and Veesler S 2009 *Org. Process Res. Dev.* **13** 1241–53
[9] Rawlings J, Miller S and Witkowski W 1993 *Ind. Eng. Chem. Res.* 1275–96
[10] Erdemir D, Lee A Y and Myerson A S 2009 *Accounts Chem. Res.* **42** 621–9
[11] Vekilov P G 2010 *Cryst. Growth Des.* **10** 5007–19
[12] Davey R J, Schroeder S L M and ter Horst J H 2013 *Angew. Rev.* **52** 2166–79
[13] Bernstein J 2011 *Cryst. Growth Des.* **11** 632–50
[14] Davey R J, Blagden N, Righini S, Alison H, Quayle M J and Fuller S 2001 *Cryst. Growth Des.* **1** 59–65
[15] Chiarella R A, Davey R J and Peterson M 2007 *Cryst. Growth Des.* **7** 1223
[16] Parveen S, Davey R J, Dent G and Pritchard R G 2005 *Chem. Commun.* 1531–33
[17] Davey R J, Dent G, Mughal R K and Parveen S 2006 *Cryst. Growth Des.* **6** 1788–96
[18] Kulkarni S A, McGarrity E S, Meekes H and ter Horst J H 2012 *Chem. Commun.* **48** 4983–85
[19] McCrone W C 1965 *Polymorphism and Chemistry of the Organic Solid State* (New York: Interscience)
[20] Sato K 1993 *J. Phys. Appl. Phys.* **26** B77–84
[21] Simone E and Nagy Z K 2015 *Crystengcomm* **17** 6538–47
[22] Simone E, Steele G and Nagy Z K K 2015 *Cryst. Eng. Comm.* **17** 9370–9
[23] Kitamura M 2002 *Thirteenth Int. Conf. Cryst. Growth conjunction with Elev. Int. Conf. Vap. Growth Ep.* 237–239 2205–14
[24] Cardew P T and Davey R J 1985 *Proc. R. Soc. London* **398** 415–28
[25] Davey R J, Cardew P T, McEwan D and Sadler D E 1986 *Proc. Eighth Int. Conf. Cryst. Growth* 79 648–53
[26] Scholl J, Bonalumi D, Vicum L and Mazzotti M 2006 *Cryst. Growth Des.* **6** 881–91
[27] Maher A, Croker D, Rasmuson A C and Hodnett K 2012 *Cryst. Growth Des.* **12** 6151–7
[28] Tian F, Zeitler J A, Strachan C J, Saville D J, Gordon K C and Rades T 2006 *J. Pharm. Biomed. Anal.* **40** 271–80
[29] O'Brien L E, Timmins P, Williams A C and York P 2004 *J. Pharm. Biomed. Anal.* **36** 335–40
[30] Smith P R, Cebula D J and Povey M J W 1994 *J. Am. Oil Chem. Soc.* **71** 1367–72
[31] Smith P R and Povey M J W 1997 *JAOCS, J. Am. Oil Chem. Soc.* **74** 169–71
[32] Klapwijk A R 2016 Disorder and additive control of solid form and morphology in continuous crystallisation *PhD Thesis* University of Bath
[33] Saleemi A N, Steele G, Pedge N I, Freeman A and Nagy Z K 2012 *Int. J. Pharm.* **430** 56–64
[34] Abu Bakar M R, Nagy Z K, Saleemi A N and Rielly C D 2008 *Cryst. Growth Des.* **9** 1378–84
[35] Jiang M *et al* 2014 *Ind. Eng. Chem. Res.* **54** 5325–36
[36] Yang Y, Song L and Nagy Z K 2015 *Cryst. Growth Des.* **15** 5839–48
[37] Hansen T B, Simone E, Nagy Z and Qu H 2017 *Org. Process Res. Dev.* **21** 855–65

[38] Simone E, Zhang W and Nagy Z K 2015 *Cryst. Growth Des.* **15** 2908–19

[39] Koswara A and Nagy Z K 2015 *9th IFAC Symp. Adv. Control Chem. Process. ADCHEM 2015 (Whistler, Canada, 7–10 June 7–10, 2015) 48 193–8*

[40] Majumder A and Nagy Z K 2015 *Cryst. Growth Des.* **15** 1129–40

[41] Lovette M A, Browning A R, Griffin D W, Sizemore J P, Snyder R C and Doherty M F 2008 *Ind. Eng. Chem. Res.* **47** 9812–33

[42] Lovette M A, Muratore M and Doherty M F 2012 *AIChE J.* **58** 1465–74

[43] Lovette M A and Doherty M F 2012 *Cryst. Growth Des.* **12** 656–69

[44] Eisenschmidt H, Voigt A and Sundmacher K 2015 *Cryst. Growth Des.* **15** 219–27

[45] Eisenschmidt H, Bajcinca N and Sundmacher K 2016 *Cryst. Growth Des.* **16** 3297–306

[46] Klapwijk A R, Simone E, Nagy Z K and Wilson C C 2016 *Cryst. Growth Des.* **16** 4349–59

[47] Yu Q, Dang L, Black S and Wei H 2012 *J. Cryst. Growth* **340** 209–15

[48] Gadewar S B and Doherty M F 2004 *J. Cryst. Growth* **267** 239–50

[49] Lemmer S and Ruether F 2012 *Chem. Eng. Sci.* **77** 143–9

[50] Gänzle M G, Haase G and Jelen P 2008 *Int. Dairy J.* **18** 685–94

[51] Garnier S, Petit S and Coquerel G 2002 *J. Cryst. Growth* **234** 207–19

[52] Parimaladevi P and Srinivasan K 2016 *Cryst. Eng. Comm.* **18** 2312–8

[53] Raghavan S L, Ristic R I, Sheen D B, Sherwood J N, Trowbridge L and York P 2000 *J. Phys. Chem.* B **104** 12256–62

[54] Jawad R, Elleman C, Vermeer L, Drake A F, Woodhead B, Martin G P and Royall P G 2012 *Pharm. Res.* **29** 511–24

[55] Salameh A K and Taylor L S 2006 *J. Pharm. Sci.* **95** 446–61

[56] Qu H, Louhi-kultanen M and Kallas J 2007 *Cryst. Growth Des.* **7** 724–29

[57] Nguyen T T H, Rosbottom I, Marziano I, Hammond R B and Roberts K J 2017 *Cryst. Growth Des.* **17** 3088–99

[58] Desiraju G R 2013 *J. Am. Chem. Soc.* **135** 9952–67

[59] Jurasekova Z, Ökten G and Domingo C 2014 *Phys. Chem. Chem. Phys.* **16** 12802–11

[60] Zembyla M, Murray B S, Radford S J and Sarkar A 2019 *J. Colloid Interface Sci.* **548** 88–99

[61] Clydesdale G, Roberts K J and Docherty R 1994 *J. Cryst. Growth* **135** 331–40

Chapter 8

Particle characterisation methods for food formulation and processing

R S Ward-Smith, R Taylor, S Cairns, M Kaszuba, D Fernandes, M Zoontjes, T Rowney, S Marsh and L Newey-Keane

8.1 Introduction

In 2016, UK Government figures [1] put the size of the UK's food and drink manufacturing industry at £29.5 billion, employing 390,000 people. Many food and drink companies use formulation processes to ensure the delivery of good-quality products made in a reproducible way.

Many raw materials are pre-treated in some way before they are used. Raw ingredients, including flour, nuts, cocoa and sugar, are often ground before incorporation, and emulsions will usually be homogenised. The effectiveness of these processes can be monitored by characterising the resulting material.

The particle size, shape and charge of food ingredients are key parameters in the final product. Particle size and shape influence how a material flows (smaller material tends to be stickier than larger material—think of the difference between icing sugar and granulated sugar), its rate of dissolution, and its mouthfeel. The human tongue is an extremely sensitive particle sizing device in its own right, detecting any particle larger than 30 μm in diameter as having a gritty or greasy texture.

The charge on a material is normally indicative of its shelf-life, with a change in pH or salt concentration often employed to induce electrostatic stabilisation, thus preventing coalescence or separation. Another option is the addition of a thickening agent, which will promote steric stabilisation, but this may also lead to a change in the product's rheological behaviour.

On occasion, there could also be contamination in our food—this can have devastating consequences, as evidenced in the Chinese milk powder scandal [2].

For all these reasons and many more, materials characterisation technologies are a key component of any food formulator's toolset. This chapter will cover many of

doi:10.1088/978-0-7503-2596-7ch8

the processes and their associated challenges in food product manufacture, and the characterisation techniques that can be used to address them.

8.2 The importance of particle size characterisation

8.2.1 Size reduction by milling

Milling, as a tool to reduce a food material's size, has been used for thousands of years; from grinding material by hand between two stones, through to wind and water mills, and more modern techniques such as ball mills and roller compactors. Milling is an energy-inefficient technique, because a great deal of energy is released in the form of heat, so it is important from an energy conservation perspective to cease milling once the desired size has been achieved. Problems are also encountered when the milling process creates a finer product than expected—this will be stickier, van der Waals forces will dominate, and it will likely dissolve.

8.2.2 Sieving

Sieving has been used as a separation technique in the food industry for thousands of years. Some of its earliest applications include the separation of wheat from chaff, and of flour from unmilled grains, and sieving is still in use for many applications to this day. Many people in the UK will still remember the advertising slogan 'Because graded grains make finer flour' [3]. The most attractive feature of this technique is its price in terms of initial outlay, which is extremely low when compared to most materials characterisation techniques.

There are various types of sieves, including punched-hole sieves and mesh sieves of many differing diameters. Sieve size extends up into the millimeter range (woven wire sieves can measure up to 125 mm) and down into the micron range (micromesh sieves can measure down to 5 μm). Sieving can be performed wet or dry, but, for most food applications, dry sieving is normally used. This is because the product being sieved would be adversely affected by the application of moisture, resulting in the swelling of flour and coffee, and the dissolution of sugar.

There are two main problems with sieving, when it is applied as a proxy for a true materials characterisation technique:

1. Sieving measures only the second largest diameter of a particle. If the material in question is composed of cylindrical or needle-shaped particles, they will pass through the sieve end-first, and will be measured according to their thickness rather than their length. This leads to incongruity with data generated by other sizing techniques such as laser diffraction or image analysis, which will often record a larger particle size. It also exposes the fallacy of the tenet that sieving is an absolute technique, which will not allow particles greater than the mesh size to pass through the mesh.
2. Sieving does not normally contain a dispersion mechanism for fine material. Often, the finer particles are adhered to the largest particles and will be

separated or sized with them. Vigorous shaking; for example, with a device such as a mechanical shaker, can provide some dispersion, but this is unlikely to provide full dispersion and is unsuitable for friable material such as that which has been freeze-dried.

8.2.3 Laser diffraction

Laser diffraction is one of the most widely-used particle size analysis techniques within the food industry. It is used to test the quality of raw materials, in research and development, and also in quality control. The major applications of diffraction within this sector are the sizing of dairy and dairy-alternative (soy, oat, pea, coconut, almond milks) products, of flavourings and colour emulsions, and of coffee, sugar, flours and chocolate.

The core principle of laser diffraction is that particles passing through a laser beam will scatter light at angles which are determined by their size. Small particles scatter light predominantly at high angles, with low intensity, and large particles scatter light predominantly at low angles, but with high intensity. A typical laser diffraction system consists of a light source, normally a laser or laser diode, to provide a source of coherent, intense light of fixed wavelength, a series of detectors to measure the light at a range of angles, and a device which passes the particles through the laser. This device may be a pump system (which could additionally contain an ultrasonic probe for sample dispersion) which can recirculate the material, or a dry powder dispersion unit. The dynamic range of the measurement is related to the angular range of the light scattering. The wavelength of the light source also influences the size range covered by the instrument, with smaller wavelengths (e.g., blue light sources) providing improved sensitivity for sub-micron materials.

In laser diffraction, particle size distributions (PSDs) are calculated by comparing a sample's scattering pattern with an appropriate optical model. Traditionally, two different models are used: the Fraunhofer approximation and Mie theory.

The Fraunhofer approximation was used in early laser diffraction instruments. It assumes that the particles being measured are opaque and will scatter light at narrow angles. As a result, it is only applicable to large particles and will give an inaccurate assessment of the fine particle fraction.

Mie theory provides a more robust solution for the calculation of PSDs from light scattering data. It predicts scattering intensities for all particles—small or large, transparent or opaque. Mie theory allows for primary scattering from the surface of the particle, with the intensity predicted by the refractive index difference between the particle and the dispersion medium. It also predicts the secondary scattering caused by light refraction within the particle. This is especially important for particles below 50 μm in diameter, as stated in the international standard for laser diffraction measurements [4].

Table 8.1. Comparison of size distribution percentiles and %>30 μm values in three chocolate types.

	Dv10 (μm)	Dv50 (μm)	Dv90 (μm)	%>30 μm
Dark	1.74	6.62	19.5	2.13
Milk	1.77	8.32	32.5	12.6
White	1.77	10.8	37.0	17.3

Case study: chocolate

Chocolate is one of the world's favourite snack foods, with over $103 billion spent on chocolate in 2017 [5]. The unique appeal of chocolate lies in its taste, aroma, mouthfeel and texture.

Several factors are considered important for a good-quality chocolate. These include [6]:

- melting temperature of 37 °C, so that it will melt in the mouth;
- shine, so it looks appealing;
- smooth texture, which gives a pleasant mouthfeel;
- snap, to provide the initial 'bite'.

Extensive consumer testing has established that particles over 30 μm in size are detected by the consumer, as they feel gritty on the tongue. However, it is not just a question of milling the ingredients to a very small size, as the increased surface area created would require the addition of more cocoa butter, dramatically increasing the cost of manufacturing. There is, therefore, a trade-off, and a very fine line is drawn between ingredients that are 'small enough' and 'too small'.

As seen in table 8.1, different types of chocolate have very different PSDs. Here we compare the 10%, 50% and 90% size distribution percentiles and the % particles greater than 30 μm (particles which can be detected by the tongue) in dark, milk and white chocolates.

The different sizes are due to their differing recipes and components. White chocolate is composed simply of sugar and cocoa butter, dark chocolate contains cocoa solids, cocoa butter and sugar, and milk chocolate contains cocoa solids, cocoa butter, sugar and milk powder.

Rheology also plays an important role in the development and processing of chocolate. As rheology is a bulk material property, it is strongly affected by the individual material properties, including particle size, shape, zeta potential, size distribution and volume fraction, of the product's ingredients. Rheology and particle properties are often used in a complementary fashion to provide an overall characterisation of food materials.

Modern rheometers can be used to analyse a material's viscosity: from water-like liquids to creams, gels and pastes. Viscoelastic characterisation, such as oscillation, can also measure the stiffness of solids and semi-solids. By applying a shear deformation or an axial movement to a sample, texture analysis (mouthfeel) can be performed to determine material properties such as viscosity, stiffness (modulus), tackiness, texture (phase angle), and melting temperatures.

Rheology is also commonly used in the food and drinks industry to ensure stability and prevent the phase separation of formulations such as smoothies, protein shakes and alcoholic suspensions. Such products require a material property known as a 'yield stress', meaning that they possess a certain amount of structure. This structure can make the material behave in a solid-like manner under rest conditions, preventing the sedimentation of particles. In the case of chocolate, different formulations or types will possess different viscosities, affecting how the chocolate will flow when enrobing a chocolate bar, how it is manipulated by a confectionary chef, or how it flows in the mouth once melted.

8.2.4 Static image analysis

Static image analysis is another particle characterisation technique that can measure both the size and shape of the material being investigated, and which enables links between particle size and particle shape to be made. Instrumentation normally consists of a light microscope and a stage, which allows a dispersion of particles to be scanned by moving the sample under the microscope. This technique is sometimes used in conjunction with an additional technique such as Raman spectroscopy, which can provide identification of the individual components of the sample—this can be invaluable in investigating potential adulteration, and also in pinpointing process changes.

Case study: wheat- and gluten-free flours

Wheat flour is used in bread production all over the world. Starch is the major component of wheat flour and serves several functions, including gelatinisation, swelling and pasting of the dough, which is related to the proportion of A- and B-type starch grains. A and B-type grains, or particles, are defined by their size, where $A > 10$ μm and $B < 10$ μm [7]. The composition of each grain type is also different, leading to different rates of digestion and hence different glycaemic indices [8]. These properties can be adjusted by controlling the relative ratio of A- and B-type starch grains [9].

The composition of the powder and particle morphology of three flour products—two standard plain flours, and one gluten-free flour—was characterised using morphologically-directed Raman spectroscopy (MDRS®). Particles were isolated by their chemical identity as defined by Raman spectroscopy and then characterised by their size and shape.

Starch was the most abundant of the components for all three brands, followed by starch-protein composite and then protein. This was followed by the trace additives, such as gamma-oryzanol, calcium carbonate and sulphate (figure 8.1).

Most protein was agglomerated with starch to form composite particles, which could affect manufacturing processes such as the milling method. The gluten-free flour (blue bars) had significantly less protein than the standard wheat flours, which is to be expected, as there should be no gluten protein present. The flour should not be devoid of protein, as this would affect its structural function as well as its manufacturing performance.

When comparing the starch component only, the two standard plain flours showed bimodality of PSD, with the modes separated at 10 μm, indicative of discrete A and B populations of starch. The gluten-free product starch PSD was monomodal but asymmetric, with a bias towards the A-type particle (>10 μm). Therefore, it is expected

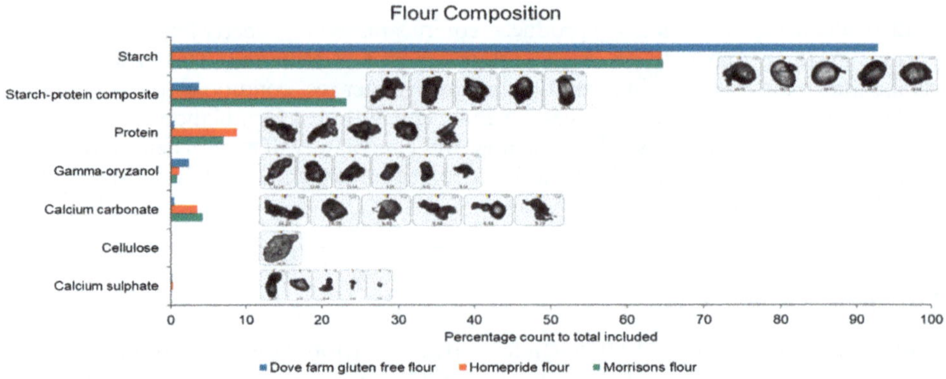

Figure 8.1. The percentage of total counted particles for each component identified by Raman spectroscopy, along with example particle images.

that the gluten-free flour will have a higher gelatinisation temperature and will swell less from water ingress. It was noted that starch particles of the gluten-free flour were notably less-elongated and more spherical than those in the standard plain flours. This may be an indication of the structural and manufacturing performance of the flour.

The relative proportion of the two starch particle types and the quantity of protein in the composition will influence the functional and nutritional properties of the dough made from the flours. Hence, the dough made from gluten-free flour is unlikely to perform equivalently to the standard flours and so may need a different preparation or recipe.

8.2.5 Homogenisation

In food processing, the performance of the product is as dependent on the type and concentration of the ingredients as it is on the efficiency of their homogenisation. For food emulsions in particular, the stability of the product will be heavily affected by the quality of the homogenisation process [10].

Briefly, homogenisation can be divided into three steps:

1. **Pre-homogenisation**: an initial dispersion step, during which the different ingredients are dissolved in their preferred phase, i.e., non-polar ingredients in the oil phase, and polar ingredients in the aqueous phase;
2. **Homogenisation**: during this step, the aim is to make both phases, oil and water, miscible using a homogeniser. This step can be further expanded into two stages: primary homogenisation, where the emulsion is directly created from the two immiscible phases, and secondary homogenisation, which involves reducing the droplet size. Depending on the process and product, one or both stages can be employed;
3. **Post-homogenisation**: finally, the emulsions may be subjected to additional steps, with the aim of ensuring long-term stability or adding specific properties. Most frequently, this involves thermal treatments such as chilling, freezing, pasteurisation or sterilisation.

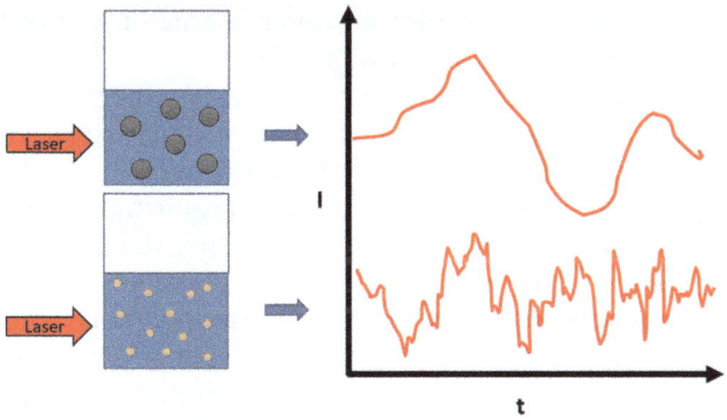

Figure 8.2. Schematic illustration showing that small particles produce more rapid fluctuations of scattered light, and the inverse for larger particles. Reproduced with permission of Malvern Panalytical.

One of the key technologies involved in investigating homogenisation is dynamic light scattering (DLS). This technique's inherent intensity-weighting and sensitivity to the presence of clusters make it an important tool for the study of the efficiency of homogenisation and other processes, as well as the quality of the final food or drink product.

8.2.6 Dynamic light scattering

DLS is a quick, cost-efficient and non-invasive technique used to determine a material's particle size at its best working capacity in the sub-micron region. The DLS principle lies in measuring the diffusion rates of randomly-moving particles as these collide with solvent molecules. This movement is also known as Brownian motion. The larger the particle, the lower its diffusion rate, and vice versa.

DLS instruments are equipped with a monochromatic coherent laser source; for instance, a He–Ne gas laser, which irradiates the sample at a specific wavelength. Once the particles are irradiated, they will scatter light at the same wavelength as the laser source, and this light will then be collected by a photon-counting device, which typically contains high-performance avalanche photodiodes (APDs). The intensity signal produced will appear to fluctuate randomly at short timescales (often less than a millisecond), due to the interference from millions of scattered waves. These intensity fluctuations can be correlated, showing that the more rapid the intensity fluctuation, the smaller the particle size [11] (figure 8.2).

In order to gain information from such short timescales, a signal comparator known as a digital correlator is required. With the aid of autocorrelation functions (ACF), seemingly random signals can be correlated, and the particle diffusion rates can be extracted from the resulting correlograms. Once the diffusion rate is known, the particle size can be determined using the Stokes–Einstein equation [12].

Historically, DLS instruments used 90° as a standard light scattering collection angle. Most samples analysed were polymer solutions which were small, weakly-scattering materials, and the 90° setup allowed for an easier internal layout of the instrument, while minimising stray light from the cell walls [13].

However, for many other applications a much wider range of sizes and concentrations must be measured. The implementation of backscatter technology allowed for a more consistent analysis of the upper-scale limit at ~1 μm–10 μm, but using backscatter alone has the drawback of requiring lower concentrations due to the increased measurement volume. The greater the measurement volume, the more likely that a photon will scatter multiple times before reaching the detector—a phenomenon known as multiple scattering. This can result in lower intensity photons being detected due to the loss of energy per scattering, which in turn will lead to lower intercept ACF and faster initial decays, but also more gradual slopes. Depending on the magnitude of multiple scattering, this will tend to produce smaller particle sizes due to the faster decays, and artificially polydisperse results arising from the randomness of multiple scattering events. This is particularly problematic for DLS, as in food analysis measurements are usually performed at higher concentrations, and dilution can fundamentally change the sample.

What rendered DLS such a versatile tool was the advent of refining the backscatter technique with the introduction of non-invasive backscattering (NIBS) technology. This technique uses an automatically-adjustable translational stage, positioned between the backscatter detector and the cuvette. If the sample is too concentrated, the stage moves the focal point of the laser source and detector closer to the cell wall, reducing multiple scattering. Conversely, for more diluted samples, in order to collect enough scattering and to avoid excessive flare from the cell wall, the stage will focus at the optical centre of the cuvette (figure 8.3).

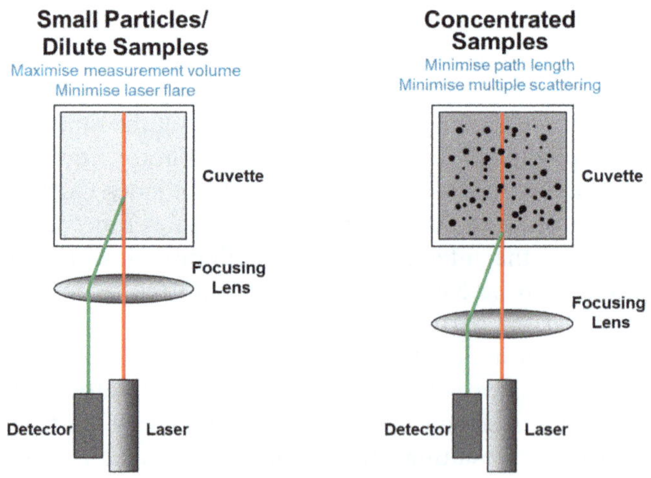

Figure 8.3. Schematic diagram illustrating the translation stage which can move as required according to sample turbidity. Reproduced with permission of Malvern Panalytical.

Case study: casein micelles in milk

One of the earliest DLS applications for food analysis was the study of casein micelles in milk [14]. Casein micelles are aggregates of several different casein proteins with calcium phosphate. The main advantage of using DLS to study casein micelles is not only to measure the PSD, but also to understand its inherent sensitivity to any changes in the sample [15].

In one of the first reports in the field, Walstra *et al* [16] investigated the existence of a protein-based layer (k-casein) on the surface of micelles, providing them with steric stability. The authors identified this layer by adding a protease (chymosin), which specifically cleaves the k-casein protein, and noticed that the hydrodynamic radius of the micelles was then reduced by 5 nm. They also noticed that as time progressed, aggregation would follow. This proved that there was a surface structure on the casein micelles that slowed the translational diffusion coefficient, making the micelles appear larger, but once the surface layer was removed, a smaller particle size was obtained, corresponding to the size of the micelle core. The aggregation that occurred upon removal of this layer suggests that it provided stability to the micelles.

Case study: KHT crystallisation in wine

More recently, in beverage research, DLS has been used as a tool for monitoring tartrate crystallisation in wine [17]. The spontaneous and unpredictable crystallisation of the winemaking by-product potassium hydrogen tartrate (KHT) is an important issue recognised by wine manufacturers, as it forms visual defects which are negatively perceived by customers.

Lankhorst *et al* [17] suggested DLS as a sensitive but simple tool to assess the early stages of KHT's crystallisation process; i.e., before it becomes visible. Although this work was performed under laboratory-controlled conditions, they found that the most common stabilisers (hydrocolloids) used to inhibit this phenomenon merely act as retardants for KHT crystallisation. Briefly, the authors described the PSD profile of a standardised wine sample, prior to the addition of ethanol and stabilisers, as predominantly sub-100 nm, and monitored how this changed as a function of time once these components were added. Crystallisation was clear when ethanol was added, but proceeded much more slowly when each stabilizer was added. The group also noticed that even though the results seemed stable after 3 and 6 days, there was a slow increase in the scattering intensity, which could be explained by longer-term crystallisation.

8.2.7 Granulation

Not all food formulation processes relate to size reduction. Some food material is granulated (coffee, gravy granules, etc), where particles are formed through mechanisms such as spray drying. In cases such as this, a technique capable of measuring large particles is appropriate.

8.2.8 Spatial filter velocimetry

Using in-line determination of powder particle size, spatial velocimetry probes provide information about a manufacturing process. The principle of measurement is evaluation of the chord length (particle size) using the extended spatial filter technique. The spatial filter is an array of optical fibres, which are illuminated by the laser. As particles pass through the measurement gap at the tip of the probe, they create shadows. The spatial frequency filter, a linear grating arrangement of optical fibres, picks up signals from the particles as they pass through the measurement gap, allowing calculation of their velocity. The time of flight detector is a single optical fibre located after the array. By combining the individual particle velocity with the time of flight, the chord length can be calculated. The technique will measure particle size within the size range 50–6000 μm, and users can view results in terms of volume and number size distributions as well as velocity (figure 8.4) [18].

Real-time monitoring of a process is achieved by writing individual measurements into a ring buffer, the size of which can be adjusted depending on the application. By continuously updating the buffer, results are presented in real-time as a rolling average and will respond quickly to any changes in the process.

Depending on the application, the user has the ability to disperse the particles within the process vessel or pipe by choosing the appropriate flow cell or in-line disperser to reliably determine the particle size in-line, even when the materials being measured are moist and sticky.

The probe is suitable for use in a range of processes such as grinding, spray-drying, mixing, separation, and the granulation or the agglomeration of instant products. It has been used for determination of the particle size of cereals, sugar, coffee, milk powder and salt.

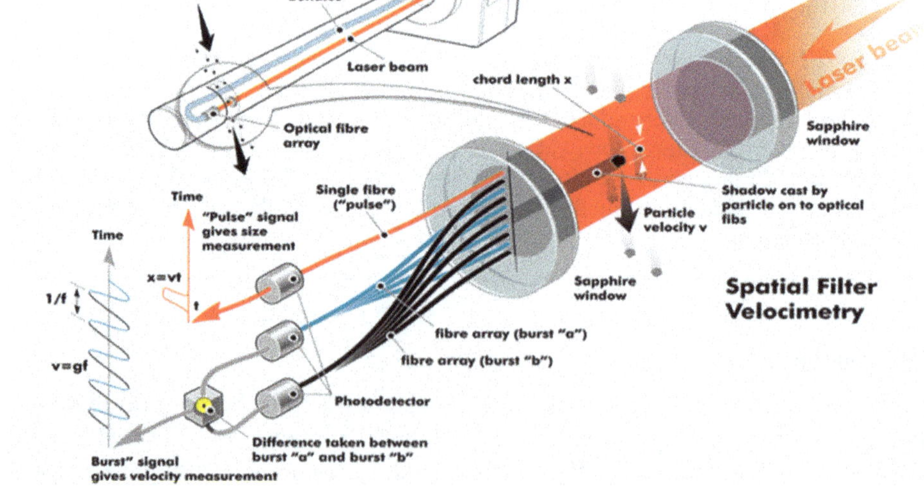

Figure 8.4. Schematic diagram of Parsum probe: extended spatial filter technique. Reproduced with permission of Malvern Panalytical.

> ***Case study: coffee***
>
> In the example below, the spatial velocimetry probe was placed at the outlet of a coffee grinder in a pilot plant. During the measurement, the gap between the rollers was changed. The probe quickly tracked the change in the process and recorded the resulting change in particle size (figure 8.5).

8.3 Optimising food formulation stability

When a food product is manufactured, a key property requiring evaluation is its stability, which impacts on the product's shelf life. Does the smaller material cream to the top? Does the large material settle to the bottom? Consumers can react strongly if the consistency of the product is not what they expect. There are two main mechanisms for optimising stability in a food product:

1. Introducing additional charge to the surface of particles increases the repulsive force between them. This is called electrostatic stabilisation.
2. Adding a (bio)polymer to the particle surface (steric stabilisation) or completely changing the medium (gelling it, such as a in a jam), reduces particle collision. In cases such as this, the properties (for example, molecular weight and branching) of the additives used are key.

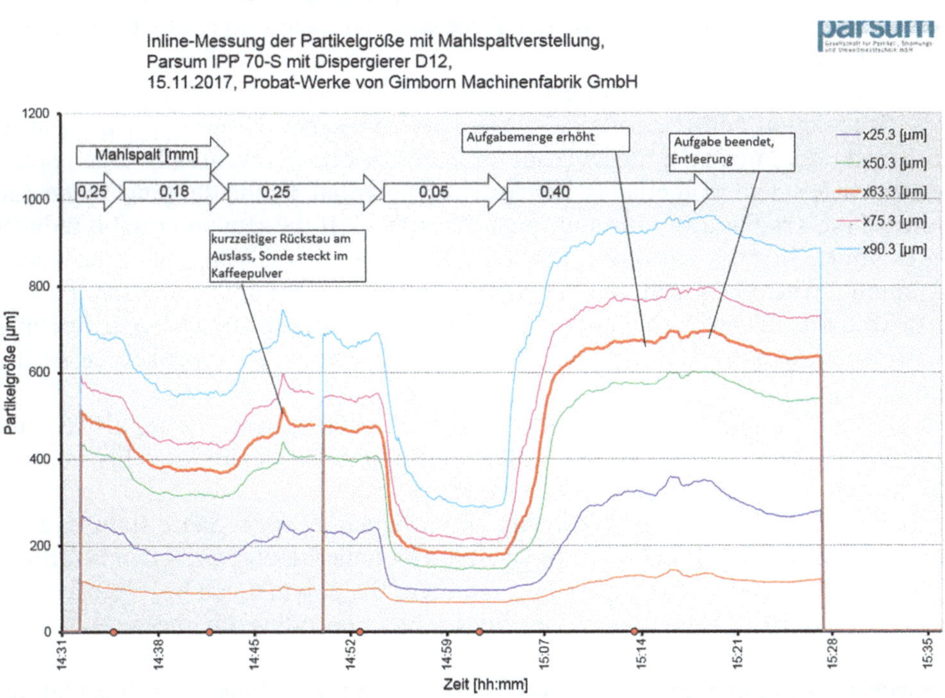

Figure 8.5. Use of spatial filter velocimetry to monitor changes in a coffee grinding process. Reproduced with permission of Malvern Panalytical.

8.3.1 Electrostatic stabilisation—zeta potential

Zeta potential is a physical property which is exhibited by any particle in suspension. It is a measure of the electrostatic repulsion between particles, and can be used as an indicator of dispersion stability. If all the particles in suspension have a large negative or a large positive zeta potential, they will tend to repel each other and tend not to aggregate/flocculate. However, if the particles have low zeta potential values, close to zero, there is no force to prevent the particles coming together and flocculating. Knowledge of a sample's zeta potential can reduce the time needed to produce trial formulations, and can also be a help in predicting long-term stability.

8.3.2 Colloidal stability and DLVO theory

The scientists Derjaguin, Verwey, Landau and Overbeek ('DVLO') developed a theory in the 1940s [19] which suggests that the stability of a system is determined by the sum of the van der Waals attractive (V_A) and electrical double layer repulsive (V_R) forces that exist between particles as they approach each other, due to the Brownian motion they are undergoing. DVLO theory proposes that an energy barrier resulting from the repulsive force prevents two particles approaching one another and adhering to one another, but if the particles collide with enough energy to overcome that barrier, the attractive force will pull them into contact, so that they adhere strongly and irreversibly.

Therefore, if the particles have a sufficiently high repulsion, the dispersion will resist flocculation and the colloidal system will be stable. However, if a repulsion mechanism does not exist, flocculation or coagulation will eventually take place.

A particle's zeta potential is its electrical potential at the slipping plane (one hydrated ion's radius away from the particle surface). If the zeta potential is reduced; for example, in high salt concentrations, there is a possibility of a 'secondary minimum', where a much weaker and potentially reversible adhesion between particles exists. These weak flocs are sufficiently stable to avoid being disrupted by Brownian motion, but may disperse under an externally-applied force such as vigorous agitation.

Therefore, to maintain the stability of the colloidal system, the repulsive forces must be dominant. So, how can colloidal stability be achieved? This is where electrostatic and steric stabilisation come in.

Each mechanism has its own benefits for particular systems. Steric stabilisation is simple, requiring just the addition of a suitable polymer. However, it can be difficult to flocculate the system subsequently if required, the polymer can be expensive, and in some cases the polymer may be undesirable, depending on the nature of the product.

Electrostatic or charge stabilisation has the benefit of stabilising or flocculating a system by simply altering the concentration of ions present. This is a reversible process and is potentially inexpensive.

8.3.3 What is the optimal zeta potential required for a stable dispersion?

The dividing line between stable and unstable aqueous suspensions is generally taken as either +30 mV or −30 mV. Particles with a zeta potential more positive than +30 mV or more negative than −30 mV are normally considered stable. However, if the particles have a different density to the dispersant, they will eventually sediment, forming a closely-packed bed.

Zeta potential is influenced by pH, ionic strength (the concentration and type of ions present) and the concentration of any charged molecules in the dispersant. The effect on the zeta potential of the pH, ionic strength of the medium, or concentration of an additive can provide useful information which can be used to give maximum stability to a product formulation.

8.3.4 Measuring zeta potential

Zeta potential can be measured using the technique of electrophoretic light scattering (ELS). Electrophoresis is the movement of a charged particle relative to the liquid in which it is suspended, under the influence of an applied electric field.

When an electric field is applied across an electrolyte, charged particles suspended in the electrolyte are attracted towards the electrode of opposite charge. Viscous forces acting on the particles tend to oppose this movement. When equilibrium is reached between these two opposing forces, the particles move with constant velocity.

Particle velocity is dependent on the strength of electric field or voltage gradient, the dielectric constant of the medium, the viscosity of the medium and the zeta potential.

8.3.5 Steric stabilisation—size exclusion chromatography

Polymer solutions have a wide array of functional properties in foods. They are included in food products primarily for thickening and gelling, but also for emulsification, stabilisation, and controlling the crystal growth of ice and sugar.

Gel formation is the phenomenon involving the association or cross-linking of the polymer chains to form a three-dimensional network that traps or immobilises the water within it to form a rigid structure that is resistant to flow [20]. Thickening occurs above a critical concentration known as overlap concentration (C^*). The polymer dispersions exhibit Newtonian behaviour below this concentration, but non-Newtonian behaviour above it. The C^* of a wide range of neutral and charged polysaccharides has been found to be dependent on the intrinsic viscosity (IV) of the formulation.

Polysaccharides that are commonly used as thickening agents include starch, various gums, and cellulose derivatives. Native polysaccharides are often depolymerised to produce different viscosity grades, and their degradation reduces their ability to thicken solutions. Depolymerisation and degradation are quantified by a change in molecular weight and molecular weight distribution. Both the molecular weight and molecular weight distribution are determined using size exclusion chromatography (SEC), which is sometimes known as gel permeation chromatography (GPC).

In SEC, particles within the sample are separated by their hydrodynamic size as they pass through a porous material, and elute from largest to smallest. After separation, each data slice is recorded as it passes through detectors. A modern multi-detector system will consist of a refractometer, viscometer and light scattering detector. At each data slice, the refractometer determines the relative concentration, and combining this with data from the other detectors, the absolute molecular weight and IV of the sample are determined.

Functionalised polysaccharides such as carboxymethylcellulose (CMC) have an additional parameter, the degree of substitution (DS), which impacts its solution rheology. With a higher DS, aggregation is minimised and stability improved. Conversely, for $DS > 0.8$, there is a decrease in the degree of crosslinking and hence a decrease in thickening [21]. CMCs with $DS \approx 0.7$ are therefore most frequently used as viscosity modifiers. Studies at Malvern Panalytical have found that by combining the IV and molecular weight data from SEC analysis to produce a Mark–Houwink plot, two CMCs of the same molecular weight with $DS = 0.7$ and 1.2 could be easily differentiated [22] (figure 8.6).

There are many parameters of polysaccharides that influence their use in the food industry. This includes molecular weight, molar mass distribution, IV and branching, which may all be characterised by SEC.

8.4 Food contamination and fortification

Food may occasionally become accidentally contaminated, or may even be deliberately adulterated for commercial gain [2]. Materials such as iron or folates may be added to make the food more nutritious. In quality control, the levels of key minerals in natural ingredients may vary seasonally or as sources change, so are important to check. The mineral levels in plant material are also characteristic of the soil in which it is grown, so can be used (with appropriate modelling in place) to

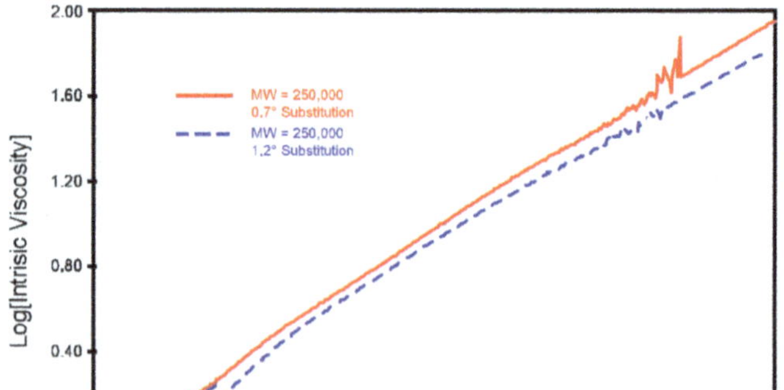

Figure 8.6. Mark–Houwink plot for carboxymethylcellulose with 0.7° and 1.2° substitutions. Reproduced with permission of Malvern Panalytical.

check that the product's origins are indeed as claimed. X-ray fluorescence (XRF) analysis is an important technique in this area.

8.4.1 X-ray fluorescence

XRF is a widely-used elemental analysis technique. X-rays are used to expel electrons from the inner shells of the atoms in a sample. Higher energy electrons will then emit photons with characteristic energies. These photons are collected and translated to a spectrum which details the elements a sample contains, and in what concentrations. XRF can be used to analyse all elements in the periodic table from beryllium onwards, from sub-ppm level up to 100% (figure 8.7).

The detection of photons can be achieved via two techniques. Historically, XRF spectrometers were based on wavelength dispersive XRF (WDXRF) technology. Excited photons were dispersed by a number of crystals, and all energies collected one-by-one. Advances over the last few decades have enabled the rise of spectrometers based on energy dispersive x-ray fluorescence (EDXRF) technology. The configuration of the optical paths does not require a crystal, and collective dispersion of the photons' energies is performed in the detector.

WDXRF is suitable for many applications, but the high power required in these spectrometers means that the technique may not be appropriate for general food applications. EDXRF uses significantly lower power for its analysis, and is generally appropriate for food applications. XRF is a non-destructive technology and can be used for the direct analysis of a wide range of sample types, including liquids, powders and solids. It is often much quicker than competitive technologies, and is well-placed for direct feedback in a production process.

In the food industry, elemental analysis can be used for various purposes, including nutrient analysis and the dosage of additives, to ensure product safety and increase production efficiency. Companies producing milk powder use XRF to dose the right additive concentration in the final product and ensure a consistently high-quality end product. XRF is frequently used to control the dosage of additives, to monitor and optimise the concentration of elements such as Na, Mg, Cl, Fe, Zn.

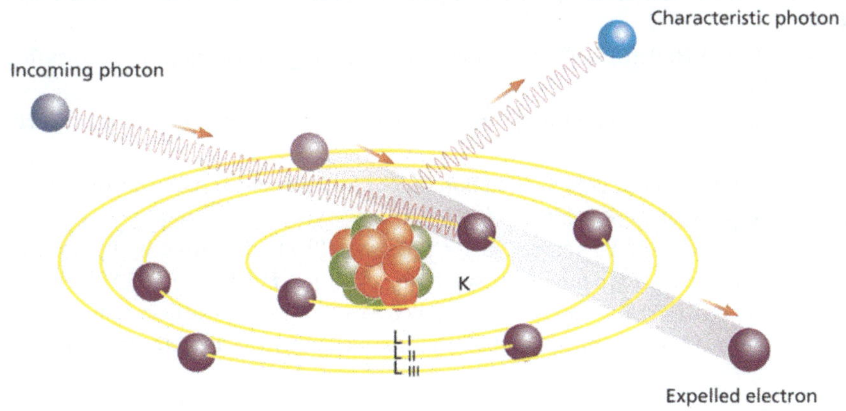

Figure 8.7. Production of characteristic radiation by XRF. Reproduced with permission of Malvern Panalytical.

8.5 Summary

This chapter has focused on many of the processes (milling, homogenisation and stabilisation) used within food formulation and processing and has covered the key analytical techniques applicable in this area. Another analytical technique which is involved in determining the safety of food through elemental analysis has also been discussed. Understanding food ingredients, their microstructure and their interactions is crucial to the effective production of safe and nutritious food products and minimising the cost of waste—ever more important for our growing population.

References

[1] https://gov.uk/government/publications/food-statistics-pocketbook-2017/food-statistics-in-your-pocket-2017-food-chain

[2] https://en.wikipedia.org/wiki/2008_Chinese_milk_scandal

[3] https://en.wikipedia.org/wiki/Homepride

[4] ISO 13320—Particle Characterisation—Laser Diffraction Method

[5] https://globenewswire.com/news-release/2018/10/22/1624439/0/en/Global-Chocolate-Market-Expected-to-Reach-USD-161-56-Billion-By-2024-Zion-Market-Research.html

[6] https://howitworksdaily.com/why-is-chocolate-so-tasty//

[7] Zhang Y, Guo Q, Feng N, Wang J-R, Wang S-J and He Z-H 2016 Characterization of A- and B-type starch granules in Chinese wheat cultivars *J. Integr. Agric.* **15** 2203–14

[8] Liu Q, Gu Z, Donner E, Tetlow I and Emes M 2007 Investigation of digestibility *in vitro* and physicochemical properties of A- and B-type starch from soft and hard wheat flour *Cereal Chem.* **84** 15–21

[9] Zeng J, Li G, Gao H and Ru Z 2011 Comparison of A and B starch granules from three wheat varieties *Molecules* **16** 10570–91

[10] McClements D J 2009 *Modern Biopolymers Science—Biopolymers in Food Emulsions* (Amsterdam: Elsevier) ch 4

[11] Chu B 1991 *Laser Light Scattering: Basic Principles and Practice* (New York: Academic)

[12] https://en.wikipedia.org/wiki/Einstein_relation_(kinetic_theory)#Stokes–Einstein_equation

[13] Xu R 2015 Light scattering: a review of particle characterization applications *Particuology* **18** 11–21

[14] Dalgleish D and Hallett F 1995 Dynamic light scattering: applications to food systems *Food Res. Int.* **28** 181–93

[15] Dalgleish D 2006 Food emulsions—their structures and structure-forming properties *Food Hydrocolloids* **20** 415–22

[16] Walstra P 1981 Effect of chymosin action on the hydrodynamic diameter of casein micelles *Biochim. Biophys. Acta* **669** 258–9

[17] Lankhorst P *et al* 2017 Prevention of tartrate crystallization in wine by hydrocolloids: the mechanism studied by dynamic light scattering *J. Agric. Food Chem.* **65** 8923–9

[18] Petrak D, Dietrich S, Eckardt G and Köhler M 2011 In-line particle sizing for real-time process control by fibre-optical spatial filtering technique (SFT) *Adv. Powder Technol.* **22** 203–8

[19] https://en.wikipedia.org/wiki/DLVO_theory
[20] Philips G O, Wedlock D J and Williams P A 1986 *Gums and Stabilizers for the Food Industry* vol 3 (Oxford: IRL Press) pp 3–5
[21] Komorowska P, Różańska S and Różański 2017 *J. Cellulose* **24** 4151
[22] Malvern Panalytical Application Note: Substitution in sodium carboxymethyl cellulose. https://malvernpanalytical.com/en/learn/knowledge-center/application-notes/AN090930SubstitutionSodiumCarboxy.html

Chapter 9

Physics in the rehydration and structure formation of recombined dairy products

Zachary J Glover and Melvin J Holmes

9.1 Introduction

Milk, the secretion of the mammalian mammary gland provides sustenance, nutritional components and confers immunological protection to the neonate. It is a remarkable evolutionary adaptation in not only providing a stable fat emulsion but also encapsulating insoluble calcium phosphate in a format readily accessible to the infant during digestion. Humans have exploited the milk of many, typically *Bovinae* or *Bovidae*, ruminant species to supply not only the fresh native form but also a multitude of derived products including cheeses and soured or fermented products. These products generate many textural and sensory profiles and benefit from significantly increased preservation capacities. Milk is ubiquitously used in foods from fresh milk, to structured fermented products and as an ingredient in an extensive array of fractionated and dried ingredients. Dairy systems can be investigated physically in many ways as they have interactions and structuring on various length scales that are interest to the food scientist and physicist alike. Milk is an oil in water emulsion (fat droplets in milk serum), a dispersion of protein (casein micelles), it can used to form fat or protein stabilised foams (whipped cream or aerated skimmed milk) and can be destabilised to form a gel, all of which have structures in the colloidal regime. Since as early as 1872, milk has been spray-dried to convert it into dry formats to increase shelf life and provide new product formulations [1].

However, in comparison with products produced from fresh milk the spray-dried equivalents have revealed sensory and textural differences which have yet to be understood. There is a current interest industrially to better understand the reconstitution and structure formation in dairy products formed from reconstituted ingredients, where functionality may have been altered from the native system due to processing. Obtaining precise knowledge of how processing affects protein

doi:10.1088/978-0-7503-2596-7ch9

functionality and the properties of reconstituted powders is crucial in the controlled production of dairy or pharmaceutically derived products.

In this chapter, we will briefly describe the compositional and functional attributes of milk and will aim to report recent developments which have aimed to identify the chemical and structural changes which occur during the drying and subsequent reconstitution processes. This will be articulated by examination of the rehydration and dissolution properties of spray-dried milk and the subsequent gelation and textural attributes in comparison with fresh-made equivalents. The use of ultrasound spectroscopy and super-resolution microscopy with quantitative image analysis methodologies will be explained, which will provide an insight into the precise novel characterisation and modelling of the onset of gelation in milk protein structures but also into the quantification of protein domain size, inter-pore distance and fractal dimensionality of the gels so formed.

9.2 Composition and nutrition

Bovine milk has a very diverse range of compositional properties due to the sources from derived species, breeds, feed, seasonality, etc. Typically, composition is in the following proportions, water (85.5%–88.7%), milk fat (2.4%–5.3%—solid at 20 °C and liquid at 37 °C), protein (3.3% with 78.8% caseins 19.4% whey), lactose (4.6%) and many trace minerals (Ca, N, P, citrate, Mg, K Na, Zn, Cl, Fe, Cu), enzymes (e.g. phosphatase and lipase) and vitamins (A, C, D, riboflavin and thiamine) and so is clearly a highly nutritious food source. Milk fats include the long chain saturated fatty acids myristic (C14), palmitic (C16) and stearic (C18) and unsaturated oleic (C18:1) and short chain fatty acids butyric (C4), caproic (C6), caprylic (C8) and capric (C10) plus some mono and diglycerides and phospholipids and cholesterol. Generic physical properties include density ρ 965 kg m^{-3}, thermal conductivity (20 °C) 0.17 J m^{-1} s^{-1} K^{-1} and a refractive index (20 °C at 589 nm) of 1.462 important in the optical measurement of milk systems [2–7].

Due to the above composition, milk is essentially an oil-in-water emulsion system combined with a colloidal suspension of casein and associated micellar aggregates. Caseins and whey proteins deliver nitrogen in approximately the same proportion of their occurrence and this is important in terms of the conformational properties of the proteins.

9.3 Chemical and functional properties

Amino acids form the smallest constituent components of the composition and structure of caseins and whey proteins within milk. Amino acids consist of a weak basic amino group and a weak acid carboxyl group with the addition of various length hydrocarbon chains. By convention, a monomer (amino acid) within an extended polymeric chain is named a residue, this provides sufficiently clear nomenclature to describe polymer composition in terms of the number of residues within a chain. With successive links and cross links through disulphide bonds, the larger defined category of proteins may be defined, e.g. as in caseins within milk. Conjugated proteins may also be formed with non-polypeptide groups, e.g. polysaccharides which increases potential for interactions with surrounding chemical species.

Shorter chain lengths of (typically < 50 amino acids linked by amide bonds) are referred to as peptides. Peptides extracted from caseins include bovine casein hydrosylate, calcium caseinate and casein hydrosylate, and are presented as food additives and dietary supplements for hypertension and high cholesterol. Polypeptide chains consist of a carboxyl C terminus and an amino N-terminus. In conformation, proteins may adopt a diverse range of intricate structures including beta sheets somewhat similar to pleated structures and alpha helices which may confer rigidity to a protein.

Casein derived from milk has been used from antiquity through modern times for the purposes of not only foodstuffs but also paints, glues, plastics, medical and dental materials. As such, significant research has been devoted to the understanding and control of milk proteins and their properties with consensus on many of the chemical and structural features. Caseins occur in four genetic forms α-S1, α-S2, β, κ with a number of phosphorylated forms shown in brackets, α-S1 (6), α-S2 (4), β (6), κ (2) [7]. Caseins combine into a structural form of a micelle facilitated through conjugation with phosphate groups esterified with serine residues binding calcium. In addition, high rigidity cyclic proline residues promote curved conformations and inhibit close-packed structures, which further expose hydrophobic residues resulting in the insoluble nature of caseins. All the caseins have approximately 20 000–25 000 molecular mass, 200 residues of which 10–40 are proline. α-S1 is composed of two hydrophobic regions containing most of the proline residues and a central phosphate polar region. α-S2 exists with N-terminus negative charge and C-terminus positive charge. Both α-S1 and α-S2 can be precipitated at low levels of calcium. β casein exhibits a high hydrophobic capacity in the C-terminal region and is highly charged at the N-terminal portion endowing the protein with amphiphilic properties and greater resilience to calcium precipitation. Intermolecular bonds, promoted by disulphide bonds require cysteine residues, these exist in both α-S2 and κ caseins but are absent in α-S1 and the β caseins. Notably, κ casein molecules may be linked by approximately 10 disulfide bridges in contrast with α-S2 exists in a dimer linked conformation. κ casein exhibits significant resistance to calcium precipitation and mediates stabilizing behaviour toward the other casein groups. κ casein comprises of para- κ casein which is hydrophobic in nature and a hydrophilic glycomacropeptide (GMP) tail and is principally located at the micellar casein surface offering a steric depth 4–10 nm. The hydrophobic properties of caseins confer stabilising potential supported by κ casein steric hindrance. The colloidal micellar system may destabilise, i.e. aggregate through the addition of Ca^{++} and be reversibly recovered through removal. This activity is obviously dependent upon salt concentration of the serum and pH. Low pH will dissolve environmental CaP initiating casein precipitation.

Due to the chemical and structural properties of caseins, a significant capacity to form colloidal particles, namely micelles, has been observed which has resulted in much study and debate as to the mechanisms for the behaviour. Proposed initial models have included the aggregation of submicelles but the consensus of recent enquiries describes the open accumulation of covalent and electrostatic bonding of protein and calcium phosphate (CaP). These structures are open, mobile and harbour significant water holding capacity. Micelles encapsulate or organize insoluble CaP in a format amenable to neonatal assimilation. Typically, micelles are of the order 50–400 nm with an average of around 250 nm diameter, as observed using a variety of optical techniques, e.g. high-resolution SEM. The structural capabilities of casein micelles have an appreciable effect on the textural and sensory attributes of derived products. These particular aspects are now being investigated in order to optimise quality attributes in food processing, in particular with the production of yogurts and other fermented products which have been formulated form dried skimmed milk powders.

In addition to the casein, approximately 20% total protein content is associated with whey or serum proteins soluble at pH4.6 comprising of α-lactalbumin (α-La) 12%, β-lactoglobulin (β-Lg) 5%, non-nitrogenic proteins 3% and proteose peptones 2%. Whey proteins have a compact globular structure and have been shown to exhibit improved emulsification capabilities in comparison to casein and derived isolates as a result of their molecular characteristics. Upon denaturation, whey proteins can offer a variety of attributes to processed milk products. In the following sections we briefly discuss the processing of milk, its effects on its constituents and the derived attributes of final products.

9.4 Milk processing

Heating can have a multitude of effects on the chemistry of the components in milk. One of the most significant is the denaturation of whey proteins [8]. The casein proteins do not have a typical tertiary structure and so are not denatured by heating, although will decompose at extreme temperatures. The whey proteins α-La and β-Lg are globular in nature, and when denatured expose a free thiol group which is usually unavailable for reaction as it is held in a hydrophobic pocket within the protein. Exposed thiol groups can lead to disulphide bridge formation between other whey proteins or with the κ-casein present at the surface of the micelle [9–12]. If whey proteins adsorb to the surface of the micelles this can lead to changes in how the micelles will aggregate following destabilisation, either through acidification in fermented dairy products, or following renneting in traditional cheese making. Depending on the characteristics of the product required, excessive heat treatments may be undesirable or confer beneficial effects to the product. β-Lg present in a gel increases the water binding capacity and alters the texture of a milk gel, and therefore a pre-heat treatment beyond standard pasteurisation is preferable in yoghurt manufacture. In cheese making, higher levels of whey protein in the cheese matrix can alter the moisture balance and provide alternative substrates for the bacterial culture to ferment, leading to different aromatic compounds being formed, which can alter the quality of the final cheese.

Spray drying provides an effective means of converting a concentrated liquid feed into a powder. Whilst the conversion of milk into powder is well established, there are continuous developments in the fractionation of milk into its constituent components. Fractionation can result either from aiming to isolate specific components with unique functionalities, or it can be to maximise the potential for liquid streams that were once essentially waste. The most obvious example being the modern use of the whey fraction that is drained during cheese making, which is an extremely versatile ingredient and, in some cases, more profitable than the cheese product. 30 years ago, the whey stream would have been used for animal feed, whereas currently a wide variety of high value whey powder products are available.

Powdered products have the benefit of an extended shelf life, reduced shipping volume and can be easily used to boost the solids content of another liquid feed. One of the current challenges surrounding recombined products is understanding of how the process of powder manufacture affects the quality and functionality of the recombined liquid feed compared to the fresh equivalent. In order to understand this fully, methods are required to track the process of powder rehydration and subsequent structure formation.

The first stage in powder manufacture is the concentration of the milk or liquid feed into a concentrated slurry that can feed into the spray drier (approx. 35% solids). The most common method of concentration is to use a multiple effect evaporator. In a single effect evaporator, the liquid is fed into a long tube called a calandria that is orientated vertically. The liquid falls down the inside surface of the calandria. The calandria tubes are encased in a sealed compartment into which concentrated steam is fed to create a heat exchanger. Heat from the steam passes through the wall of the calandria and into the falling liquid. The calandria are kept under vacuum at a negative pressure to lower the boiling point of the fluid. Lowering the boiling point has the advantage of being more energy efficient and preventing excessive thermal damage to the milk components. Thermal damage during processing is likely to lead to fouling, i.e. accumulation of surface components, which would require more intensive cleaning, and reduce the heat transfer efficiency. As the liquid falls and heats up in the calandria it will reach its boiling point causing some amount of water (and volatile components) to evaporate. The concentrated feed keeps falling and is separated from the vapour in a cyclone separator which minimises vapour entrainment [7, 13].

The principle behind a multiple effect evaporator is that the concentrated feed is now fed into a second calandria. This stage is operated at a higher vacuum; therefore, the boiling point of the fluid is reduced further. The heat source for the second stage is the vapour that has been removed in the previous stage. This process can then be repeated a number of times, each effect operating at a lower pressure than the previous one for up to seven effects [7, 13]. Efficiencies of the process can be increased by recompressing the vapour from the evaporator either thermally or mechanically. Thermal vapour recompression, mixes atmospheric steam with the vapour from the evaporator, leading to a compression of the vapour and increase in temperature, maximizing the potential for the vapour to be used in subsequent effects. Mechanical vapour recompression increases the temperature of the vapour from the evaporator using a pump to physically compress the vapour [7, 13]. A two-effect evaporator with a thermocompression requires 0.25 kg steam per 1 kg of water removed compared to 0.6 kg without recompression. A five-effect evaporator requires 0.20 kg of steam per 41 kg water removed compared to 0.40 kg without recompression. Concentration is limited by viscosity, as the heat transfer coefficient decreases and fouling becomes more likely, this typically goes to 48% solids for skimmed milk and 52% solids for whole milk [7, 13]

One main advantage of working under vacuum is the reduction in temperature, below the denaturation temperature of the whey proteins (65 °C). However, although the evaporators are run at a lower pressure to reduce to temperatures that the feed is exposed to, this stage of manufacturing is where most of the thermal damage occurs to the milk components due to the higher temperatures and long exposure times. Also, temperature variation occurs between proteins directly contacting the calandria surface to the bulk, despite the high degree of fluid/vapour flow.

An alternative method of concentration that has fewer thermal effects is membrane filtration. Successive filtration operations with membranes of different pore size facilitates effective fractionation of the various components in milk. Microfiltration (MF) utilities a membrane with pore sizes of >0.2 μm at 1 bar pressure. MF can be used to remove smaller particles and microorganisms and control the level of fat in whey streams. Ultrafiltration (UF) operates with pore sizes between 3–300 nm at 1–10 bar pressure. UF can separate molecules and is primarily used for concentrating proteins in milk and whey and is therefore useful in standardising protein contents in liquid feeds. Nanofiltration (NF) has pore sizes of 1–10 nm at 20–40 bar pressure. NF can remove monovalent ions and is of use for partial desalination of whey or UF permeate. Reverse osmosis (RO) has pore sizes of 0.1–1 nm at 30–60 bar pressure. RO can be used to remove water from a liquid feed, so can be used for concentration or obtaining clean water from cleaning streams. The main disadvantage of filtration-based concentrations methods is the increased operating costs [7, 13].

Concentrated feeds may be used directly for product formulation or be further treated to remove liquid content through spray drying for example. The main advantage to spray drying is that a large volume of material can be processed with lower temperature exposure in comparison with other techniques such as roller drying. The material is atomised into the top of a spray drying tower where there is hot turbulent air. Individual droplets begin to lose moisture from the surface which can lead to the formation of a case. The rate of moisture loss then depends on the rate at which moisture can migrate through the drying droplet and reach the surface. Temperatures inside the drying tower can reach 200 °C, but due to evaporative cooling/latent heat, the temperature of the actual particle should not exceed 90 °C. The dry particles are separated from the hot air flow and fines in a cyclone separator. Alterations in drier design can impact the quality of the powder produce. For example, the size of the tower will affect the time it takes a droplet to fall through and pass through the drying tower. The nozzle design and number of nozzles in the tower is important for the size and uniformity of the droplets that will be released into the tower and the viscosity of feed than can be atomised [7, 13].

9.5 Milk powder

There are several parameters that affect the quality of the final powder that relate to the ability of the powder to rehydrate and form a liquid feed that has the desired functionality. The powder must have high wettability, i.e. not behave hydrophobically to ensure easier rehydration and mixing during processing. Wettability can be increased by agglomeration, whereby the dry powder is moistened in steam to increase cohesive properties and then re-dried on a fluidised-bed drier. This increases the effective size of the particles but also introduces capillary channels between the powdered particles. Agglomeration reduces the surface area: volume ratio, therefore there is a lower surface energy required to wet the agglomerate than the same volume of individually wetted powder particles. The capillary channels draw liquid into the structure and increase the wetting rate. Powder particles can be instantized through the addition of an emulsifier to the surface, which reduces the surface energy of wetting. In addition, dependent on the final application, e.g. coffee, the powder may need to be stable under high temperatures or acidic conditions [14].

There are several stages to rehydration once the powder has been added to a liquid. Initially, the powder particles must disperse in the liquid. Dispersion is followed by swelling of the powder particles. The individual powder particles then begin to disintegrate into their constituent components [15–17]. An important step at this stage is the rehydration of the individual components. The system will not have reached equilibrium once the powder particles have disintegrated [15, 18–23]. Particle size information is not sufficient to determine whether the system has reached an equilibrium state. The equilibrium point is important in milk due to the complex mineral balance, and the fact that the mineral balance contributes to a structural role in the casein micelles.

9.6 Structure formation

Once the product has been recombined it can then be used for product manufacture. Producing a product often requires controlled destabilisation of the system in order to lead to structure formation. For many dairy products this begins by destabilising the casein micelles through acidification or renneting, either collapsing or cleaving the stabilising κ-casein layer. Micelles can then aggregate, and over time extend longer range interactions until a network spanning system, a gel, is produced.

It is likely that alterations in the mineral balance of the recombined product lead to changes in the functional behaviour of the milk system. The degree of thermal exposure the product received over its entire processing history will contribute further to the altered behaviour of the recombined liquid. The presence of denatured whey proteins leads to changes in the microstructure of a gel, its mechanical properties and water binding capacity. The use of recombined milk compared to fresh milk leads to changes in microstructure and rheology in derived gels the route of which is not explained solely through differences in denatured whey proteins and instead is likely to lie in alterations in the mineral balance of the recombined liquid.

Ultrasound spectroscopy has been applied to material and food characterisation, in bulk measurements and under dynamic processing conditions [24–28].

9.7 Ultrasound techniques

Ultrasound, namely acoustic frequencies above human audible frequencies (16 kHz) consist of pressure waves propagating through an intervening medium and comprises of longitudinal waves in the direction of propagation which is parallel to the incident plane. In solid materials transverse waves can be supported which are perpendicular or at other angles to the incident plane, although in fluids and gases these quickly decay (evanescent waves). The primary quantities to describe ultrasonic waves are the speed of propagation and attenuation, i.e. the decay in the initial signal amplitude per unit length propagation. In the former case of an acoustic wave propagating in a single phase medium, Newton first derived the isothermal speed v (ms^{-1}) (equation (9.1)) by considering the pressure differential ΔP on a compressed column of gas having density ρ (kg m^{-3}) and compressibility κ (Pa^{-1}). Newton incorrectly assumed that this was the adiabatic velocity which was later demonstrated by Laplace which is the case provided the adiabatic condition (equation (9.2)) is satisfied.

Here τ (W m^{-1} K^{-1}) denotes the thermal conductivity of the medium, C_V (J kg^{-1} K^{-1}) the specific heat capacity at constant volume and f (Hz) the frequency of the pressure perturbation. Here we note the relation $v = \lambda f$ (λ wavelength m) thus with a given wave speed we may determine the acoustic wavelength e.g. water $\lambda = 1.48$ mm.

Laplace equation:

$$v^2 = \frac{1}{\rho\kappa}. \tag{9.1}$$

Adiabatic condition:

$$\frac{2f\tau}{v^2\rho C_V} \ll 1. \tag{9.2}$$

In one dimension, the attenuation or amplitude fluctuation $A(x,t)$ at the spatial position x units form the origin at time t may be represented by the exponential function shown in equation (9.3) where $\omega = 2\pi f$ is the angular frequency, $i = \sqrt{-1}$ is the imaginary number and $k_C = k_C' + i\alpha$ is the complex compressional wavenumber.

$$A(x,\, t) = A_0 e^{i(\omega t - k_C x)}. \tag{9.3}$$

If the wave amplitude decreases during propagation, i.e. is attenuated then the parameter α (Nepers or dB) is the attenuation coefficient often expressed at particular frequencies with units α [dB MHz^{-1} cm^{-1}]. k_c' is the real part of the acoustic wavenumber defined as in equation (9.4).

$$k_C' = \frac{2\pi}{\lambda} = \frac{\omega}{v} = \frac{2\pi f}{\nu}. \tag{9.4}$$

Using these basic quantities and the facts that low power ultrasound is non-destructive, can be applied non-invasively and furthermore can probe the interior of solid or opaque materials, ultrasonic methods have been used extensively in many applications. Notably, medical imaging has benefited hugely from the advantages of ultrasound techniques. The real-time scans of an in-utero foetus or damaged or diseased internal organ are now routine and safe procedures. Also, in the the case of high-power focussed ultrasound, non-invasive ablation is achievable in conditions such as kidney stones. With appropriate selection of ultrasonic frequency or wavelength, ultrasonic microscopy can be also achieved providing the potential to interrogate the cellular level [29, 30].

Elsewhere in other fields, extensive investigations and uses have been developed in material testing and the characterisation of physical properties such as density, compressibility, elastic moduli and viscosity and in the monitoring of phase state, the initiation of crystal nucleation or evaluating enzymic activity [31–38].

In the food industry, ultrasonic techniques are used to characterise fats, oils and emulsions [39–43], and provide methods to acquire and utilise processing and quality markers [39–43]. Precise measurement of water and air (velocity and attenuation) under a variety of temperature and pressure conditions provides repeatable reference and calibration standards with which to analyse materials to a high degree of precision [44–48]. For an extensive research list refer to [49]. In terms of emulsions and particle sizing, ultrasound has been researched since the seminal work of Strutt [50–52], who first introduced the elastic scattering of a single spherical droplet in a continuous media subject to an incoming plane acoustic wave. This theory was further extended to include viscosity and thermal conduction using appropriate pressure and temperature coupling, i.e. Navier–Stokes and energy conservation by Epstein and Carhart [53], and subsequently by Allegra and Hawley [54], to address solid particles. The model provides a linearized solution of the compressional φ, thermal y and shear c fields both in the continuous and dispersed domains by application of appropriate boundary conditions at the interface of the droplet. This is the well-established theoretical model ECAH which permits the calculation of the attenuation α [Np m^{-1} or dB m^{-1}] due to a particle of a known radius r and volume fraction ϕ at each frequency f (Hz) of the incoming wave. The spherical harmonic solutions, consisting of radial spherical Hankel functions h_n and Legendre polynomials P_n of the exterior compressional field take the form as shown in equation (9.5)

$$\varphi = \sum_{n=0}^{\infty} i^n(2n + 1)A_n h_n(k_C r)P_n(\cos\theta) \tag{9.5}$$

where k_c denotes the more complete compressional wavenumber given in equation (9.6) and

$$k_C = \frac{\omega}{v} + i\alpha = \frac{\omega}{v} + i\frac{\eta\omega^2}{2\rho v^3}\left[\frac{4}{3} + \frac{\mu}{\eta} + \frac{(\gamma - 1)\tau}{\eta C_p}\right]. \tag{9.6}$$

Here v is the velocity in the phase, ω the angular frequency, η is the shear viscosity (N s m^{-1}), ρ is the density (kg m^{-3}), μ is the bulk viscosity (N s m^{-1}), γ is the ratio of specific heats, τ is the thermal conductivity (W m^{-1} K^{-1}), C_p is the specific heat at constant pressure (J kg^{-1} K^{-1}) and A_n are the scattering coefficients determined from solution of the boundary value problem.

Under specific model assumptions, e.g. point-particle, dilute systems with non-overlapping thermal fields, Lloyd and Berry [55] integrated the intrinsic absorption and scattering from the single particle solution to account for an ensemble of droplets and provided a multiple scattering theory, thus enabling determination of an effective wavenumber B defined as in equation (9.7). The zero A_0 scattering coefficient is a thermal monopole and A_1 a viscous dipole term. For full details and model assumptions, refer to [34–36, 49, 56].

$$\left(\frac{B}{k_C}\right)^2 = 1 - \frac{3i\phi}{k_C^3a^3}(A_0 + 3A_1) - \frac{27\phi^2}{k_C^6a^6}\left(A_0A_1 + \frac{10}{3}A_0A_2 + 2A_1^2 + 11A_1A_2 + \frac{230}{21}A_2^2\right). \tag{9.7}$$

The real and imaginary components of the effective wavenumber B now supply the equivalent dispersion quantities in equation (9.8);

$$\mathfrak{Re}(B) = \frac{2\pi}{\lambda} = \frac{\omega}{v},\ \mathfrak{Im}(B) = \alpha. \tag{9.8}$$

The scattering contribution attributed to the ensemble of particles is characterised by the excess attenuation (equation (9.9)) determined by subtracting the attenuation from the pure continuous phase and from the dispersed droplets in proportion to their associated volume fractions, i.e.

$$\alpha_{\text{excess}} = \alpha_{\text{total}} - (1 - \phi)\alpha_{\text{continuous}} - \phi\alpha_{\text{dispersed}}. \tag{9.9}$$

Experimentally, measured attenuation spectra at each frequency f_i ($i = 1,\ldots,n$ frequencies) can be evaluated and minimised against model predictions, e.g. using a sum-squared residual fit (SSD) as in equation (9.10).

$$SSD = \sum_{i=1}^{n}[\alpha_T(f_i) - \alpha_E(f_i)]^2 \tag{9.10}$$

(T—theoretical, E—experimental) thus providing particle size and volume fraction estimates toward the particle size and volume fraction distributions. Instruments such as the Malvern Ultrasizer MSV may provide acoustic spectra in the range 1–120 MHz to with precision ±0.1 dB in^{-1} and at volume fractions from 0.5 up to 50% dependent upon physical properties and particle size determination in the range 0.01 μm to 1000 μm [34].

Many studies have utilised these approaches for particle sizing and in the investigation of novel emulsion systems. For brevity and in the context of structural evaluation of dairy products, herein we shall present recent results which highlight the capability of ultrasound to monitor the rehydration of casein micelles, skimmed milk powder and in the reversible aggregation through shear. This is attained through examination of the acoustic spectrum and attenuation measurements. These aims bid to optimise product processing and quality.

In figure 9.1(a), the attenuation spectra of 0.8% (w/w) casein dispersions are shown highlighting the reductions in attenuation due to increasing hydration. A time period of 0–20 h with intervals of approximately 30 min is presented. As hydration increases, scattering from the initial aggregates diminishes as the large scale structures fragment followed by micelle swelling to reduce disparities in densities between the dispersed and continuous phases. This latter aspect reduces the scattering potential. This process is typical considered to be a relatively fast process, e.g. approximately 1 h. Ultrasound spectroscopy herein indicates that this process is in fact a significantly longer process. The initial stage is dominated by fragmentation of the micelle aggregates and then due to the hydrophobic potential of the micelles during uptake of the continuous phase. Figure 9.1(b) shows the non-dimensional product of attenuation and wavelength of the initial and final state after 17 h hydration. Initial mean particle size (diameter) over volume (also or DeBroukere mean) D_{43} estimate is 1.7 μm and final prediction size is 50 nm. The final sizes of the particles in all casein samples to date have been predicted in the size range 50–100 nm which is in agreement with published findings. Figure 9.1(c) shows a comparison of micellar casein dispersion, whey protein isolate emulsion and 10% skimmed milk powder fully hydrated. In the latter case, attenuation contributions from the relevant fractions of the skimmed milk powder may be evaluated. Figure 9.1(d) shows reversible switching on and off aggregation at pH 4.6 (isoelectric point) of the

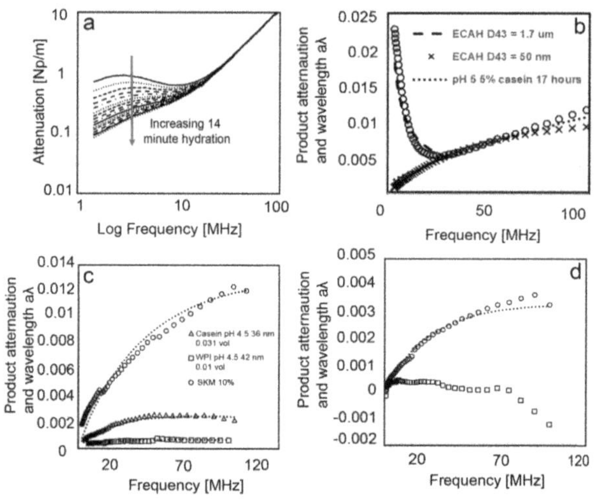

Figure 9.1. (a) The attenuation spectra of 0.8% (w/w) casein dispersion showing the reductions in attenuation due to increasing hydration. Time period 0–20 h with intervals of approximately 30 min; (b) shows the non-dimensional product of attenuation and wavelength of the initial and final state after 17 h hydration. Initial mean particle size (diameter) over volume (also or DeBroukere mean) D_{43} estimate is 1.7 μm and final prediction size is 50 nm; (c) shows a comparison of micellar casein dispersion, whey protein isolate emulsion and 10% skimmed milk powder fully hydrated; and (d) shows reversible switching on and off aggregation at pH 4.6 (isoelectric point) of the dispersion through the application of shear, clearly showing the increased scattering phenomenon when particles are evenly distributed throughout the sample. Data for figure 9.1 was acquired by Qinya Yu and included in her Master's thesis alongside her contribution to [57].

Table 9.1. Physical properties of water and casein at 25 °C used in the numerical calculations to solve ECAH model [34, 45, 47, 56].

	Water	Casein
Ultrasound velocity (m s^{-1})	1497	1563
Density (kg m^{-3})	997	1076
Specific heat capacity (J kg^{-1} K^{-1})	4177	3818
Thermal conductivity (W m^{-1} K^{-1})	0.611	0.521
Thermal expansivity (K^{-1})	2.1×10^{-4}	7.5×10^{-4}
Attenuation exponent (MHz^{-2})	2.0	1.4
Attenuation coefficient (Np m^{-1})	0.023	4.02

dispersion through the application of shear, clearly showing the increased scattering phenomenon when particles are evenly distributed throughout the sample. All theoretical predictions are performed using equations (9.7)–(9.10) but requires the detailed solution of the ECAH model, for details refer to [34]. The thermo-physical properties of the continuous and dispersed phases used to make all the necessary ECAH model calculations are shown in table 9.1 [57].

Hydration time is a critical feature to ensure stability of measurements and clearly also a significant factor when considering product formulation and processing. Ultrasound spectroscopy can quantitatively assess hydration, agglomeration, dispersion and discern differences between milk protein constituents. These capabilities will provide a valuable and insightful tool to monitor the rehydration of milk powders and inform product technologists of specific process and quality measures to ensure equivalence between dairy derived products produced from fresh and dried ingredients. In conjunction with complementary methods such as the optical methods described herein, a complete understanding of powder to protein gels can be realised.

9.8 Structure characterisation

A suitable methodology is required order to be able to determine how various processes have affected food microstructure. Such a method must be able to describe the native microstructures, and the structures formed following different processes or formulations. If the point at which a reconstituting powder has reached equilibrium can be determined with ultrasound spectroscopy as the results above suggest, it would then be possible to compare the structures formed versus the fresh equivalent. Differences on a nano or micro scale have been shown to relate to changes on the macro scale. A methodology based on precise spatial features can be used to optimise recombination processes to minimise differences between the fresh and recombined products or explore novel structures that are possible using recombined ingredients.

Optical microscopy has been widely used to investigate the structures of food products and colloidal materials. Confocal microscopy has provided a powerful tool for the food scientist to be able to explore multi-component materials, which can be optically sectioned, giving an indication of the three-dimensionality of the microstructure [58–65]. The resolution of optical microscopy methods has been limited by the diffraction of light to approximately 250 nm. Many components and interactions of interest in the colloidal regime fall below this length scale. Higher resolutions have been achieved with non-optical microscopy such as transmission electron microscopy (TEM) [66, 67] and scanning electron microscopy (SEM) [65, 68, 69]. The disadvantage to these techniques is the sample preparations methodologies required, which would often lead to the fat phase being lost from the sample or requiring the sample to be fixed, which has the potential to lead to artefacts in the images [65], in such cases cryo-SEM may offer an alternative if the sample can be frozen [70].

The development of super-resolution microscopy has opened a new avenue to be able to investigate food structures down to the nanoscale, without requiring freezing, drying or fixation. Samples can be imaged under conditions that are of industrial relevance [71, 72].

Stimulated emission depletion (STED) microscopy shared the Nobel prize in chemistry in 2014 for overcoming the diffraction limit of light [73]. STED is based on a confocal microscope set up where a laser is used to excite a fluorophore in the sample [74]. The fluorophore transitions to an excited state, lose some energy to reach the lowest vibrational energy state of the excited state and then emits a photon of fluorescent light to return to its ground state. The fluorescent light is collected by the objective lens and passes to a detector. By scanning the laser across the sample, the image is built up pixel by pixel. The use of confocal pinholes in the microscope's optical path allows out of focus light to be blocked out, only 'in-focus' light is collected, which is from a relatively thin (600 nm) section. Images can therefore be acquired by changing the focal position to build up a 3D image. The resolution of confocal microscopy is still diffraction limited to a resolution of 250 nm in x and y [75]. STED microscopy utilises a second laser, called the depletion beam. The depletion beam passes through a phase plate which shapes the beam into a torus, with zero intensity at the centre. The depletion beam is overlaid with the excitation beam. As the lasers are scanned across the sample the excitation beam excites the fluorophores in the sample as before, however, wherever the depletion beam interacts with an exited fluorophore it forces that molecule to emit fluorescent light at the same wavelength as the depletion beam through stimulated emission. The photons from the fluorophores that have undergone stimulated emission is filtered out. Therefore, light is only collected from the central region of the torus where the fluorescent light is not depletion, as this is a smaller area the pixel size is smaller and so the final image has a higher resolution [75, 76]. The practical resolution of STED is approximately 30 nm and it has been used recently to resolve structures in dairy gels to below 100 nm, which is on the order of a single casein micelle.

9.9 Quantitative image analysis

To extract the maximum level of information from super-resolution microscopy images must be quantitatively analysed. The microstructures of many food products are far too complex to be subjectively analysed by eye, particularly when the features are irregular and fractal in nature. Methods of analysis based on 2D spatial correlation analysis have recently been applied to analyse images of dairy gels consisting of a protein network or protein network with embedded fat droplets. Autocorrelation analysis has been used to determine the typical length of protein domains in a gel or size of fat droplets, the inter-pore distance in the gel and the fractal dimension of the network. Cross-correlation analysis has been used to assess the relative spatial distribution of protein around fat droplets, determining the typical distance between a fat droplet interface and the maximum local protein intensity. Separation distances have been measured to be below the diffraction limit of light, a measurement only possible with a super-resolution technique that has not stripped the fat from the sample. 2D spatial correlation methods of image analysis have so far been used to significantly differentiate the dairy gel microstructures of samples produced from fresh or reconstituted milk, heat treated and non-heat treated milks, gels induced from acid or rennet and whole milk samples that have either been homogenised or not.

The 2D spatial correlation analyses are based on either the auto or cross-correlation image or either one image or a pair of images, respectively. The 2D spatial correlation image (G,a,b) is defined as:

$$G(a, b) = \sum_{x=1}^{M} \sum_{y=1}^{N} I_1(x, y) \cdot I_2(x - a, y - b) \tag{9.11}$$

where $I_n(x,y)$ are the intensity values in the two images of $M \times N$ pixels. If the images are the same the autocorrelation image is generated, if they are different the cross-correlation image is generated. a and b are spatial lags from coordinates x and y [71, 72, 77].

The Wiener–Khinchin theorem allows the autocorrelation image to be calculated in a computationally efficient way based on the power spectrum image $S_{I1,I2}$ [78],

$$S_{I1,I2}(kx, ky) = \mathcal{F}[I_1(x, y)] \cdot \mathcal{F}^*[I_2(x, y)] \tag{9.12}$$

$$G(a, b) = \mathcal{F}^{-1}[S(I_1, I_2)] \tag{9.13}$$

where (kx, ky) are co-ordinates in frequency space, $\mathcal{F}[I_1(x, y)]$ is the Fourier transform of Image 1, $\mathcal{F}^*[I_2(x, y)]$ is the conjugate of the Fourier transform of Image 2, $G(a, b)$ is the spatial correlation image and $\mathcal{F}^{-1}$ is the inverse Fourier transform [71, 77].

The correlation images can be normalised which can remove any variation in pixel intensity that may have arisen in the acquisition system. The normalised correlation image, $g(a, b)$, can be defined as [71, 77, 79–81]:

$$g(a, b) = \frac{\mathcal{F}^{-1}[\mathcal{F}(I_1 - \langle I_1 \rangle) \cdot \mathcal{F}^*(I_2 - \langle I_2 \rangle)]}{\sigma_{I1} \cdot \sigma_{I2}}. \tag{9.14}$$

The auto and cross-correlation images generated from images of isotropic systems can be radially averaged and interpreted. It has been shown that a stretched exponential model can be fit to the radial decay of an autocorrelation image to extract the typical length, ξ, of the protein domain or the size of fat droplets. The form factor, β, can provide information on the shape of the building blocks of the network [82].

$$p(r) = \mathrm{C}e^{-(\frac{r}{\xi})^{\beta}}. \tag{9.15}$$

The stretched exponential model has been adapted to include a dampened oscillator, λ, which extracts the inter-pore distance in microscopy images of porous materials, where r and r_1 are spatial displacement and the starting spatial location, respectively [71].

$$p(\mathrm{r}) = Ce^{-(\frac{r}{\xi})^{\beta}}\cos\left(\frac{2\pi(r - r_1)}{\lambda}\right). \tag{9.16}$$

The fractal dimension of an image can be determined from the radial average of the power spectrum image. When plotted on a log-log plot the gradient of the slope relates to the fractal dimension:

$$D_f = 4 - (\beta/2) \tag{9.17}$$

where β is the gradient of the decay [71, 77, 83, 84].

A two-channel microscopy image of two different components such as fat and protein can be analysed through a 2D spatial cross-correlation [85, 86]. For isotropic systems the cross-correlation image can be radially averaged. Cross-correlation analysis has been applied to whole milk gels, containing fat droplets dispersed in the protein matrix. For systems where the fat is associated with the protein network it has been shown that the maximum point of the radially averaged cross-correlation curve is equal to the fat droplet radius and the distance to the point of maximum intensity of the proximal protein distribution. The fat droplet radius can be determined by through autocorrelation analysis of the fat channel image and extraction of parameter ξ. From the cross-correlation curve and autocorrelation of the single channel a separation distance from the fat droplet interface to maximum protein intensity can be determined [77]. Figure 9.2 shows the process involved in the autocorrelation and cross-correlation analysis for two copies of one image or two different images, respectively. In this example the autocorrelation is used to find the size of the fat droplets by fitting with equation (9.16), alternatively the protein network could be characterised through autocorrelation analysis using equation (9.17).

The cross-correlation analysis can be applied globally to pairs of two-channel images, which gives the ensemble average of all sizes of fat droplets in the image. The analysis can be run in a region-specific manner, in order to focus on individual fat droplets. Both the distance to the protein maxima, and the mean protein intensity can be determined in radial segments and plotted as a function of angle around an individual droplet. Region specific analysis allows smaller length scales to be probed in a manner that is both visual and intuitive on a subjective level but backed up with quantitative information [77].

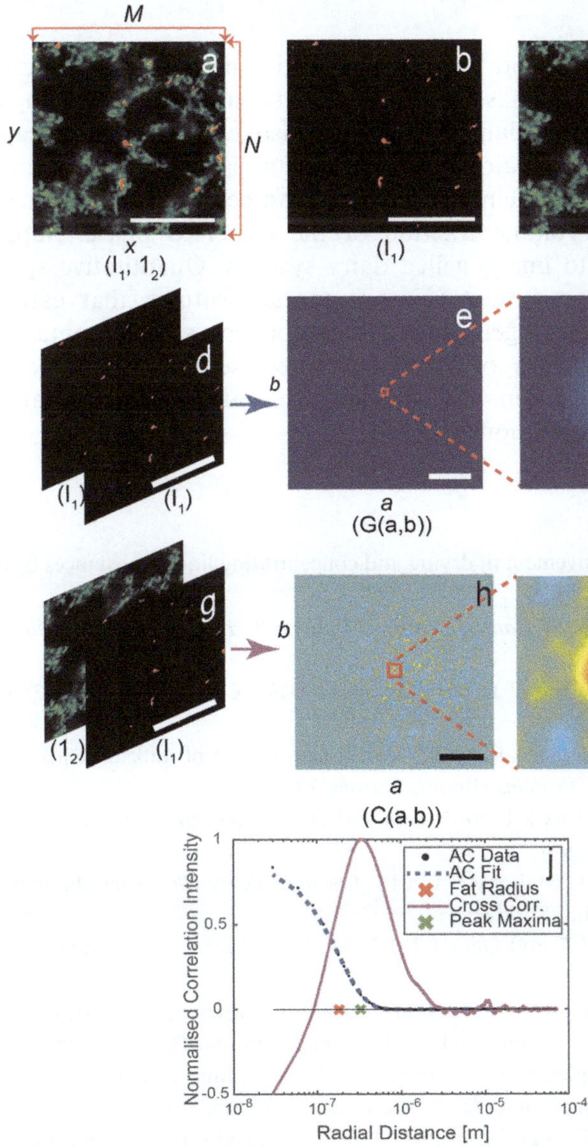

Figure 9.2. Workflow for the autocorrelation and cross-correlation analysis of two channel microscopy images. Two colour microscopy image of acid induced homogenised whole milk gel acquired with confocal microscopy for image I_1, with the fat shown in red and with STED for image I_2 where the protein is shown in green (a). Confocal image showing the fat channel in *a* (b). Autocorrelation image calculated from the red confocal channel in *a* (c). Zoom in on the centre of the autocorrelation image *c* (d). STED image showing the protein channel in image *a* (e). Cross-correlation image calculated from *b* and *c* (f). Zoom in on the centre of the cross-correlation image *f* (g). Plot of the normalised radially averaged autocorrelation decay from image *c*, shown in black points, to which a stretched exponential is fit (equation (9.16)), shown in blue. The normalised radially averaged cross-correlation decay calculated from image *f* is shown in magenta. The fat droplet radius, ξ, (equation (9.16)), is shown in red and the peak maxima of the cross-correlation curve is shown in green (h). Scalebar 5 μm, where not indicated on figure. Reprinted from [77], copyright 2019, with permission from Elsevier.

9.10 Summary

A combination of ultrasound spectroscopy and super-resolution microscopy with quantitative image analysis has been presented in relation to monitoring rehydration and structure formation in dairy systems. Ultrasound spectroscopy can be used to monitor the rehydration of dairy powders and establish when the system has reached equilibrium. The acoustic attenuation spectra can be used to determine the particle size distribution and volume fraction of the dispersed phase. Super-resolution imaging can be used to image gelled dairy systems. Quantitative spatial correlation-based analysis can extract relevant spatial features that can be used to characterise dairy protein gels and two-component system containing fat and protein. The two branches of techniques presented provide a complementary overview of the rehydration and structure formation processes at appropriate temporal and spatial resolutions.

References

[1] Percy S R 1872 Improvement in drying and concentrating liquid substances by atomizing *US Patent* 125406

[2] Fox P F 2013 *Advanced Dairy Chemistry Volume 3: Lactose, Water, Salts and Vitamins* (Berlin: Springer)

[3] Fox P F and McSweeney P L 2007 *Advanced Dairy Chemistry Volume 2: Lipids* (Berlin: Springer)

[4] Hambraeus L and Lönnerdal B 2003 Nutritional aspects of milk proteins *Advanced Dairy Chemistry Volume 1: Proteins* (Berlin: Springer) 605–45

[5] McSweeney P L and Fox P F 2013 *Advanced Dairy Chemistry: Volume 1A: Proteins: Basic Aspects* (Berlin: Springer)

[6] McSweeney P L and O'Mahony J A 2015 *Advanced Dairy Chemistry: Volume 1B: Proteins: Applied Aspects* (Berlin: Springer)

[7] Walstra P, Wouters J T and Geurts T J 2005 *Dairy Science and Technology* (Boca Raton, FL: CRC Press)

[8] Anema S G and Li Y 2003 Association of denatured whey proteins with casein micelles in heated reconstituted skim milk and its effect on casein micelle size *J. Dairy Res.* **70** 73–83

[9] Corredig M and Dalgleish D G 1996 Effect of temperature and pH on the interactions of whey proteins with casein micelles in skim milk *Food Res. Int.* **29** 49–55

[10] Nair P K, Dalgleish D G and Corredig M 2013 Colloidal properties of concentrated heated milk *Soft Matter. R. Soc. Chem.* **9** 3815–24

[11] Van Vliet T, Lakemond C M and Visschers R W 2004 Rheology and structure of milk protein gels *Curr. Opin. Coll. Interf. Sci.* **9** 298–304

[12] Vasbinder A J, Alting A C, Visschers R W and de Kruif C G 2003 Texture of acid milk gels: formation of disulfide cross-links during acidification *Int. Dairy J.* **13** 29–38

[13] Bylund G 2003 *Dairy Processing Handbook* (Tetra Pak Processing Systems AB)

[14] Dhanalakshmi K, Ghosal S and Bhattacharya S 2011 Agglomeration of food powder and applications *Crit. Rev. Food Sci. Nutr.* **51** 432–41

[15] Fang Y, Selomulya C, Ainsworth S, Palmer M and Chen X D 2011 On quantifying the dissolution behaviour of milk protein concentrate *Food Hydrocoll.* **25** 503–10

[16] Kim E H-J, Chen X D and Pearce D 2002 Surface characterization of four industrial spray-dried dairy powders in relation to chemical composition, structure and wetting property *Colloids Surf.* B **26** 197–212

[17] Mimouni A, Deeth H C, Whittaker A K, Gidley M J and Bhandari B R 2009 Rehydration process of milk protein concentrate powder monitored by static light scattering *Food Hydrocoll.* **23** 1958–65

[18] Fang Y, Selomulya C and Chen X D 2007 On measurement of food powder reconstitution properties *Drying Technol.* **26** 3–14

[19] Fang Y, Selomulya C and Chen X D 2010 Characterization of milk protein concentrate solubility using focused beam reflectance measurement *Dairy Sci Technol.* **90** 253–70

[20] Gaiani C, Banon S, Scher J, Schuck P and Hardy J 2005 Use of a turbidity sensor to characterize micellar casein powder rehydration: Influence of some technological effects *J. Dairy Sci.* **88** 2700–6

[21] Gaiani C, Morand M, Sanchez C, Tehrany E A, Jacquot M and Schuck P *et al* 2010 How surface composition of high milk proteins powders is influenced by spray-drying temperature *Coll. Surfaces B: Biointerf* **75** 377–84

[22] Gaiani C, Schuck P, Scher J, Desobry S and Banon S 2007 Dairy powder rehydration: influence of protein state, incorporation mode, and agglomeration *J. Dairy Sci.* **90** 570–81

[23] Hussain R, Gaiani C and Scher J 2012 From high milk protein powders to the rehydrated dispersions in variable ionic environments: a review *J. Food Eng.* **113** 486–503

[24] Al-Aufi Y, Hewakandamby B, Dimitrakis G, Holmes M, Hasan A and Watson N 2019 Thin film thickness measurements in two phase annular flows using ultrasonic pulse echo techniques *Flow Meas. Instrum.* 66 67–78

[25] Corona E, Garcia-Perez J V, Alvarez-Arenas T E G, Watson N, Povey M J and Benedito J 2013 Advances in the ultrasound characterization of dry-cured meat products *J. Food Eng.* **119** 464–70

[26] Escrig J, Woolley E, Rangappa S, Simeone A and Watson N 2019 Clean-in-place monitoring of different food fouling materials using ultrasonic measurements *Food Control* **104** 358–66

[27] Meyer S, Rajendram V S and Povey M J 2006 Characterization of reconstituted milk powder by ultrasound spectroscopy *J. Food Qual.* **29** 405–18

[28] Watson N, Povey M, Reynolds G, Xu B and Ding Y 2014 Acoustic emission monitoring from a lab scale high shear granulator—A novel approach *Int. J. Pharm.* **465** 262–74

[29] Jenkins P and Briggs G 1991 Acoustic microscopy *Engineering Materials Advisory Services Ltd, Fatigue Crack Measurement: Techniques and Applications(UK)* 147–72

[30] Parker N, Nelson P and Povey M 2010 A versatile scanning acoustic platform *Meas. Sci. Technol.* **21** 045901

[31] Cheeke J D N 2002 *Fundamentals and Applications of Ultrasonic Waves* (Boca Raton, FL: CRC Press)

[32] Holmes M, Parker N and Povey M 2011 Temperature dependence of bulk viscosity in water using acoustic spectroscopy *J. Phys.: Conf. Series* 012011

[33] Holmes M, Southworth T, Watson N and Povey M 2014 Enzyme activity determination using ultrasound *J. Phys.: Conf. Series* **498** 012003

[34] Povey M J W 1997 *Ultrasonic Techniques for Fluids Characterization* (New York: Academic)

[35] Povey M J 2006 Acoustic methods for particle characterisation *KONA Powder Part. J. Hosokawa Powder Technology Foundation* **24** 126–33

[36] Povey M J 2013 Ultrasound particle sizing: a review *Particuology* **11** 135–47

[37] Povey M J 2014 Crystal nucleation in food colloids *Food Hydrocoll.* **42** 118–29

[38] Povey M and Challis R 2006 Applications of ultrasound to analysis/quantitation of dairy lipids *Advanced Dairy Chemistry Volume 2: Lipids* (Berlin: Springer) 709–23

[39] Chanamai R and McClements D J 1998 Ultrasonic attenuation of edible oils *J. Am. Oil Chem. So. (Wiley Online Library)* **75** 1447–8

[40] Coupland J N and McClements D J 2001 Droplet size determination in food emulsions: comparison of ultrasonic and light scattering methods *J. Food Eng.* **50** 117–20

[41] Hipp A K, Storti G and Morbidelli M 2002 Acoustic characterization of concentrated suspensions and emulsions. 1. Model analysis *Langmuir* **18** 391–404

[42] Hipp A K, Storti G and Morbidelli M 2002 Acoustic characterization of concentrated suspensions and emulsions. 2. Experimental validation *Langmuir* **18** 405–12

[43] McClements D and Povey M 1989 Scattering of ultrasound by emulsions *J. Phys. D: Appl. Phys.* **22** 38

[44] Bilaniuk N and Wong G S 1993 Speed of sound in pure water as a function of temperature *J. Acous. Soc. Am. ASA* **93** 1609–12

[45] Del Grosso V and Mader C 1972 Speed of sound in pure water *J. Acous. Soc. Am. ASA* **52** 1442–6

[46] Kaye G W C and Laby T 1995 *Tables of Physical and Chemical Constants* 16th edn (Harlow: Longman)

[47] Vance S and Brown J M 2010 Sound velocities and thermodynamic properties of water to 700 MPa and 10 to 100° C *J. Acous. Soc. Am. ASA* **127** 174–80

[48] Wong G S and Zhu S 1995 Speed of sound in seawater as a function of salinity, temperature, and pressure *J. Acous. Soc. Am. ASA* **97** 1732–6

[49] Challis R, Povey M, Mather M and Holmes A 2005 Ultrasound techniques for characterizing colloidal dispersions *Rep. Prog. Phys.* **68** 1541

[50] Strutt J W (Baron Rayleigh) 1945 *The Theory of Sound* (New York: Dover)

[51] Strutt H J W 1871 On the scattering of light by small particles *Phil. Mag. Series 4* **41** 447–54

[52] Strutt J 1871 Investigation of the disturbance produced by a spherical obstacle on the waves of sound *Proc. of the London Mathematical Society* **1** 253–83

[53] Epstein P S and Carhart R R 1953 The absorption of sound in suspensions and emulsions. I. Water fog in air *J. Acous. Soc. Am. ASA* **25** 553–65

[54] Allegra J and Hawley S 1972 Attenuation of sound in suspensions and emulsions: Theory and experiments *J. Acous. Soc. Am. ASA* **51** 1545–64

[55] Lloyd P and Berry M 1967 Wave propagation through an assembly of spheres: IV. Relations between different multiple scattering theories *Proc. Phys. Soc.* **91** 678

[56] Alba F, Higgs D, Jack R and Kippax P 1998 Ultrasound spectroscopy: A sound approach to sizing of concentrated particulates *Handbook on Ultrasonic and Dielectric Characterization Techniques for Suspended Particulates* (Westerville, OH: American Ceramic Society)

[57] Francis M, Glover Z, Yu Q, Povey M and Holmes M 2019 Acoustic characterisation of pH dependant reversible micellar casein aggregation *Coll. Surf. A: Physicochem. Eng. Asp.* **568** 259–65

[58] Auty M, Fenelon M, Guinee T, Mullins C and Mulvihill D 1999 Dynamic confocal scanning laser microscopy methods for studying milk protein gelation and cheese melting *Scanning* **21** 299–304

[59] Auty M, O'Kennedy B, Allan-Wojtas P and Mulvihill D 2005 The application of microscopy and rheology to study the effect of milk salt concentration on the structure of acidified micellar casein systems *Food Hydrocoll.* **19** 101–9

[60] Auty M A, Twomey M, Guinee T P and Mulvihill D M 2001 Development and application of confocal scanning laser microscopy methods for studying the distribution of fat and protein in selected dairy products *J. Dairy Res.* **68** 417–27

[61] Dubert-Ferrandon A, Niranjan K and Grandison A S 2006 A novel technique for differentiation of proteins in the development of acid gel structure from control and heat treated milk using confocal scanning laser microscopy *J. Dairy Res.* **73** 423–30

[62] Lucey J, Munro P and Singh H 1998 Rheological properties and microstructure of acid milk gels as affected by fat content and heat treatment *J. Food Sci.* **63** 660–4

[63] Michalski M, Cariou R, Michel F and Garnier C 2002 Native vs. damaged milk fat globules: membrane properties affect the viscoelasticity of milk gels *J. Dairy Sci.* **85** 2451–61

[64] Ong L, Dagastine R, Kentish S and Gras S 2010 The effect of milk processing on the microstructure of the milk fat globule and rennet induced gel observed using confocal laser scanning microscopy *J. Food Sci.* **75** E135–45

[65] Ong L, Dagastine R R, Kentish S E and Gras S L 2011 Microstructure of milk gel and cheese curd observed using cryo scanning electron microscopy and confocal microscopy *LWT-Food Sci. Technol.* **44** 1291–302

[66] Andoyo R, Guyomarc'h F, Burel A and Famelart M-H 2015 Spatial arrangement of casein micelles and whey protein aggregate in acid gels: Insight on mechanisms *Food Hydrocoll.* **51** 118–28

[67] Marchin S, Putaux J-L, Pignon F and Léonil J 2007 Effects of the environmental factors on the casein micelle structure studied by cryo transmission electron microscopy and small-angle x-ray scattering/ultrasmall-angle x-ray scattering *J. Chem. Phys.* **126** 01B617

[68] Dalgleish D G and Corredig M 2012 The structure of the casein micelle of milk and its changes during processing *Ann. Rev. Food Sci. Technol. Ann. Rev.* **3** 449–67

[69] Martin A H, Goff H D, Smith A and Dalgleish D G 2006 Immobilization of casein micelles for probing their structure and interactions with polysaccharides using scanning electron microscopy (SEM) *Food Hydrocoll.* **20** 817–24

[70] Awad S, Hassan A and Muthukumarappan K 2005 Application of exopolysaccharide-producing cultures in reduced-fat Cheddar cheese: texture and melting properties *J. Dairy Sci.* **88** 4204–13

[71] Glover Z J, Ersch C, Andersen U, Holmes M J, Povey M J and Brewer J R *et al* 2019 Super-resolution microscopy and empirically validated autocorrelation image analysis discriminates microstructures of dairy derived gels *Food Hydrocoll.* **90** 62–71

[72] Glover Z J, Francis M J, Bisgaard A H, Andersen U, Johansen L B and Povey M J *et al* 2020 Dynamic moisture loss explored through quantitative super-resolution microscopy, spatial micro-viscosity and macroscopic analyses in acid milk gels *Food Hydrocoll.* **101** 105501

[73] Hell S W and Wichmann J 1994 Breaking the diffraction resolution limit by stimulated emission: stimulated-emission-depletion fluorescence microscopy *Opt. Lett. Opt. Soc. Am.* **19** 780–2

[74] Willig K, Keller J, Bossi M and Hell S W 2006 STED microscopy resolves nanoparticle assemblies *New J. Phys.* **8** 106

[75] Hell S W 2009 Microscopy and its focal switch *Nat. Methods* **6** 24

[76] Hell S W 2003 Toward fluorescence nanoscopy *Nat. Biotechnol.* **21** 1347

[77] Glover Z J, Bisgaard A H, Andersen U, Povey M J, Brewer J R and Simonsen A C 2019 Cross-correlation analysis to quantify relative spatial distributions of fat and protein in super-resolution microscopy images of dairy gels *Food Hydrocoll.* **97** 105225

[78] Robertson C and George S C 2012 Theory and practical recommendations for autocorrelation-based image correlation spectroscopy *J. Biomed. Opt. Inter. Soc. Opt. Photon.* **17** 080801

[79] Costantino S, Comeau J W, Kolin D L and Wiseman P W 2005 Accuracy and dynamic range of spatial image correlation and cross-correlation spectroscopy *Biophys. J.* **89** 1251–60

[80] Kolin D L and Wiseman P W 2007 Advances in image correlation spectroscopy: measuring number densities, aggregation states, and dynamics of fluorescently labeled macromolecules in cells *Cell Biochem. Biophys.* **49** 141–64

[81] Mir S M, Baggett B and Utzinger U 2012 The efficacy of image correlation spectroscopy for characterization of the extracellular matrix *Biomedical Opt. Exp. Opt. Soc. Am.* **3** 215–24

[82] Ako K, Durand D, Nicolai T and Becu L 2009 Quantitative analysis of confocal laser scanning microscopy images of heat-set globular protein gels *Food Hydrocoll.* **23** 1111–9

[83] Marangoni A G, Acevedo N, Maleky F, Peyronel F, Mazzanti G and Quinn B *et al* 2012 Structure and functionality of edible fats *Soft Matter. Roy Soc. Chem.* **8** 1275–300

[84] Super B J and Bovik A C 1991 Localized measurement of image fractal dimension using Gabor filters *J. Vis. Commun. Image Repres.* **2** 114–28

[85] Hebert B, Costantino S and Wiseman P W 2005 Spatiotemporal image correlation spectroscopy (STICS) theory, verification, and application to protein velocity mapping in living CHO cells *Biophys. J.* **88** 3601–14

[86] Petersen N and Wiseman P *et al* 1999 Analysis of membrane protein cluster densities and sizes *in situ* by image correlation spectroscopy *Faraday Discuss. R. Soc. Chem.* **111** 289–305

Chapter 10

Non-invasive sensing for food reassurance

Xiaobo Zou, Xiaowei Huang, Xiaodong Zhai, Jiong Shi and Megan Povey

Adapted from [1] with permission of The Royal Society of Chemistry.

Consumers and governments are concerned about safety, authenticity and quality of food commodities; consequently, non-invasive sensing techniques are in development for rapid analysis. In this chapter, a wide range of techniques: optical, acoustical, electrical, to nuclear magnetic, x-ray, biosensor, microwave and terahertz, are considered for food detection according to their physical principles. Some of these techniques are now in a period of transition between experimental and applied utilization. With continued innovation and attention to key challenges, such non-invasive sensors and biosensors could open up new avenues in the field of portable and wearable wireless sensing devices and connecting with mobile networks, and thus find considerable use in a wide range of food assurance applications.

10.1 Introduction

In a previous article, we reviewed the field of non-invasive sensing for food quality [1]. In this chapter, we update that article, reviewing the latest innovations in food sensing. The food industry (including suppliers, manufactures and supermarkets) is required to provide and confirm the authenticity and point of origin of food products and their components, in order to protect consumer rights and prevent fraudulent or deceptive practices such as food adulteration. However, common methods for food reassurance, such as chemical reactions, immunological and enzymatic techniques, and DNA and protein based assays, are generally applied in laboratories [2–4]. Even though these methods are usually capable of detecting low levels of adulteration, they are expensive, invasive, sophisticated, laborious, and technically demanding [5, 6].

Recently, promising non-invasive (NIV) and rapid sensing techniques combined with multivariate data analysis techniques have been developed for food reassurance. This is a rapidly expanding area of research as indicated by the number of recent publications in NIV technologies (figure 10.1).

doi:10.1088/978-0-7503-2596-7ch10

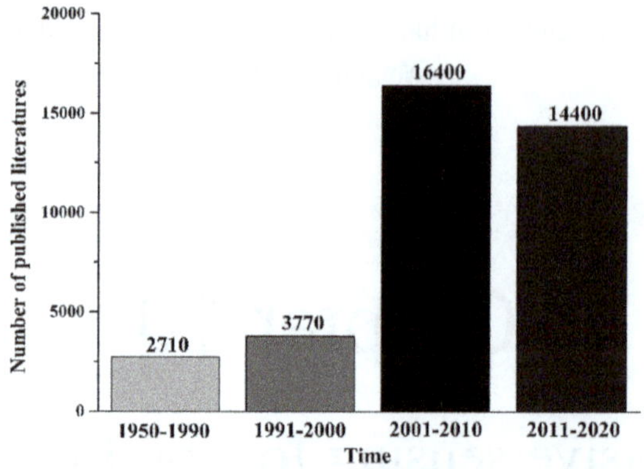

Figure 10.1. Number of published literatures on non-invasive sensing of food since 1950 (data was taken on August 21, 2019 from Google scholar using 'non-invasive sensing' and 'food' in the advanced search option). Reproduced from [1] by permission of The Royal Society of Chemistry.

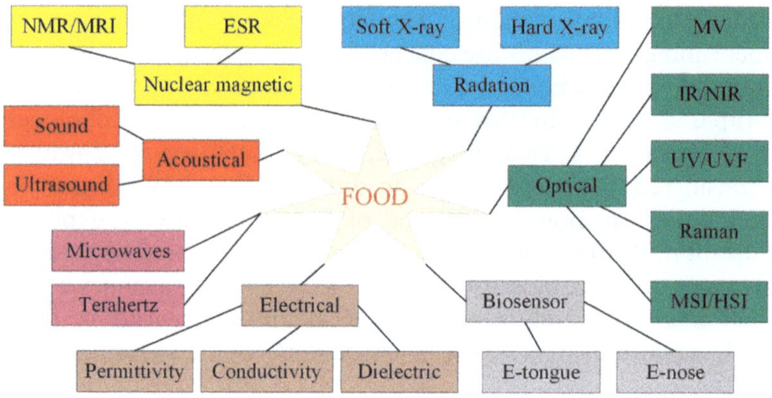

Figure 10.2. Non-invasive sensing techniques in food assurance. Reproduced from [1] by permission of The Royal Society of Chemistry.

In this chapter, we organize, non-invasive techniques and chemometrics for food assurance according to their physical principles rather than their application, because some techniques can be used to determine different food quality attributes. The advantages, limitations, and future trends of the techniques will be provided.

10.2 Techniques employed for non-invasive food assurance

There are many non-invasive sensing techniques in food, as shown in figure 10.2. Among the techniques used, the NIV technique is mainly based on physical methods of analysis excepting biosensors. Some NIV techniques such as machine visions (MVs) can provide external information such as color, shape, size, and surface defects of food products [7], and some NIV techniques such as ultrasound can provide internal quality such as firmness and crispiness, taste, aroma, and internal defects [8, 9]. Some NIV techniques such as electronic nose [10] and hyperspectral imaging [11–14] can give quality attributes such as freshness, safety, nutritional value and health-promoting properties, chemical residues. Looking at the enormous amount of literature produced during the last 15 years, one group of optical methods, visible/near/mid-infrared spectroscopy and imaging, and the other group of acoustic methods, ultrasound spectroscopy and imaging, may be considered the most researched non-invasive techniques for food assurance.

10.2.1 Optical techniques

Many physical methods deploy electromagnetism (figure 10.3). The light may be absorbed, scattered or reflected. These effects are quantified by measuring the diffuse reflection (I_{fd}), transmission (I_t), and emission (I_e) of light and contain structural and chemically related information.

10.2.1.1 Machine vision (MV)

The external appearance of food is a very important quality for both consumer and producer. MV, or computer vision, which can give the external characteristics of food, consists of a digital camera connected to a computer and software for further image analysis. MV in visible light regions can recognize size [15], shape [16], color [17, 18], and texture [19] of food. Some MV systems are also able to inspect these objects in parts of the electromagnetic spectrum invisible to humans, such as x-ray

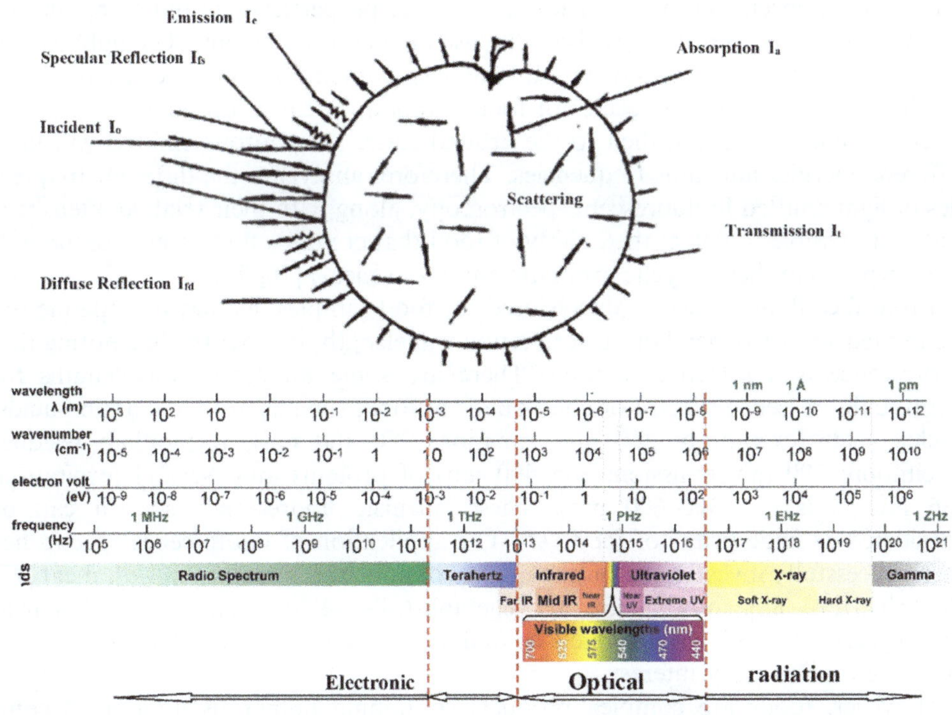

Figure 10.3. The light–food interaction and the electromagnetic spectrum (adapted from Google image) [1]. Reproduced by permission of the Royal Society of Chemistry.

[20], ultraviolet (UV) [21], near-infrared (NIR) [22], and infrared (IR) [23] [24–26]. Extensive reviews of MV in food can also be found in [27–30]. The food industry is now ranked among the top 10 industries using MV [30, 31].

10.2.1.2 Vis/NIR spectroscopy

Vis/NIR (near infrared) light can penetrate up to a few centimeters, depending on the wavelength into biological tissue, as the absorption by water is relatively low in the Vis/NIR range compared to the mid-infrared ranges [32]. In addition, the scattering of the light in this region is significantly larger, enabling light to diffuse in the sample volume and then reemitted at the tissue boundaries [33–35]. Therefore, Vis/NIR spectroscopy can offer finger-print spectra for food assurance but requires advanced chemo metric techniques to extract needed information. Vis/NIR spectroscopy has been used successfully to validate many foods such as fruit [36, 37], vegetables [38, 39], meat [40, 41], fish [42], vines [43], milk [44], starch [45, 46], oil [47–49], cereal [50], transgenic food [51], etc. And it has also been successful in giving some physical-chemical parameters of food, such as freshness [52, 53], water content [54, 55], soluble solids contents [56, 57], acidity [58–60], firmness [60–62], texture [63–65], which are related to quality and safety of foods. There are also many recent reviews [49, 51, 66–75].

10.2.1.3 UV–visible fluorescence spectroscopy

Fluorescence spectroscopy can determine several properties (e.g. function, composition, and nutrition) of food without the use of chemical reagents. UV light excites electrons in certain compounds in food and causes them to emit lower energy light. As the electrons of the molecules of food compounds can drop down into any of several vibrational levels, then to the ground state, the photons emitted will have different energies and thus frequencies. Therefore, analyzing the different frequencies of light emitted in fluorescent spectroscopy, along with their relative intensities, makes it possible to detect the quality of food that contains fluorescent compounds (e.g. tryptophan, heterocyclic aromatic amines in meat) [76, 77].

Front face fluorescence is always used for food samples, as they are opaque. As the excited species depend on the excitation wavelength, it is worthwhile noting that fluorescence is a selective method. There are some important wavelengths for intrinsic fluorescence spectra of compounds in food, such as aromatic-amino-acids, nucleic-acids (excitation: 250 nm, emission: 280–480 nm), tryptophan residues (excitation: 290 nm, emission: 305–400 nm) of proteins and NADH (excitation: 336 nm, emission: 360–600 nm). These intrinsic fluorescence spectra can be considered as fingerprints of food [77]. The application of fluorescence spectra has been successfully investigated on the quality of animal (i.e. dairy [78, 79], meat [77, 80], fish [31, 81, 82], and egg [83]) and vegetable (oils [84], cereal [85], sugar [86], fruit and vegetable [79, 87–89]) products as well as the identification of bacteria [78, 90, 91] of agro-alimentary interest.

However, foods are complex products containing numerous fluorescent compounds. In such cases, the signals of the different chromophores may overlap, and it becomes more complicated for food assurance.

10.2.1.4 Hyperspectral imaging and multispectral imaging

Spectral imaging (i.e. hyperspectral and multispectral imaging) can rapidly and non-invasively monitor physical and morphological characteristics, as well as intrinsic chemical and molecular information of a food product. During the last decade, spectral imaging has developed rapidly as an effective method for safety and quality inspection of food; due to its ability to integrate conventional imaging and spectroscopy giving both spatial and spectral information for an object [92, 93]. Hyperspectral images are usually used as fundamental datasets ('Groundtruth') to determine optimal wavebands that can then be used by a multispectral imaging solution for a particular application [94]. 400–1000 nm light sources are usually used for fruit and vegetables, because they are generally cheaper than those with a 900–1700 nm light source and are less influenced by water [95, 96].

Combining different image acquisition methods (e.g. reflectance/transmittance and reflectance/fluorescence) can give the hyperspectral/multispectral imaging systems more inspection capabilities than systems that use only a single imaging mode [97]. Such image acquisition combinations along with data fusion techniques are likely to expand in the future for building multitasking food inspection systems. There are many recently published reviews dealing with the application of Vis/NIR hyperspectral/multispectral imaging for classification and grading [98, 99], defect [100–102] and disease [103] detection, distribution visualization of chemical attributes [104, 105], and for inspection of meat [106, 107], fruit [93, 108, 109] vegetables [93, 110, 111] and fish [112, 113].

10.2.1.5 Raman spectroscopy

Raman spectroscopy may provide detailed chemical information on a sample without a need for labeling and is almost insensitive to polar materials such as water [114] [115]. It is a non-invasive technique that yields reliable results for solid and liquid multicomponent food samples [114]. In a typical Raman analysis of food, the spectral fingerprint can be used to identify a molecule of interest in a complex food matrix as the assignment of Raman-scattering bands to corresponding vibrational modes of a molecule is unique [116, 117].

Raman spectroscopy may be combined with FT-IR for the study of molecular vibrations and structure of the chemical composition of heterogeneous foods and food ingredients [118]. Raman spectroscopy also has been successfully applied for characterizing olives [119] [120], juices [121], fruits [122] [123] [124], and analyzing wax [125], fatty acid [126], protein [127], etc in food, and investigating structural and texture [128]. Raman spectroscopy also enhances the vibrational spectrum of molecules adsorbed on or near metal particles and/or surfaces. There are recent and potential advances in surface-enhanced Raman spectroscopy or surface-enhanced Raman scattering (SERS), which is a surface-sensitive technique that enhances Raman scattering by molecules adsorbed on rough metal surfaces or by nanostructures such as plasmonic-magnetic silica nanotubes. The enhancement factor can be as much as 10^{10} to 10^{11}, which means the technique may detect single molecules. Such applications are highlighted in sections that deal with (a) the detection of food-borne pathogenic microorganisms [129–132] and (b) the detection of food contaminants and adulteration [133], concentrated specifically on antibiotics [134], drugs [135], hormones [129], melamine [136–139], and pesticides [140–142] as reviewed in references [128, 129, 143–145].

10.2.1.6 Summary of optical techniques

Optical radiation propagates well in air, and optical spectroscopy can be used without contact with food. The interaction between the optical radiation and the food samples under test can be assessed by measuring either the reflected, transmitted, scattered or diffused light, or several of these parameters. Optical instrumentation is quite inexpensive and easy to miniaturize, allowing the development of portable devices suitable for industrial applications. The typical detection results of optical techniques are shown in figure 10.4, taking meat as an example (adapted from the literature [146–150]).

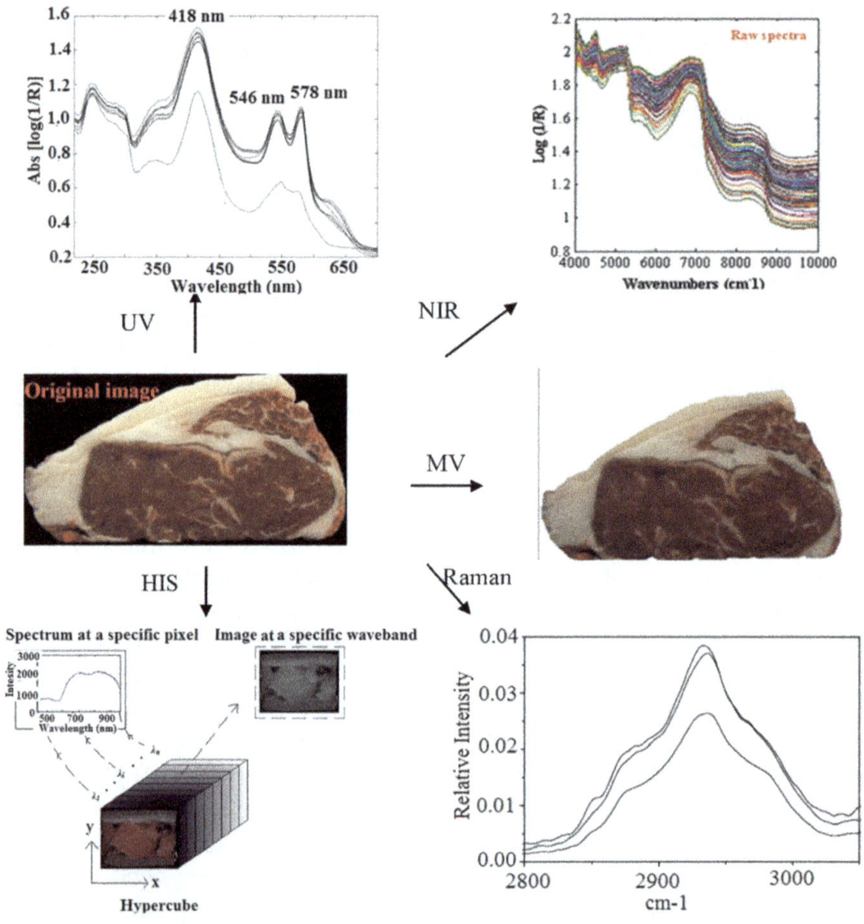

Figure 10.4. A typical application of optical techniques used in meat quality inspection. Reproduced from [1] by permission of the Royal Society of Chemistry.

10.2.2 Acoustical techniques

Quality and acceptability of food products are often assessed based on the sounds produced during crushing or biting of the food [151]. Sound is transmitted through a food product to obtain the natural resonant frequencies of the product. It is used to monitor the composition and physicochemical properties of food components and products during processing and storage, which is crucial for controlling the food properties and improving its quality.

From figure 10.5, by tuning frequency, sound can be utilized in many industrial applications including food [152]. Sound propagates through food materials as mechanical waves causing alternating compressions and decompressions [153]. These sound waves have characteristic wavelength, velocity, frequency, pressure and period. The interaction of sound waves with matter alters both the velocity and attenuation of the sound waves via absorption and/or scattering mechanisms [154].

For acoustic techniques, there are four main measurement instruments, microphone passive measurement technique, pulse-echo ultrasound technique, pitch and catch technique, and resonance techniques, as shown in figure 10.6.

Acoustical techniques have been utilized for non-invasive analysis and monitoring of fresh vegetables and fruits in both pre- and postharvest, cheese during processing, commercial cooking oils, bread and cereal products, bulk and emulsified fat based food products, raw and fermented meat products, fish and poultry, food gels, aerated and frozen foods. Other applications include the detection of honey adulteration and assessment of the aggregation state, size and type of protein [155].

Scanning acoustic microscopy and ultrasound imaging are the two main acoustic imaging techniques to capture images from detected objects. In a scanning acoustic microscope, a focused ultrasound beam is coupled into the sample and the returning signal is analyzed in a variety of ways, including simple pulse echo [156, 157]. Ultrasonic imaging (ultrasonography) is a medical technology that has been used as a diagnostic technique for a long time. This technique has found several applications in food technology, like the use of high contrast images to detect defects in food packaging seals using pulse–echo ultrasound [156].

Passive recording of the sounds made when eating food can also provide information on the crispy/crunchiness of food. Such techniques may also be combined with a force-displacement technique such as the Stable Micro Systems food tester [158–161].

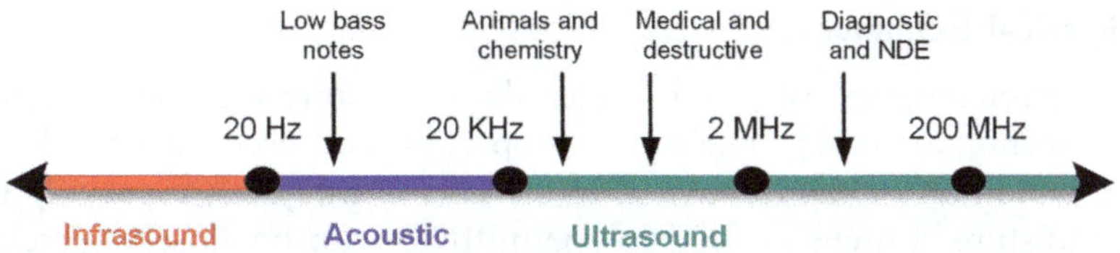

Figure 10.5. Approximate frequency ranges corresponding to sound, with rough guide to some applications. Reproduced from [1] by permission of the Royal Society of Chemistry.

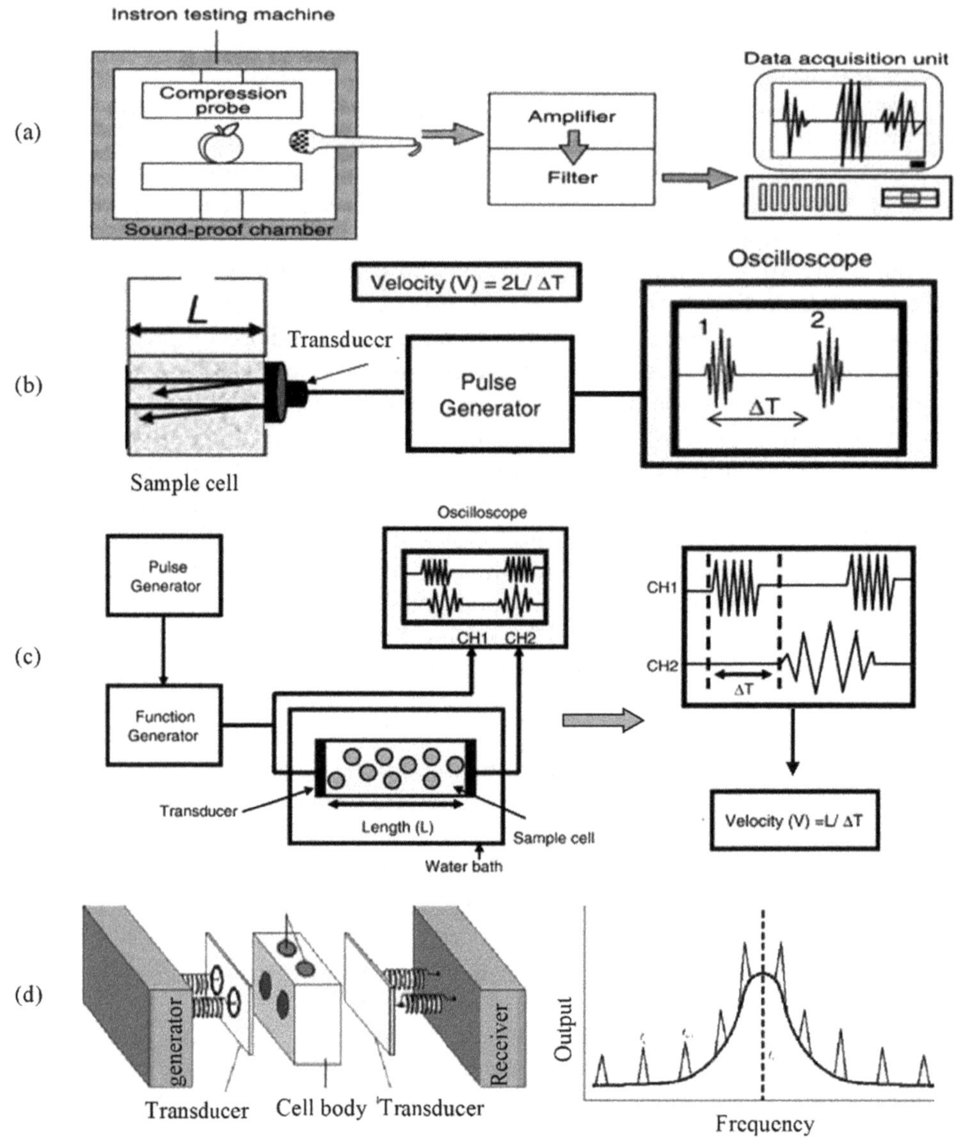

Figure 10.6. Acoustic techniques, (a) microphone, (b) pulse–echo, (c) pitch and catch, (d) resonance techniques. Reproduced from [1] by permission of the Royal Society of Chemistry.

10.2.3 Electrical techniques

Electrical characteristics of food materials is a long-standing nondestructive measuring technique [162]. Dielectric properties are used in the detection of processing condition or the quality of food. This technique is especially useful for detecting moisture content in foods as permittivity can be closely correlated with water mobility. Bound water has only a small impact on the microwave dielectric factor. During microwave thawing of food, it is very important to detect the rates and uniformity of heating in partially frozen food material by using electrical methods [163]. As the measurement is rapid and does not destroy the substance, it is therefore suitable for on-line assurance of food quality [77, 164].

As most of food tissue is biomass, when an electric current passes through food, it passes through the extracellular fluid (ECF) or through both the ECF and intracellular fluid (ICF) compartments [165]. The current pathway is generally represented as two parallel branches: one through the ECF and the other through the capacitive membrane and the ICF compartment. The extracellular pathway is considered to be purely resistive, whereas the intracellular pathway including the capacitive effect of the cell membrane is an impedance (i.e. complex), with a resistive part (real) and a capacitive part (imaginary), resulting in the magnitude of the intracellular impedance being frequency dependent [166].

Conductivity plays a fundamental role in food assurance measurements [167–171]. The conductivity of a food is generally measured by passing a known current at constant voltage through a known volume of the material and by determining resistance. The electrical conductivity of foods has been found to increase linearly with temperature, and water/ionic content.

Electrical permittivity is another parameter used to characterize the dielectric property of food. It depends on dielectric constant ε', which is related to capacitance of a substance and its ability to store electrical energy, and the dielectric loss factor ε'', related to energy losses when the food is subjected to an alternating electrical field, i.e. dielectric relaxation and ionic conduction [172]. The permittivity can be related to chemical composition, physical structure, frequency, and temperature, with moisture content being the dominant factor [173].

Various factors influence the dielectric properties of food materials such as frequency of the applied alternating electric field, moisture content, bulk density, temperature [77, 163]. At constant temperature, dielectric constant (ε') and loss factor (ε'') of food materials increase with decreasing frequency. The penetration depth (PD) decreases with increasing frequency, temperature and moisture content [163, 173]. At higher microwave frequencies only salty foods show an increase in ε'' with temperature but at lower microwave frequencies there is a general increase in ε'' [173]. Both ε' and ε'' of various foods increase with increasing moisture content. For fruits and vegetables [174, 175], ε' increases with temperature, whereas ε'' increases with increasing temperature and ε' increases with temperature at lower frequencies, but decreases with temperature at the higher frequencies. For meat [176], both ε' and ε'' increase with decreasing frequency at constant temperature; however, ε' decreases and ε'' increases with increasing temperature at constant frequency. For fish, a sharp increase in dielectric properties is observed around the freezing point [177]. Both ε' and ε'' increase with increased water content at constant temperature; ε' and ε'' of lean tuna are larger than those of fatty tuna. ε' of marinated catfish and shrimp, generally decrease with increasing temperature, whereas ε'' increases with temperature. Both ε' and ε'' of fish meal increase non-linearly with moisture content and also increase with temperature in a relatively linear manner [178–180].

The interest in dielectric and electric properties of food materials has historically been associated with the design of electrical equipment. As shown in figure 10.7(a),

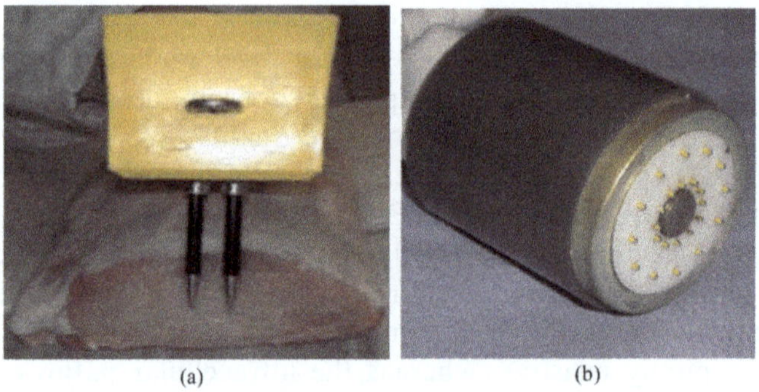

Figure 10.7. A small device for impedance measurement with (a) two electrodes and (b) a multidirectional sensor. Reproduced from [1] by permission of the Royal Society of Chemistry.

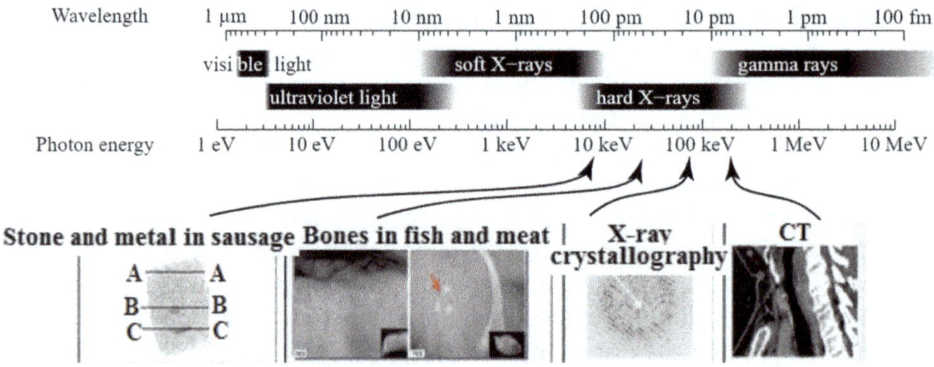

Figure 10.8. Different applications use different parts of the x-ray spectrum (adapted from Wikipedia, http://en.wikipedia.org/wiki/X-ray, author: Ulflund, we added some soft x-ray applications). Bottom image reproduced from [1] by permission of the Royal Society of Chemistry.

simple detection devices only need a low frequency generator, a couple of electrodes and a view meter. It is a convenient method for evaluating food quality, especially for detecting moisture content in foods. Figure 10.7(b) shows a circular multi-electrode sensor based on anisotropy index methods which was successfully used in meat aging detection [168]. Jha *et al* [163] have given a detailed review which covers theoretical aspects of different electrical properties, their measurement techniques, applications of dielectric properties in agriculture/food processing sector for quality and safety assessment in food processing. The electrical and dielectrical apparatus has been applied for determination of various characteristics of food such as frost sensitiveness, chilling and freezing tolerance, moisture content, seeds germination, and so forth [77, 165, 166, 169, 170, 180–186].

10.2.4 X-ray techniques

As shown in figure 10.8, the wavelengths of x-radiation (composed of x-rays) are shorter than those of UV rays and longer than those of gamma rays. X-rays have a wavelength in the range of 0.01 to 10 nm, corresponding to frequencies in the range 30 PHz to 30 EHz and energies in the range 100 eV to 100 keV. X-rays with photon energies above 5–10 keV (below 0.2–0.1 nm wavelength) are called hard x-rays, while those with lower energy are called soft x-rays [187, 188].

X-rays are widely used to image the inside of objects by using their penetrating ability [189]. The FDA (US Food and Drug Administration) recommends that the maximum dose permitted for irradiation of spice is 30 kGy. Different applications use different parts of the x-ray spectrum, as shown in figure 10.8. Applications in food mainly use soft x-rays.

X-rays that are elastically or inelastically scattered from a sample provide valuable information about the internal characteristics of a sample such as the molecular structure, density and atomic number [190, 191]. There are four techniques for quality and safety control of food based on x-rays. Firstly, the transmission imaging measurement, such as two-dimensional radiography well known from medicine or the line scan method producing images when the product passes through a vertical plane of x-rays, is used for foreign bodies detection in food [187, 192–194]; secondly, x-ray microtomography for microstructural measurement [195–197]; thirdly, the x-ray fluorescence spectroscopy measurement, such as the use of x-ray fluorescence spectra for determination of trace elements in food [198–201]; and the fourth, the small angle measurement for the structure analysis of food component, such as the structure of proteins, allergens, and starches [202–206]. Among them, the x-rays transmission imaging measurement techniques are widely used in the food industry.

Conventional x-ray transmission imaging measurement technique, as shown in figure 10.9(a), has long been used for online detection of foreign bodies in food products, relying on the absorption contrast between the foreign body and food product. A dense area, such as contaminant, will reduce the energy even further. As the x-ray exits the product, and then reaches a sensor. The sensor then converts the energy signal into an image of the interior of the food product [207]. The x-rays transmission inspection system, as shown in figure 10.9(b) and (c), mainly comprises a computer-controlled x-ray generator (i.e. x-ray source tube), a line-scanning sensor for x-ray detection, conveying belt, stepping motor, image-acquisition card and computer.

Foreign bodies are a major reason for consumer complaints in the food industry. For the meat industry, the principle of operation of x-ray measurement systems is that various components of muscle-lean meat, fat, and bone-have different properties. The relative density is the critical property, as lean tissue has a consistent density of 1.07 to 1.08, whereas fat varies depending on the temperature. Skin has a similar density to lean meat, which means that collagen measurement is hardly possible. As materials attenuate x-rays depending on their energy, use can be made of the selective attenuation of one, two, or more energy levels; that is, mono, dual (DEXA), or multiple energy x-ray absorptiometry (MEXA) [208–210]. Some internal disorders with negative effects on quality should be detectable by x-ray techniques [39, 194, 211, 212]. Because the method is primarily dependent on the density of the tissue, not the chemical composition, the selectivity is limited. X-rays have not yet been used to their fullest potential in a range of application areas, especially the agricultural and food industries. McFarlane *et al* examined constraints

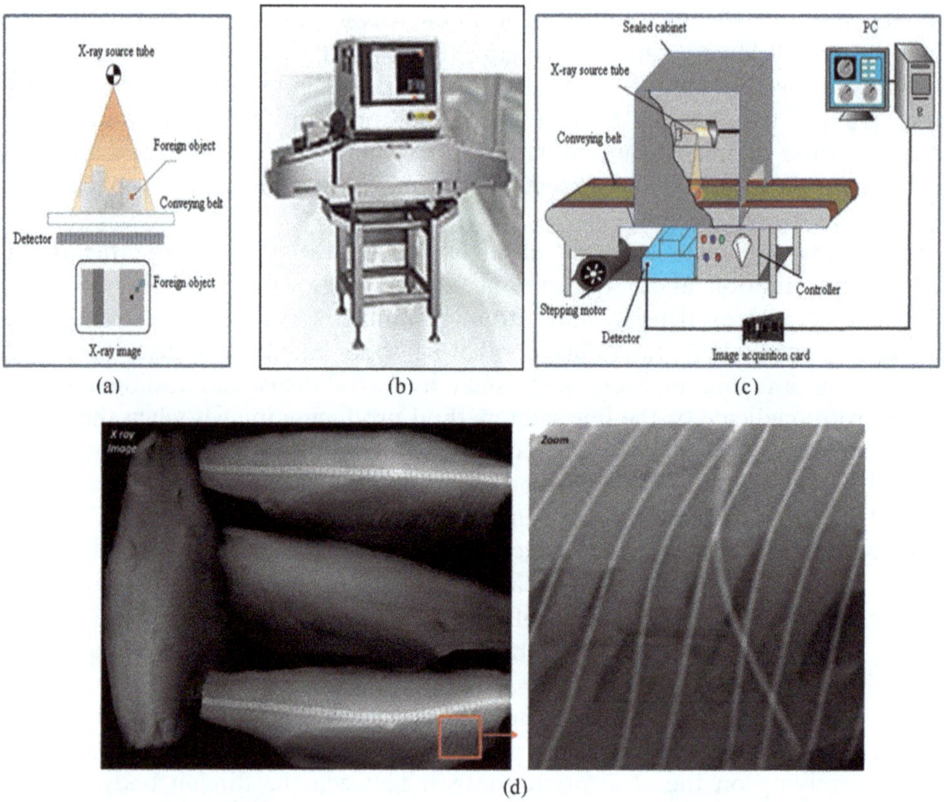

Figure 10.9. Principle of x-rays transmission radiography (a) and soft x-ray inspection system (b) picture, (c) schematic diagram, (d) x-ray image of fish fillets in which the fish bones are detectable. Reproduced from [1] by permission of the Royal Society of Chemistry.

on foreign body detection in food by using Compton scattered x-rays. The detection of a 4 mm glass fragment in water, instant coffee and muesli using Compton scatter was demonstrated by experiment [213]. Mery *et al* developed an x-ray MV approach to automatically detect fish bones in fish fillets (figure 10.9(d)) [189].

10.2.5 NMR and MRI

10.2.5.1 Nuclear magnetic resonance

NMR is based on the emission and absorption of energy in the radiofrequency range of the electromagnetic spectrum [214]. All nuclei that contain odd numbers of protons or neutrons can be observed with NMR. With the development of large magnets for medical applications, a variety of new types of magnets and radio frequency coils have become available for use in engineering applications of NMR [215]. The most commonly measured nuclei are ^{1}H and ^{13}C, although nuclei from the isotopes of many other elements can be observed (^{23}Na, ^{31}P, etc). NMR is sensitive to the existence of mobile water, oil, and sugar, which are major components of agricultural materials. Therefore, owing to its non-invasive nature,

NMR has a high potential for use in internal-quality evaluation of agricultural products whenever a component of the product contains an NMR sensitive nuclei that is correlated with quality, although the size of the magnets required is an obstacle to the wide application of NMR imaging techniques in the food industry.

(1) ^{1}H NMR is a better spectroscopic method for assessing water holding capacity (WHC), intra muscular fat (IMF) and total water content in porcine muscles than visual, fluorescent and near infrared (NIR) reflectance spectrophotometry. As water protons are easily visible with ^{1}H NMR, this technique is obviously especially useful for studying water in food products. The parameters linked to vegetables, fruit, meat and fish products are numerous: water content, WHC, water distribution (i.e. water state) [216], water mobility, etc.

(2) ^{31}P NMR is mainly applied to post mortem evaluation in meat and fish muscle. ^{31}P NMR is used in the prediction of fish freshness, of WHC in rabbit muscle, in association with ^{1}H MAS and NMR [217].

(3) ^{13}C NMR spectroscopy has been successfully used for detection of sugars and acids and that ^{13}C resonances of amino acids can be used as fingerprint for the monitoring of wines. Proton and ^{13}C NMR data have been used for the differentiation of white wines coming from three German regions [214, 217, 218].

(4) ^{23}Na NMR is performed to quantify salt content in meat and fish products. For instance, in association with ^{23}Na MRI, it is a rapid and reliable alternative for optimizing and understanding industrial salting processes in the fish industry [219].

10.2.5.2 Magnetic resonance imaging

Similarity to the NMR, MRI is also a nondestructive, nonintrusive spectroscopic technique based on the interaction of electromagnetic radiation in the radiofrequency range with matter. MRI is particularly suitable for biological materials (given that protons are abundant therein), mainly in water, but also in fat, oil, or salt, and it allows one to distinguish these components [220].

MRI has been used successfully in the past to measure several quality parameters of fresh fruit and vegetables, including the presence of internal defects, such as chilling injury in micro-tom tomato [221], water core development in apples [222], postharvest ripening of tomato [223], microporosity in fruit [214]. MRI has also been used to measure physical properties, such as size, shape and volume, and has been correlated with firmness, soluble solids, or acid content [224–226]. MRI has been also used successfully to measure several quality parameters of meat and meat products. An original experimental approach has been developed based on the quantitative, local, dynamic and *in situ* analysis deformation and water content of chicken during cooking [227]. MRI has been used to detect the water distribution [228], viscoelastic properties [229] and muscle structure [230] in meat.

Compared to other techniques such as x-ray imaging, the MRI image acquisition speed is relatively low, and is strongly determined by the required image quality and whether 1D or 2D imaging is performed [215]. In figure 10.10, CT and MRI images of a healthy apple and one with severe water core symptoms, a disorder characterized by water-soaked regions in the fruit, are compared.

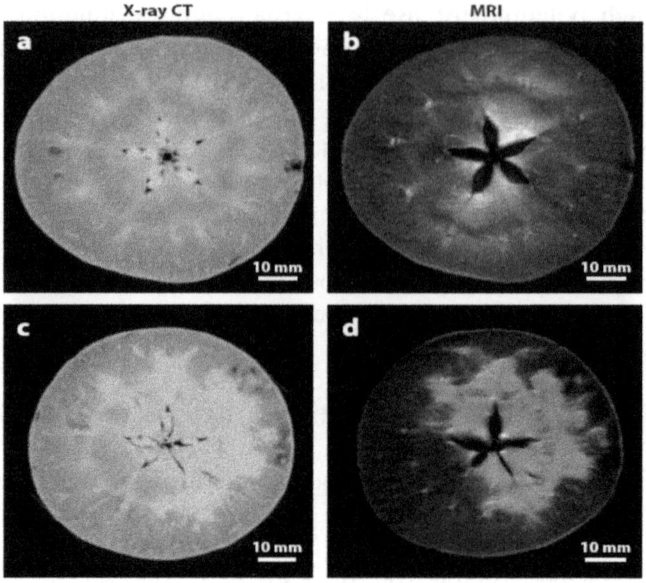

Figure 10.10. X-ray CT (left) and MRI (right) images of Ascara apples without (a, b) and with (c, d) water core. Both techniques are capable of detecting the water core region inside the fruit; however, the contrast in the MRI images is better due to the particular pulse sequence used. Figure is reprinted from [195], copyright 2014, with permission from Elsevier. Abbreviations: CT, computed tomography; MRI, magnetic resonance imaging [217].

10.2.6 Microwaves and terahertz waves

10.2.6.1 Microwaves

In North America, only four microwave and three radio frequencies are permitted by the Federal Communications Commission (FCC) for dielectric heating applications (Decareau 1985). The allotted microwave frequencies are 915, 2450 and 5800 MHz, and the radio frequencies are 13.56, 27.12, and 40.68 MHz [231]. The interest in the dielectric properties of food has been principally for predicting heating rates describing the behavior of materials. The relevant quality attributes that can be assessed with the structure, composition (particularly water content), water state and water distribution in the product, and particularly water content profiles [77]. In the high frequency range the dielectric properties of food are closely correlated with water state and water content. Dielectric relaxation spectroscopy determines the response of the molecular motion of polar molecules in the sample to a weak external alternating electric field [232]. As the frequency of the electric field is increased, it reaches a frequency called the relaxation frequency when the polar molecule can no longer rotate with the electric field. Dielectric properties change significantly around this relaxation frequency [233].

The rare industrial applications currently using microwave are based either on cavity, antenna or probe measurements. The basic configuration of this method uses two antennas, namely one transmitting and one receiving, and the meat streaming on a conveyor belt was set between the two non-contacting antennas [234].

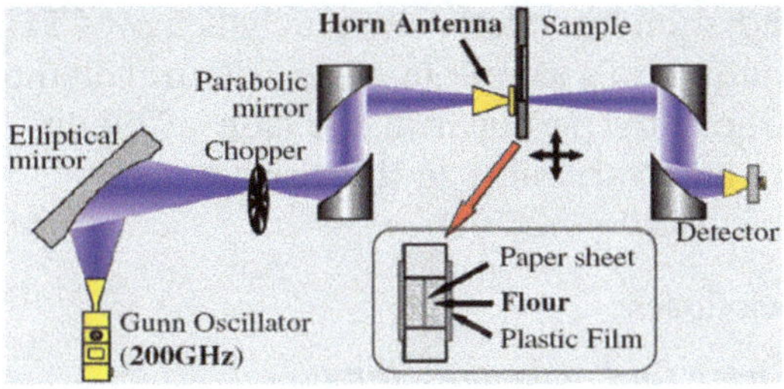

Figure 10.11. Transmission CW THz imaging system based on the parabolic mirrors operating at 0.2 THz using a pyramidal horn antenna located in front of the sample (adapted from Kim *et al* [238] with permission of Springer).

Applications of microwave sensors in other sectors of the food industry could be adapted for the meat industry, particularly for measuring water content. Work has been carried out over the last two decades by the US Department of Agriculture on water content, water state and density of grain and seed [235], and meat [233].

10.2.6.2 Terahertz

Terahertz (THz) radiation, or THz wave, is an electromagnetic wave that lies between mid-infrared and microwave radiation. Unlike x-rays, terahertz waves are not harmful, so there are no exposure worries for practitioners or patients. Moreover, THz waves have penetrative properties in various low-moisture materials, due to their low photon energy, and sub-millimeter spatial resolution. In particular, the low photon energy is suitable for food inspection with active substances [236, 237]. Due to the high absorbance of THz radiation by water, the most widespread application of THz spectroscopy would seem to be for quantification of moisture content in foods [238]. At the same time, THz radiation interacts only very weakly with materials composed of non-polar molecules, such as plastics and ceramics and is reflected by most metals [238, 239]. This indicates the THz technology has the potential for detection of some foreign material in food [240]. A transmission continuous-wave (CW) THz imaging system using reflecting mirrors is illustrated schematically in figure 10.11 [241].

The developed system was employed for simultaneous determination of the sugar and alcohol content of commercial alcoholic beverages independent of other properties such as color, organic matter content, carbonation and flavor [242 243]. Common packaging materials made from cardboard and polymers are transparent to THz radiation. This makes THz spectroscopy and imaging attractive tools for quality validation of packaged products [244]. THz spectroscopy was also used to detect pesticide in food powders (sticky rice, sweet potato, and lotus root) [245], antibiotics in milk [245] and egg powder [246], and characterization of optical properties in vegetable oil (sunflower, peanut, soybean and rapeseed oil) [247].

Although THz waves are useful for food inspection, few researches have been reported, since THz waves still have many drawbacks [248]. One of the limitations of THz spectroscopy for moisture content detection is that it is not suitable for high moisture products with thickness greater than 1 mm. This is due to the high absorption of THz radiation by water. Another challenge of THz spectroscopy is the effect of physical variations in a sample (e.g. particle size) on the refractive index of a material. This is particularly pertinent in the case of quality monitoring of fresh produce which shows high variability in this respect.

10.2.7 Biosensors techniques

10.2.7.1 Electronic nose and electronic tongue

The electronic nose (e-nose) was developed in order to mimic human olfaction that functions as a non-separative mechanism, i.e. an odor stimulus that generates a characteristic fingerprint (or smell-print) from the sensor array. The sensor array consists of broadly tuned (non-specific) sensors made of a variety of odor-sensitive biological or chemical materials. The patterns or fingerprints are used to construct a database and train a pattern recognition system so that unknown odors can subsequently be classified and identified. The detection system of e-noses usually consists of an array of chemical gas sensors or biosensors (i.e. sensors that incorporate a biological sensing element). For the chemical gas sensors, a variety of different sensor types have been developed, to which three types of materials are commonly used: metal oxide semiconductors (MOS) [249], conducting polymers composites and intrinsically conducting polymers. Recently, new technologies such as optical sensors [250], gas sensitive field effect transistors [251] and quartz microbalance (QMB) sensors [252] and ion mobility spectrometry (IMS) [253] have entered this field.

Similarly to an e-nose, electronic tongue (e-tongue) can be also considered as analytical instruments which simulate the human senses. The e-tongue devices are composed of sensors array coupled to chemometric processing to characterize complex liquid samples. In fact, the e-tongue can be thought of as an analog to both olfaction and taste senses and can recognize the flavor of liquid or liquefied products. It can also be used for quantitative detection of a variety of dissolved compounds, including volatile substances that are responsible for odors but originate from either the liquid or solid phases.

E-nose and e-tongue are composed of three elements: (i) sampler component, (ii) array of chemical sensors with different selectivity, and (iii) software with the appropriate algorithm to process the signal and get the results [254]. Pre-processing of multivariate signals in sensor arrays represents an essential part of the measuring system. Again, similar to biology, one of the most often used data-processing methods is an artificial neural network (ANN) the algorithms of which are based on modeling of learning and recognition processes in the human brain [255]. More traditional data processing techniques such as principal component analysis (PCA), linear discriminate analysis (LDA), partial least squares (PLS), functional discriminate analysis (FDA), cluster analysis (CA), and fuzzy logic are also widely used for processing of the data from multi-sensor systems [254, 256].

The strengths of both the e-nose and e-tongue include high sensitivity and the fact that they are easy to build, cost-effective and provide a short time of analysis. Therefore, these devices are becoming more and more popular as objective automated nondestructive techniques to classify and recognize a large variety of foods, such as meat [257, 258], vegetables [259], fish [260, 261], juice [262, 263], beverages [264], oil [265] and wines [266].

10.2.7.2 Colorimetric sensor array

Recently, a new artificial olfaction technology based on colorimetric sensor array has been used for odors and volatile organic compounds (VOCs). This colorimetric sensor array was proposed by using thin films of chemically responsive dyes on porous membranes. The array of multiple dyes, whose colors change based on the full range of intermolecular interactions, provides enormous discriminatory power among odorants in a simple device that can be easily digitally imaged.

For the colorimetric sensor array, it naively relies on bond formation, acid–base interactions, hydrogen-bonding, dipolar and multipolar interactions, p–p molecular complexation, Van der Waals and physical adsorption interactions between sensor and analyte. This is believed to be a fundamental flaw in the development of chemical sensors with both high sensitivity and high selectivity [267, 268].

A colorimetric sensor array system appears in figure 10.12. Gas streams (N_2) containing the vapors of interests were generated by flowing nitrogen through the sample chamber in a thermo-stated, glass-fritted bubbler. Digital mass-flow controllers controlled nitrogen flow speed. The 'before' image was first acquired on the flatbed scanner; an array was then exposed to a flowing stream of N_2 containing the analytes of interest, and the array was then scanned again after equilibration. Experiments were run until full equilibration was demonstrated by comparison of repeated scans. Color-difference maps were obtained from the scanned RGB images by digitally subtracting the image before exposure to analytes from the image after exposure, using a 314-pixel average from the center of each pigment spot, as shown in (figure 10.12) [269].

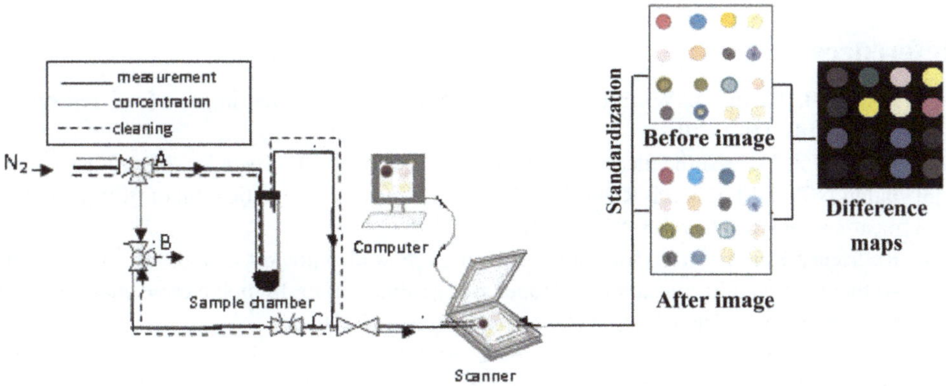

Figure 10.12. Colorimetric measurement system diagram: A, B, C: three-way valve. Three measurement phases occur: concentration, measurement and cleaning. The electro-valves, controlled by a computer program, guided the N2 through different circuits depending on the measurement phase. Reproduced from [1] by permission of the Royal Society of Chemistry.

This colorimetric sensor array is not sensitive to humidity and temperature due to the hydrophobicity of the sensor materials and sensors plate [269, 270]. It is superior to traditional MOS gas sensor techniques in the analysis of solid and liquid samples aroma. Normally, advanced chemometric techniques are needed to extract information from these color-difference maps. The colorimetric sensor array described here has been used successfully to detect many foods such as evaluation of meat [271, 272] and fish freshness [260, 273, 274], discrimination of wine [275, 276], vinegar [277–279], coffee [280], sugar [281] and tea [282, 283].

10.3 Conclusions and future trends

Nondestructive evaluation techniques inform manufacturers and regulators about product properties such as: discontinuities and separations; structure; dimensions and metrology; physical and mechanical properties; composition and chemical analysis; stress and dynamic response. As computers become ever more powerful, non-invasive assessment and evaluation of food samples becomes more popular.

Understanding the best way to evaluate food samples is of vital importance to the food scientist. For external attributes such as color, size, and absence of external defects, techniques such as MV and, spectral imaging, are now widely used on commercial sorting lines. For internal quality attributes, there are also several techniques available, such as spectroscopy, ultrasound, x-ray, and nuclear magnetic resonance. These techniques measure a complex signal that requires to be related to the quality attribute of interest via chemometric techniques. Some techniques such as NIR spectroscopy require frequent recalibration. For flavor of food, the emerging technology of electronic nose and electronic tongue based on biosensors are also available commercially. The success of techniques for food assurance often depends critically on how their measurement principle mimics the way humans assess a particular property. Future sensor designs may, therefore, preferably be based on biomimetic principles.

References

[1] Zou X B, Huang X W and Povey M 2016 Non-invasive sensing for food reassurance *Analyst* **141** 1587–610

[2] Ballin N Z 2010 Authentication of meat and meat products *Meat Sci.* **86** 577–87

[3] Ballin N Z and Lametsch R 2008 Analytical methods for authentication of fresh vs. thawed meat—a review *Meat Sci.* **80** 151–58

[4] Rodríguez-Ramírez R, González-Córdova A F and Vallejo-Cordoba B 2011 Review: Authentication and traceability of foods from animal origin by polymerase chain reaction-based capillary electrophoresis *Anal. Chim. Acta* **685** 120–6

[5] Esslinger S, Riedl J and Fauhl-Hassek C 2014 Potential and limitations of non-targeted fingerprinting for authentication of food in official control *Food Res. Int.* **60** 189–204

[6] Huck-Pezzei V A, Seitz I, Karer R, Schmutzler M, De Benedictis L and Wild B *et al* 2014 Alps food authentication, typicality and intrinsic quality by near infrared spectroscopy *Food Res. Int.* **62** 984–90

[7] Aghbashlo M, Hosseinpour S and Ghasemi-Varnamkhasti M 2014 Computer vision technology for real-time food quality assurance during drying process *Trends Food Sci. Technol.* **39** 76–84
[8] Chandrapala J, Oliver C, Kentish S and Ashokkumar M 2012 Ultrasonics in food processing—Food quality assurance and food safety *Trends Food Sci. Technol.* **26** 88–98
[9] Oliveri P, Di Egidio V, Woodcock T and Downey G 2011 Application of class-modelling techniques to near infrared data for food authentication purposes *Food Chem.* **125** 1450–6
[10] Canhoto O, Pinzari F, Fanelli C and Magan N 2004 Application of electronic nose technology for the detection of fungal contamination in library paper *Inter. Biodet. Biodeg.* **54** 303–9
[11] Cheng J-H, Sun D-W, Zeng X-A and Pu H-B 2014 Non-destructive and rapid determination of TVB-N content for freshness evaluation of grass carp (Ctenopharyngodon idella) by hyperspectral imaging *Innov. Food Sci. Emer. Technol.* **21** 179–87
[12] Cheng J-H and Sun D-W 2015 Rapid and non-invasive detection of fish microbial spoilage by visible and near infrared hyperspectral imaging and multivariate analysis *LWT – Food Sci. Technol.* **62** 1060–8
[13] Edelman G J, Gaston E, van Leeuwen T G, Cullen P J and Aalders M C G 2012 Hyperspectral imaging for non-contact analysis of forensic traces *Forensic Sci. Int.* **223** 28–39
[14] Fernández de la Ossa M Á, Amigo J M and García-Ruiz C 2014 Detection of residues from explosive manipulation by near infrared hyperspectral imaging: A promising forensic tool *Forensic Sci. Int.* **242** 228–35
[15] Atas M, Yardimci Y and Temizel A 2012 A new approach to aflatoxin detection in chili pepper by machine vision *Comput. Electron. Agric.* **87** 129–41
[16] Igathinathane C, Pordesimo L O and Batchelor W D 2009 Major orthogonal dimensions measurement of food grains by machine vision using Image *J Food Research Int.* **42** 76–84
[17] Dowlati M, Mohtasebi S S, Omid M, Razavi S H, Jamzad M and de la Guardia M 2013 Freshness assessment of gilthead sea bream (Sparus aurata) by machine vision based on gill and eye color changes *J. Food Eng.* **119** 277–87
[18] Yagiz Y, Balaban M O, Kristinsson H G, Welt B A and Marshall M R 2009 Comparison of Minolta colorimeter and machine vision system in measuring colour of irradiated Atlantic salmon *J. Sci. Food Agric.* **89** 728–30
[19] Kerr I, Shi F, Brown N, Jackson M and Parkin R 2004 Quality inspection of food packaging seals using machine vision with texture analysis *Proc. Inst. Mech. Eng. B J. Engin. Manu.* **218** 1591–9
[20] Nansen C, Herrman T and Swanson R 2010 Machine vision detection of bonemeal in animal feed samples *Appl. Spectrosc.* **64** 637–43
[21] Yang C C, Chao K, Kim M S, Chan D E, Early H L and Bell M 2010 Machine vision system for on-line wholesomeness inspection of poultry carcasses *Poul. Sci.* **89** 1252–64
[22] Lee D-J, Schoenberger R, Archibald J and McCollum S 2008 Development of a machine vision system for automatic date grading using digital reflective near-infrared imaging *J. Food Eng.* **86** 388–98
[23] Teena M, Manickavasagan A, Mothershaw A, El Hadi S and Jayas D S 2013 Potential of machine vision techniques for detecting fecal and microbial contamination of food products: a review *Food Bioprocess. Technol.* **6** 1621–34

[24] Alfatni M S M, Shariff A R M, Abdullah M Z, Marhaban M H B and Saaed O M B 2013 The application of internal grading system technologies for agricultural products—review *J. Food Eng.* **116** 703–25
[25] Ying Y, Zhang W, Jiang Y and Zhao Y 2000 Application of machine vision technique in automatic harvesting and processing of agricultural products *Nongye Jixie Xuebao/Trans. Chin. Soc. Agricul. Mach.* **31** 112–5
[26] Brosnan T and Sun D W 2002 Inspection and grading of agricultural and food products by computer vision systems—a review *Comput. Electron. Agric.* **36** 193–213
[27] Chen Y R, Chao K L and Kim M S 2002 Machine vision technology for agricultural applications *Comput. Electron. Agric.* **36** 173–91
[28] Davies E R 2009 The application of machine vision to food and agriculture: a review *Imaging Sci J.* **57** 197–217
[29] Gumus B, Balaban M O and Unlusayin M 2011 Machine vision applications to aquatic foods: a review *Turk. J. Fish. Aquat. Sci.* **11** 167–76
[30] Janardhana S, Jaya J, Sabareesaan K J and George J 2013 IEEE computer aided inspection system for food products using machine vision – a review *2013 Int. Conf. on Current Trends in Engineering and Technology (ICCTET)* 29–33
[31] Ottavian M, Fasolato L, Facco P and Barolo M 2013 Foodstuff authentication from spectral data: toward a species-independent discrimination between fresh and frozen–thawed fish samples *J. Food Eng.* **119** 765–75
[32] Dianov E M 2008 Bi-doped optical fibers: a new active medium for NIR lasers and amplifiers – art. no. 68900H ed M J F Digonnet, S Jiang, J W Glesener and J C Dries *Optical Components and Materials V. Proc. of the Society of Photo-Optical Instrumentation Engineers* (SPIE) 68902008 H8900
[33] Hemmer E, Venkatachalam N, Hyodo H, Hattori A, Ebina Y and Kishimoto H *et al* 2013 Upconverting and NIR emitting rare earth based nanostructures for NIR-bioimaging *Nanoscale* **5** 11339–61
[34] Luo S, Zhang E, Su Y, Cheng T and Shi C 2011 A review of NIR dyes in cancer targeting and imaging *Biomaterials* **32** 7127–38
[35] Peng M, Dong G, Wondraczek L, Zhang L, Zhang N and Qiu J 2011 Discussion on the origin of NIR emission from Bi-doped materials *J. Non-Cryst. Solids* **357** 2241–5
[36] Nicolai B M, Beullens K, Bobelyn E, Peirs A, Saeys W and Theron K I *et al* 2007 Nondestructive measurement of fruit and vegetable quality by means of NIR spectroscopy: a review *Postharvest Biol. Technol.* **46** 99–118
[37] Magwaza L S, Opara U L, Nieuwoudt H, Cronje P J R, Saeys W and Nicolai B 2012 NIR spectroscopy applications for internal and external quality analysis of citrus fruit – a review *Food Bioprocess. Technol.* **5** 425–44
[38] Nicolaie B M, Beullens K, Bobelyn E, Peirs A, Saeys W and Theron K I *et al* 2007 Nondestructive measurement of fruit and vegetable quality by means of NIR spectroscopy: a review *Postharvest Biol. Technol.* **46** 99–118
[39] Butz P, Hofmann C and Tauscher B 2005 Recent developments in noninvasive techniques for fresh fruit and vegetable internal quality analysis *J. Food Sci.* **70** R131–141
[40] Mlcek J, Rop O, Sustova K, Simeonovova J and Gal R 2010 Potentials of NIR spectroscopy in meat industry *Chem. Listy* **104** 855–60
[41] Prevolnik M, Candek-Potokar M and Skorjanc D 2004 Ability of NIR spectroscopy to predict meat chemical composition and quality – a review *Czech J. Anim. Sci.* **49** 500–10

[42] Liu D, Zeng X-A and Sun D-W 2013 NIR spectroscopy and imaging techniques for evaluation of fish quality – a review *Appl. Spectrosc. Rev.* **48** 609–28

[43] Sethuramasamyraja B, Sachidhanantham S, Yen M and Wample R (ed) 2007 *Interpolation of wine grape quality indicators (Anthocyanin and Brix) and development of differential harvest attachment* (Minneapolis, MN: American Society of Agricultural and Biological Engineers)

[44] Ruzickova J and Sustova K 2006 Determination of selected parameters of quality of the dairy products by NIR spectroscopy *Czech J. Food Sci.* **24** 255–60

[45] McGlone V A, Clark C J and Jordan R B 2007 Comparing density and VNIR methods for predicting quality parameters of yellow-fleshed kiwifruit (Actinidia chinensis) *Postharvest Biol. Technol.* **46** 1–9

[46] Dai N, German M A, Matsevitz T, Hanael R, Swartzberg D and Yeselson Y *et al* 2002 LeFRK2, the gene encoding the major fructokinase in tomato fruits, is not required for starch biosynthesis in developing fruits *Plant Sci.* **162** 423–30

[47] Schulz H, Baranska M, Quilitzsch R, Schutze W and Losing G 2005 Characterization of peppercorn, pepper oil, and pepper oleoresin by vibrational spectroscopy methods *J. Agric. Food Chem.* **53** 3358–63

[48] Borges G B C, Rohwedder J J R and Bortoni E C 2012 Evaluation of the use of near infrared spectroscopy (NIR) in on-line monitoring of power transformer insulation oil *AASRI Conf. on Power and Energy Systems* 2 56–61

[49] Nunes C A 2014 Vibrational spectroscopy and chemometrics to assess authenticity, adulteration and intrinsic quality parameters of edible oils and fats *Food Res. Int.* **60** 255–61

[50] Qiu Z, Feng L, Li W and He Y (ed) 2005 *Application of Near Infrared Spectroscopy in Nondestructive Quality Inspection of Agricultural Products* (Bellingham, WA: International Society for Optical Engineering)

[51] Alishahi A, Farahmand H, Prieto N and Cozzolino D 2010 Identification of transgenic foods using NIR spectroscopy: a review *Spectrochim. Acta A* **75** 1–7

[52] Bae Y M, Cho S I, Kim Y Y, Park T S and Hwang K Y 2006 Estimation of freshness of beef using near-infrared spectroscopy *Trans. ASABE* **49** 557–61

[53] Chen Y-R, Huffman R W and Park B 1996 Changes in the visible/near-infrared spectra of chicken carcasses in storage *J. Food Process Eng* **19** 121–34

[54] Kestens V, Charoud-Got J, Bau A, Bernreuther A and Emteborg H 2008 Online measurement of water content in candidate reference materials by acousto-optical tuneable filter near-infrared spectrometry (AOTF-NIR) using pork meat calibrants controlled by Karl Fischer titration *Food Chem.* **106** 1359–65

[55] Thorvaldsson K and Skjoldebrand C 1998 *Method and instrument for measuring local water content inside food Chalmers Tekniska Hogskola, Goteborg, Sweden Report* 0346-718X

[56] Alamar M C, Bobelyn E, Lammertyn J, Nicolai B M and Molto E 2007 Calibration transfer between NIR diode array and FT-NIR spectrophotometers for measuring the soluble solids contents of apple *Postharvest Biol. Technol.* **45** 38–45

[57] Cen H Y, He Y and Huang M 2006 Measurement of soluble solids contents and pH in orange juice using chemometrics and vis-NIRS *J. Agric. Food Chem.* **54** 7437–43

[58] Shao Y, He Y and Feng S 2007 Measurement of yogurt internal quality through using Vis/NIR spectroscopy *Food Res. Int.* **40** 835–41

[59] Karoui R, Pillonel L, Schaller E, Bosset J O and De Baerdemaeker J 2007 Prediction of sensory attributes of European Emmental cheese using near-infrared spectroscopy: a feasibility study *Food Chem.* **101** 1121–9
[60] Gomez A H, He Y and Pereira A G 2006 Non-destructive measurement of acidity, soluble solids and firmness of satsuma mandarin using Vis/NIR-spectroscopy techniques *J. Food Eng.* **77** 313–9
[61] Shao Y N, He Y, Gomez A H, Pereir A G, Qiu Z J and Zhang Y 2007 Visible/near infrared spectrometric technique for nondestructive assessment of tomato 'Heatwave' (Lycopersicum esculentum) quality characteristics *J. Food Eng.* **81** 672–8
[62] Ma G, Ying Y, Fu X, Lu H, Yu H and Liu Y (ed) 2005 *Prediction of Chufa MT-Firmness using FT-NIR Spectroscopy* (Bellingham, WA: International Society for Optical Engineering)
[63] Srisawas W, Jindal V K and Thanapase W 2007 Relationship between sensory textural attributes and near infrared spectra of cooked rice *J. Near Infrared Spectrosc.* **15** 333–40
[64] Bruun S W, Sondergaard I and Jacobsen S 2007 Analysis of protein structures and interactions in complex food by near-infrared spectroscopy. 2. hydrated gluten *J. Agric. Food Chem.* **55** 7244–51
[65] Garcia-Rey R M, Garcia-Olmo J, De Pedro E, Quiles-Zafra R and de Castro M D L 2005 Prediction of texture and colour of dry-cured ham by visible and near infrared spectroscopy using a fiber optic probe *Meat Sci.* **70** 357–63
[66] Schmitt S, Garrigues S and de la Guardia M 2014 Determination of the Mineral composition of foods by infrared spectroscopy: a review of a Green Alternative *Crit. Rev. Anal. Chem.* **44** 186–97
[67] Usha K and Singh B 2013 Potential applications of remote sensing in horticulture – A review *Sci. Horticult.* **153** 71–83
[68] Teixeira dos Santos C A, Lopo M, RNMJ P and Lopes J A 2013 A review on the applications of portable near-infrared spectrometers in the agro-food industry *Appl. Spectrosc.* **67** 1215–33
[69] Lopez A, Arazuri S, Garcia I, Mangado J and Jaren C 2013 A review of the application of near-infrared spectroscopy for the analysis of potatoes *J. Agric. Food Chem.* **61** 5413–24
[70] Cattaneo T M P and Holroyd S E 2013 The use of near infrared spectroscopy for determination of adulteration and contamination in milk and milk powder: updating knowledge *J. Near Infrared Spectrosc.* **21** 341–9
[71] Shao X, Ning Y, Liu F, Li J and Cai W 2012 Application of near-infrared spectroscopy in micro inorganic analysis *Acta Chim. Sinica* **70** 2109–14
[72] Weeranantanaphan J, Downey G, Allen P and Sun D-W 2011 A review of near infrared spectroscopy in muscle food analysis: 2005–2010 *J. Near Infrared Spectrosc.* **19** 61–104
[73] Rodriguez-Saona L E and Allendorf M E 2011 Use of FTIR for rapid authentication and detection of adulteration of food *Ann. Rev. Food Sci. Technol.* **22011** 467–83
[74] Kehlenbeck V 2011 Use of near infrared spectroscopy for in- and off-line performance determination of continuous and batch powder mixers: opportunities & challenges *Eleventh Int. Congress on Engineering and Food (ICEF11)* **1** 2015–22
[75] Wang J, Wang J-q, Bu D-p, Guo W-j, Shen J-s and Wei H-y *et al* 2009 Application of near-infrared spectroscopy to quality detection of milk and its products *Spectrosc. Spectral Anal.* **29** 1281–5

[76] Karoui R and Blecker C 2011 Fluorescence spectroscopy measurement for quality assessment of food systems—a review *Food Bioproc. Technol.* **4** 364–86

[77] Damez J-L and Clerjon S 2013 Quantifying and predicting meat and meat products quality attributes using electromagnetic waves: an overview *Meat Sci.* **95** 879–96

[78] Chen X, Wu X, Gan M, Xu F, He L and Yang D *et al* 2015 Rapid detection of *Staphylococcus aureus* in dairy and meat foods by combination of capture with silica-coated magnetic nanoparticles and thermophilic helicase-dependent isothermal amplification *J. Dairy Sci.* **98** 1563–70

[79] Lefcourt A M, Kim M S and Chen Y R 2005 A transportable fluorescence imagining system for detecting fecal contaminants *Comput. Electron. Agric.* **48** 63–74

[80] Jordan G, Thomasius R, Schröder H, Wulf J S, Schlüter O and Sumpf B *et al* 2009 Non-invasive mobile monitoring of meat quality *J Verbr Lebensm.* **4** 7–14

[81] Hassoun A and Karoui R 2015 Front-face fluorescence spectroscopy coupled with chemometric tools for monitoring fish freshness stored under different refrigerated conditions *Food Control* **54** 240–9

[82] Karoui R, Thomas E and Dufour E 2006 Utilisation of a rapid technique based on front-face fluorescence spectroscopy for differentiating between fresh and frozen–thawed fish fillets *Food Res. Int.* **39** 349–55

[83] Abels J P and Ludescher R D 2003 Native fluorescence from juvenile stages of common food storage insects *J. Agric. Food Chem.* **51** 544–9

[84] Bai Z, Wang B, Chen H and Wang M 2015 Spatial wettability patterning of glass microchips for water-in-oil-in-water (W/O/W) double emulsion preparation *Sens. Actuators* B **215** 330–6

[85] Liu Z-Y, Zhu T, Gu E-d, Liu Y and Song C-y 2008 Study on fluorescence spectra of Chinese rice wine *Spectrosc. Spectral Anal.* **28** 2409–12

[86] Švecová B, Bordovská M, Kalvachová D and Hájek T 2015 Analysis of Czech meads: sugar content, organic acids content and selected phenolic compounds content *J. Food Compos. Anal.* **38** 80–8

[87] Huang S-J, Lin C-P and Tsai S-Y 2015 Vitamin D2 content and antioxidant properties of fruit body and mycelia of edible mushrooms by UV-B irradiation *J. Food Compos. Anal.* **42** 38–45

[88] Pao S, Davis C L, Friedrich L M and Parish M E 2002 Utilization of fluorogenic assay for rapid detection of *Escherichia coli* in acidic fruit juice *J. Food Prot.* **65** 1943–8

[89] Pérez-Ruiz T, Martínez-Lozano C, Tomás V and Martín J 2001 Flow injection determination of methamidophos using online photo-oxidation and fluorimetric detection *Talanta* **54** 989–95

[90] Chaumpluk P and Chaiprasart P 2013 Fluorescence biosensor based on *N*-(2-Aminoethyl) glycine peptide nucleic acid for a simple and rapid detection of *Escherichia coli* in fresh-cut mango *Int. Mango Symp.* 992 551–60

[91] Thepwiwatjit N, Thattiyaphong A, Limsuwan P, Tuitemwong K and Tuitemwong P 2014 Rubpy dye-doped silica nanoparticles as signal reporter in a dot fluorescence immunoassay strip *J. Nanomater*

[92] Lee K-S, Cohen W B, Kennedy R E, Maiersperger T K and Gower S T 2004 Hyperspectral versus multispectral data for estimating leaf area index in four different biomes *Remote Sens. Environ.* **91** 508–20

[93] Nicolai B M, Beullens K, Bobelyn E, Peirs A, Saeys W and Theron K I *et al* 2007 Nondestructive measurement of fruit and vegetable quality by means of NIR spectroscopy: a review *Postharvest Biol. Technol.* **46** 99–118
[94] Qin J, Chao K, Kim M S, Lu R and Burks T F 2013 Hyperspectral and multispectral imaging for evaluating food safety and quality *J. Food Eng.* **118** 157–71
[95] Wu D and Sun D-W 2013 Advanced applications of hyperspectral imaging technology for food quality and safety analysis and assessment: a review — Part I: fundamentals *Innov. Food Sci. Emer. Technol.* **19** 1–14
[96] Wu D and Sun D-W 2013 Advanced applications of hyperspectral imaging technology for food quality and safety analysis and assessment: a review — Part II: applications *Innov. Food Sci. Emer. Technol.* **19** 15–28
[97] Ariana D P and Lu R 2008 Quality evaluation of pickling cucumbers using hyperspectral reflectance and transmittance imaging: Part I. Development of a prototype *Sens. Instrum. Food Qual. Saf.* **2** 144–51
[98] Jin J, Tang L, Hruska Z and Yao H 2009 Classification of toxigenic and atoxigenic strains of Aspergillus flavus with hyperspectral imaging *Comput. Electron. Agric.* **69** 158–64
[99] Jiang L, Zhu B, Tao Y and Da-Wen S 2010 Hyperspectral image classification methods *Hyperspectral Imaging for Food Quality Analysis and Control* (San Diego, CA: Academic) pp 79–98
[100] Cho B-K, Kim M S, Baek I-S, Kim D-Y, Lee W-H and Kim J *et al* 2013 Detection of cuticle defects on cherry tomatoes using hyperspectral fluorescence imagery *Postharvest Biol. Technol.* **76** 40–9
[101] Mehl P M, Chen Y-R, Kim M S and Chan D E 2004 Development of hyperspectral imaging technique for the detection of apple surface defects and contaminations *J. Food Eng.* **61** 67–81
[102] Ariana D P and Lu R 2010 Evaluation of internal defect and surface color of whole pickles using hyperspectral imaging *J. Food Eng.* **96** 583–90
[103] Zhang M, Qin Z, Liu X and Ustin S L 2003 Detection of stress in tomatoes induced by late blight disease in California, USA, using hyperspectral remote sensing *Int. J. Appl. Earth Obs. Geoinf.* **4** 295–310
[104] Zou X, Shi J, Hao L, Zhao J, Mao H and Chen Z *et al* 2011 *In vivo* noninvasive detection of chlorophyll distribution in cucumber (*Cucumis sativus*) leaves by indices based on hyperspectral imaging *Anal. Chim. Acta* **706** 105–12
[105] Schaeffer B A, Morrison J M, Kamykowski D, Feldman G C, Xie L and Liu Y *et al* 2008 Phytoplankton biomass distribution and identification of productive habitats within the Galapagos Marine Reserve by MODIS, a surface acquisition system, and *in situ* measurements *Remote Sens. Environ.* **112** 3044–54
[106] ElMasry G and Sun D-W 2010 Meat quality assessment using a hyperspectral imaging system ed S Professor Da-Wen *Hyperspectral Imaging for Food Quality Analysis and Control* (San Diego, CA: Academic) pp 175–240
[107] Díaz R, Cervera L, Fenollosa S, Ávila C and Belenguer J 2011 Hyperspectral system for the detection of foreign bodies in meat products *Procedia Eng.* **25** 313–6
[108] Ye X, Sakai K, Sasao A and Asada S-I 2008 Potential of airborne hyperspectral imagery to estimate fruit yield in citrus *Chemometr. Intell. Lab. Syst.* **90** 132–44
[109] Lu R and Da-Wen S 2008 Quality evaluation of fruit by hyperspectral imaging *Computer Vision Technology for Food Quality Evaluation* (Amsterdam: Academic) pp 319–48

[110] Huang M, Wan X, Zhang M and Zhu Q 2013 Detection of insect-damaged vegetable soybeans using hyperspectral transmittance image *J. Food Eng.* **116** 45–9

[111] Qin J and Lu R 2008 Measurement of the optical properties of fruits and vegetables using spatially resolved hyperspectral diffuse reflectance imaging technique *Postharvest Biol. Technol.* **49** 355–65

[112] Garcia-Allende P B, Conde O M, Mirapeix J, Cobo A and Lopez-Higuera J M 2008 Quality control of industrial processes by combining a hyperspectral sensor and Fisher's linear discriminant analysis *Sens. Actuators* B **129** 977–84

[113] Sivertsen A H, Chu C-K, Wang L-C, Godtliebsen F, Heia K and Nilsen H 2009 Ridge detection with application to automatic fish fillet inspection *J. Food Eng.* **90** 317–24

[114] Schrader B, Schulz H, Andreev G N, Klump H H and Sawatzki J 2000 Non-destructive NIR-FT-Raman spectroscopy of plant and animal tissues, of food and works of art *Talanta* **53** 35–45

[115] Celedon A and Aguilera J M 2002 Applications of microprobe Raman spectroscopy in food science *Food Sci. Technol. Int.* **8** 101–8

[116] Qin J, Chao K and Kim M S 2011 Investigation of Raman chemical imaging for detection of lycopene changes in tomatoes during postharvest ripening *J. Food Eng.* **107** 277–88

[117] Daarre A, Jaffiol R, Richard A and Tchtio P 2002 Raman hyperspectral imaging applied to chemical co-localization in diluted samples of perylene-doped nanotubes *Chem. Phys. Lett.* **366** 274–8

[118] Thygesen L G, Lokke M M, Micklander E and Engelsen S B 2003 Vibrational micro-spectroscopy of food. Raman vs. FT-IR *Trends Food Sci. Technol.* **14** 50-7

[119] Muik B, Lendl B, Molina-Diaz A, Ortega-Calderon D and Ayora-Canada M J 2004 Discrimination of olives according to fruit quality using Fourier transform Raman spectroscopy and pattern recognition techniques *J. Agric. Food Chem.* **52** 6055–60

[120] Guzman E, Baeten V, Fernandez Pierna J A and Garcia-Mesa J A 2012 A portable Raman sensor for the rapid discrimination of olives according to fruit quality *Talanta* **93** 94–8

[121] Camerlingo C, Zenone F, Delfino I, Diano N, Mita D G and Lepore M 2007 Investigation on clarified fruit juice composition by using visible light micro-Raman spectroscopy *Sensors* **7** 2049–61

[122] Seidler-Lozykowska K, Baranska M, Baranski R and Krol D 2010 Raman analysis of caraway (*Carum carvi* L.) single fruits. Evaluation of essential oil content and its composition *J. Agric. Food Chem.* **58** 5271–5

[123] Nikbakht A M, Hashjin T T, Malekfar R and Gobadian B 2011 Nondestructive determination of tomato fruit quality parameters using Raman spectroscopy *J. Agricul. Sci. Technol.* **13** 517–26

[124] Trebolazabala J, Maguregui M, Morillas H, de Diego A and Manuel Madariaga J 2013 Use of portable devices and confocal Raman spectrometers at different wavelength to obtain the spectral information of the main organic components in tomato (Solanum lycopersicum) fruits *Spectrochim. Acta A* **105** 391–9

[125] Prinsloo L C, du Plooy W and van der Merwe C 2004 Raman spectroscopic study of the epicuticular wax layer of mature mango (*Mangifera indica*) fruit *J. Raman Spectrosc.* **35** 561–7

[126] Afseth N K, Segtnan V H, Marquardt B J and Wold J P 2005 Raman and near-infrared spectroscopy for quantification of fat composition in a complex food model system *Appl. Spectrosc.* **59** 1324–32

[127] Zhao Y, Ma C Y, Yuen S N and Phillips D L 2004 Study of acetylated food proteins by Raman spectroscopy *J. Food Sci.* **69** C206–13
[128] Herrero A M 2008 Raman spectroscopy for monitoring protein structure in muscle food systems *Crit Rev Food Sci.* **48** 512–23
[129] Craig A P, Franca A S and Irudayaraj J 2013 Surface-enhanced Raman spectroscopy applied to food safety *Ann. Rev. Food Sci. Technol.* **4** 369–80
[130] Harz M, Rosch P, Peschke K D, Ronneberger O, Burkhardt H and Popp J 2005 Micro-Raman spectroscopic identification of bacterial cells of the genus Staphylococcus and dependence on their cultivation conditions *Analyst* **130** 1543–50
[131] Krause M, Roesch P, Radt B and Popp J 2008 Localizing and identifying living bacteria in an abiotic environment by a combination of Raman and fluorescence microscopy *Anal. Chem.* **80** 8568–75
[132] Yang H and Irudayaraj J 2003 Rapid detection of foodborne microorganisms on food surface using Fourier transform Raman spectroscopy *J. Mol. Struct.* **646** 35–43
[133] Lee S, Lohumi S, Lim H-S, Gotoh T, Cho B-K and Kim M S *et al* 2015 Development of a detection method for adulterated onion powder using Raman spectroscopy *J. Facul. Agricul. Kyushu Univ.* **60** 151–6
[134] Lu X, Al-Qadiri H M, Lin M and Rasco B A 2011 Application of mid-infrared and Raman spectroscopy to the study of bacteria *Food Bioprocess. Technol.* **4** 919–35
[135] Zhang Y, Yu W, Pei L, Lai K, Rasco B A and Huang Y 2015 Rapid analysis of malachite green and leucomalachite green in fish muscles with surface-enhanced resonance Raman scattering *Food Chem.* **169** 80–4
[136] Cheng Y and Dong Y 2011 Screening melamine contaminant in eggs with portable surface-enhanced Raman Spectroscopy based on gold nanosubstrate *Food Control* **22** 685–9
[137] Hu Y, Feng S, Gao F, Li-Chan E C Y, Grant E and Lu X 2015 Detection of melamine in milk using molecularly imprinted polymers-surface enhanced Raman spectroscopy *Food Chem.* **176** 123–9
[138] Li X, Lee H K, Phang I Y, Lee C K and Ling X Y 2014 Superhydrophobic-oleophobic Ag nanowire platform: an analyte-concentrating and quantitative aqueous and organic toxin surface-enhanced Raman scattering sensor *Anal. Chem.* **86** 10437–44
[139] Okazaki S, Hiramatsu M, Gonmori K, Suzuki O and Tu A T 2009 Rapid nondestructive screening for melamine in dried milk by Raman spectroscopy *Forensic Toxicol.* **27** 94–7
[140] Fan Y, Lai K, Rasco B A and Huang Y 2015 Determination of carbaryl pesticide in Fuji apples using surface-enhanced Raman spectroscopy coupled with multivariate analysis *Lwt-Food Sci Technol.* **60** 352–7
[141] Lin L, Wu R-m, Liu M-h, Wang X-b and Yan L-Y 2015 Surface-enhanced Raman spectroscopy analysis of thiabendazole pesticide *Spectrosc. Spectral Anal.* **35** 404–8
[142] Nguyen T H D, Zhang Z, Mustapha A, Li H and Lin M 2014 Use of graphene and gold nanorods as substrates for the detection of pesticides by surface enhanced Raman spectroscopy *J. Agric. Food Chem.* **62** 10445–51
[143] Herrero A M 2008 Raman spectroscopy a promising technique for quality assessment of meat and fish: a review *Food Chem.* **107** 1642–51
[144] Yang D and Ying Y 2011 Applications of Raman spectroscopy in agricultural products and food analysis: a review *Appl. Spectrosc. Rev.* **46** 539–60
[145] Zheng J and He L 2014 Surface-enhanced Raman spectroscopy for the chemical analysis of food *Compreh. Rev. Food Sci. Food Safety* **13** 317–28

[146] Chen K, Sun X, Qin C and Tang X 2010 Color grading of beef fat by using computer vision and support vector machine *Comput. Electron. Agric.* **70** 27–32
[147] Huang L, Zhao J, Chen Q and Zhang Y 2014 Nondestructive measurement of total volatile basic nitrogen (TVB-N) in pork meat by integrating near infrared spectroscopy, computer vision and electronic nose techniques *Food Chem.* **145** 228–36
[148] Estévez M, Kylli P, Puolanne E, Kivikari R and Heinonen M 2008 Fluorescence spectroscopy as a novel approach for the assessment of myofibrillar protein oxidation in oil-in-water emulsions *Meat Sci.* **80** 1290–6
[149] Herrero A M, Cambero M I, Ordóñez J A, de la Hoz L and Carmona P 2008 Raman spectroscopy study of the structural effect of microbial transglutaminase on meat systems and its relationship with textural characteristics *Food Chem.* **109** 25–32
[150] Chen Q, Zhang Y, Zhao J and Hui Z 2013 Nondestructive measurement of total volatile basic nitrogen (TVB-N) content in salted pork in jelly using a hyperspectral imaging technique combined with efficient hypercube processing algorithms *Anal. Methods* **5** 6382–8
[151] Demattè M L, Pojer N, Endrizzi I, Corollaro M L, Betta E and Aprea E *et al* 2014 Effects of the sound of the bite on apple perceived crispness and hardness *Food Qual. Preference* **38** 58–64
[152] Mohammadi V, Ghasemi-Varnamkhasti M, Ebrahimi R and Abbasvali M 2014 Ultrasonic techniques for the milk production industry *Measurement* **58** 93–102
[153] Povey M J W 1997 *Ultrasonic Techniques for Fluids Characterization* (New York: Academic)
[154] Povey M J W 2007 Acoustic techniques to characterize food microstructure *Understanding and Controlling the Microstructure of Complex Foods* ed D J McClements (Cambridge: Woodhead Publishing) 311–33
[155] Awad T S, Moharram H A, Shaltout O E, Asker D and Youssef M M 2012 Applications of ultrasound in analysis, processing and quality control of food: A review *Food Res. Int.* **48** 410–27
[156] Povey M J W, Watson N and Parker N G 2013 Ultrasonic and acoustic microscopy: principles and applications to food microstructures *Food Microstructures* ed V J M Groves (Cambridge: Woodhead Publishing) 192–222
[157] Povey M J W 2014 Crystal nucleation in food colloids *Food Hydrocoll.* **42** Part 1 118–29
[158] Chen J, Karlsson C and Povey M 2005 Acoustic Envelope Detector for crispness assessment of biscuits *J. Texture Stud.* **36** 139–56
[159] Povey M and Harden C 1981 An application of the ultrasonic pulse echo technique to the measurement of crispness of biscuits *J. Food Technol.* **16** 167–75
[160] Povey M J W 2007 Acoustic techniques to characterize food microstructure Understanding and Controlling the Microstructure of Complex *Foods* **149** 311–33
[161] Varela P, Chen J, Fiszman S and Povey M J W 2006 Crispness assessment of roasted almonds by an integrated approach to texture description: texture, acoustics, sensory and structure *J Chemometr.* **20** 311–20
[162] So N 1973 Electrical properties of agricultural products *Trans. ASAE* **16** 384–400
[163] Jha S N, Narsaiah K, Basediya A L, Sharma R, Jaiswal P and Kumar R *et al* 2011 Measurement techniques and application of electrical properties for nondestructive quality evaluation of foods-a review *J Food Sci Tech Mys.* **48** 387–411
[164] Picard B, Lebret B, Cassar-Malek I, Liaubet L, Berri C and Le Bihan-Duval E *et al* 2015 Recent advances in omic technologies for meat quality management *Meat Sci.* **109** 18–26

[165] Zywica R, Pierzynowska-Korniak G and Wojcik J 2005 Application of food products electrical model parameters for evaluation of apple puree dilution *J. Food Eng.* **67** 413–8

[166] Damez J L and Clerjon S 2007 Modelling the electrical properties of meat mesostructure during ageing *53rd Int. Congress of Meat Science and Tech.* **1** 215–6

[167] Krauss S, Schnitzler W H, Grassmann J and Woitke M 2006 The influence of different electrical conductivity values in a simplified recirculating soilless system on inner and outer fruit quality characteristics of tomato *J. Agric. Food Chem.* **54** 441–8

[168] Garcia-Breijo E, Barat J M, Torres O L, Grau R, Gil L and Ibanez J *et al* 2008 Development of a puncture electronic device for electrical conductivity measurements throughout meat salting *Sens. Actuat. A - Physical.* **148** 63–7

[169] Bozkurt H and Icier F 2010 Electrical conductivity changes of minced beef-fat blends during ohmic cooking *J. Food Eng.* **96** 86–92

[170] Yang Y, Li Q, Yu X, Chen X and Wang Y 2014 A novel method for determining peroxide value of edible oils using electrical conductivity *Food Control* **39** 198–203

[171] Zhu S, Luo Y, Feng L and Bao Y 2015 Establishment of kinetic models based on electrical conductivity and global stability index for predicting the quality of allogynogenetic crucian carps (carassius auratus gibelio) during chilling storage *J. Food Process. Preserv.* **39** 167–74

[172] Damez J-L, Clerjon S, Abouelkaram S and Lepetit J 2007 Dielectric behavior of beef meat in the 1–1500 kHz range: simulation with the Fricke/Cole–Cole model *Meat Sci.* **77** 512–9

[173] Ryynänen S 1995 The electromagnetic properties of food materials: a review of the basic principles *J. Food Eng.* **26** 409–29

[174] Sosa-Morales M E, Tiwari G, Wang S, Tang J, Garcia H S and Lopez-Malo A 2009 Dielectric heating as a potential post-harvest treatment of disinfesting mangoes, Part II: Development of RF-based protocols and quality evaluation of treated fruits *Biosystems Eng.* **103** 287–96

[175] Sosa-Morales M E, Tiwari G, Wang S, Tang J, Garcia H S and Lopez-Malo A 2009 Dielectric heating as a potential post-harvest treatment of disinfesting mangoes, Part I: Relation between dielectric properties and ripening *Biosystems Eng.* **103** 297–303

[176] VN T, SS S 1987 Dielectric properties of beef, beef liver, chicken and salmon at frequencies from 100 to 2,500 MHz *J Microw Power* **22** 29–33

[177] C L, N S 1999 Dielectric properties of tuna at 2,450 and 915 MHz as a function of temperature *J. Jpn Soc. Food Sci. Technol.* **20** 42–5

[178] Martinsen Ø G, Grimnes S and Mirtaheri P 2000 Non-invasive measurements of post-mortem changes in dielectric properties of haddock muscle—a pilot study *J. Food Eng.* **43** 189–92

[179] Basaran P, Basaran-Akgul N and Rasco B A 2010 Dielectric properties of chicken and fish muscle treated with microbial transglutaminase *Food Chem.* **120** 361–70

[180] Sosa-Morales M E, Valerio-Junco L, López-Malo A and García H S 2010 Dielectric properties of foods: Reported data in the 21st Century and their potential applications *LWT - Food Science and Technology.* **43** 1169–79

[181] Corwin D L and Plant R E 2005 Applications of apparent soil electrical conductivity in precision agriculture *Comput. Electron. Agric.* **46** 1–10

[182] Legrand A, Leuliet J C, Duquesne S, Kesteloot R, Winterton P and Fillaudeau L 2007 Physical, mechanical, thermal and electrical properties of cooked red bean (*Phaseolus vulgaris* L.) for continuous ohmic heating process *J. Food Eng.* **81** 447–58

[183] Damez J-L, Clerjon S, Abouelkaram S and Lepetit J 2008 Electrical impedance probing of the muscle food anisotropy for meat ageing control *Food Control* **19** 931–9

[184] Zhao X, Yin J, Chen J, Song G, Wang L and Zhu H *et al* 2009 Inhibitory effects and mechanisms of intestinal electrical stimulation on gastric tone, antral contractions, pyloric tone, and gastric emptying in dogs *Am. J. Physiol. - Reg. Integ. Comp. Physiol.* **296** R36–42

[185] Jin Y, Yang N, Duan X, Wu F, Tong Q and Xu X 2015 Determining total solids and fat content of liquid whole egg products via measurement of electrical parameters based on the transformer properties *Biosystems Eng.* **129** 70–7

[186] Zywica R and Banach J K 2015 Simple linear correlation between concentration and electrical properties of apple juice *J. Food Eng.* **158** 8–12

[187] Zwiggelaar R, Bull C R and Mooney M J 1996 X-ray simulations for imaging applications in the agricultural and food industries *J. Agric. Eng. Res.* **63** 161–70

[188] Warning A, Verboven P, Nicolaï B, van Dalen G and Datta A K 2014 Computation of mass transport properties of apple and rice from X-ray microtomography images *Innov. Food Sci. Emer. Technol.* **24** 14–27

[189] Mery D, Lillo I, Loebel H, Riffo V, Soto A and Cipriano A *et al* 2011 Automated fish bone detection using X-ray imaging *J. Food Eng.* **105** 485–92

[190] Leboutet H and Aucouturier J 1985 Theoretical evaluation of induced radioactivity in food products by electron—or X-ray beam sterilization *Rad. Phys. Chem. (1977)* **25** 233–42

[191] Bull C R, Zwiggelaar R and Speller R D 1997 Review of inspection techniques based on the elastic and inelastic scattering of X-rays and their potential in the food and agricultural industry *J. Food Eng.* **33** 167–79

[192] Nielsen M S, Lauridsen T, Christensen L B and Feidenhans'l R 2013 X-ray dark-field imaging for detection of foreign bodies in food *Food Control* **30** 531–5

[193] Sjögren E, Abrahamsson B, Augustijns P, Becker D, Bolger M B and Brewster M *et al* 2014 *In vivo* methods for drug absorption—Comparative physiologies, model selection, correlations with *in vitro* methods (IVIVC), and applications for formulation/API/excipient characterization including food effects *Eur. J. Pharm. Sci.* 57 99–151

[194] Toyofuku N and Haff R P 2012 Computer vision for foreign body detection and removal in the food industry *Computer Vision Technology in the Food and Beverage Industries* ed D-W Sun (Cambridge: Woodhead Publishing) pp 181–205

[195] Herremans E, Verboven P, Defraeye T, Rogge S, Ho Q T and MLATM H *et al* 2014 X-ray CT for quantitative food microstructure engineering: the apple case *Nucl. Instrum. Methods Phys. Res., Sect.* B **324** 88–94

[196] Laverse J, Mastromatteo M, Frisullo P and Del Nobile M A 2011 X-ray microtomography to study the microstructure of cream cheese-type products *J. Dairy Sci.* **94** 43–50

[197] Lim K S and Barigou M 2004 X-ray micro-computed tomography of cellular food products *Food Res. Int.* **37** 1001–12

[198] Nielson K K, Mahoney A W, Williams L S and Rogers V C 1991 X-ray fluorescence measurements of Mg, P, S, CI, K, Ca, Mn, Fe, Cu, and Zn in fruits, vegetables, and grain products *J. Food Compos. Anal.* **4** 39–51

[199] Jastrzębska A, Brudka B, Szymański T and Szłyk E 2003 Determination of phosphorus in food samples by X-ray fluorescence spectrometry and standard spectrophotometric method *Food Chem.* **83** 463–7

[200] Perring L, Andrey D, Basic-Dvorzak M and Hammer D 2005 Rapid quantification of iron, copper and zinc in food premixes using energy dispersive X-ray fluorescence *J. Food Compos. Anal.* **18** 655–63

[201] Otaka A, Hokura A and Nakai I 2014 Determination of trace elements in soybean by X-ray fluorescence analysis and its application to identification of their production areas *Food Chem.* **147** 318–26

[202] McFarlane N J B, Bull C R, Tillett R D, Speller R D, Royle G J and Johnson K R A 2000 The potential for Compton scattered X-rays in food inspection: the effect of multiple scatter and sample inhomogeneity *J. Agric. Eng. Res.* **75** 265–74

[203] Almeida A P G, Braz D, Barroso R C and Lopes R T 2007 Low-angle X-ray scattering properties of irradiated spices *Nucl. Instrum. Methods Phys. Res., Sect.* A **580** 764–7

[204] Blazek J and Gilbert E P 2011 Application of small-angle X-ray and neutron scattering techniques to the characterisation of starch structure: a review *Carbohyd. Polym.* **85** 281–93

[205] Petoukhov M V and Svergun D I 2013 Applications of small-angle X-ray scattering to biomacromolecular solutions *Int. J. Biochem. Cell Biol.* **45** 429–37

[206] Khandelwal M and Windle A H 2014 Small angle X-ray study of cellulose macromolecules produced by tunicates and bacteria *Int. J. Biol. Macromol.* **68** 215–7

[207] Chen Q, Zhang C, Zhao J and Ouyang Q 2013 Recent advances in emerging imaging techniques for non-destructive detection of food quality and safety *TrAC, Trends Anal. Chem.* **52** 261–74

[208] Xia Y, Ergun D L, Wacker W K, Wang X, Davis C E and Kaul S 2014 Relationship between dual-energy X-ray absorptiometry volumetric assessment and X-ray computed tomography–derived single-slice measurement of visceral fat *J. Clin. Densitometry* **17** 78–83

[209] Grégoire O, Cleland M R, Mittendorfer J, Dababneh S, Ehlermann D A E and Fan X *et al* 2003 Radiological safety of food irradiation with high energy X-rays: theoretical expectations and experimental evidence *Radiat. Phys. Chem.* **67** 169–83

[210] Solanki R B, Prasad M, Sonawane A U and Gupta S K 2012 Probabilistic safety assessment for food irradiation facility *Ann. Nucl. Energy* **43** 123–30

[211] Demaurex G and Sallé L 2014 Detection of physical hazards ed Y M Lelieveld *Food Safety Management* (San Diego, CA: Academic) pp 511–33

[212] Edwards M 2014 Food hygiene and foreign bodies *Hygiene in Food Processing* ed H L Lelieveld, J T Holah and D Napper 2nd edn (Cambridge: Woodhead Publishing) pp 441–64

[213] McFarlane N J B, Bull C R, Tillett R D, Speller R D and Royle G J 2001 SE—Structures and environment: time constraints on glass detection in food materials using Compton scattered X-rays *J. Agric. Eng. Res.* **79** 407–18

[214] Musse M, De Guio F, Quellec S, Cambert M, Challois S and Davenel A 2010 Quantification of microporosity in fruit by MRI at various magnetic fields: comparison with X-ray microtomography *Magn. Reson. Imaging* **28** 1525–34

[215] Kim S-M, Chen P, McCarthy M J and Zion B 1999 Fruit internal quality evaluation using on-line nuclear magnetic resonance sensors *J. Agric. Eng. Res.* **74** 293–301

[216] Bertram H C, Kohler A, Bocker U, Ofstad R and Andersen H J 2006 Heat-induced changes in myofibrillar protein structures and myowater of two pork qualities. A combined FT-IR spectroscopy and low-field NMR relaxometry study *J. Agric. Food Chem.* **54** 1740–6

[217] Bertram H C, Jian Zhi H, Rommereim D N, Wind R A and Andersen H J 2004 Dynamic high-resolution 1H and 31P NMR spectroscopy and 1H T2 measurements in postmortem rabbit muscles using slow magic angle spinning *J. Agric. Food Chem.* **52** 2681–8

[218] Brescia M A, Caldarola V, De Giglio A, Benedetti D, Fanizzi F P and Sacco A 2002 Characterization of the geographical origin of Italian red wines based on traditional and nuclear magnetic resonance spectrometric determinations *Anal. Chim. Acta* **458** 177–86

[219] Bertram H C, Holdsworth S J, Whittaker A K and Andersen H J 2013 Salt diffusion and distribution in meat studied by 23Na nuclear magnetic resonance imaging and relaxometry *J. Agric. Food Chem.* **53** 7814–8

[220] BMN I, Defraeye T, De Ketelaere B, Herremans E, MLAT H and Saeys W *et al* 2014 Nondestructive measurement of fruit and vegetable quality *Annual Rev. Food Sci. Technol.* **5** 285–312

[221] Tao F, Zhang L, McCarthy M J, Beckles D M and Saltveit M 2014 Magnetic resonance imaging provides spatial resolution of chilling injury in micro-tom tomato (*Solanum lycopersicum* L.) fruit *Postharvest Biol. Technol.* **97** 62–7

[222] Melado-Herreros A, Muñoz-García M-A, Blanco A, Val J, Fernández-Valle M E and Barreiro P 2013 Assessment of watercore development in apples with MRI: effect of fruit location in the canopy *Postharvest Biol. Technol.* **86** 125–33

[223] Musse M, Quellec S, Cambert M, Devaux M-F, Lahaye M and Mariette F 2009 Monitoring the postharvest ripening of tomato fruit using quantitative MRI and NMR relaxometry *Postharvest Biol. Technol.* **53** 22–35

[224] Abbott J A, Lu R, Upchurch B L and Stroshine R L 1997 Technologies for non-destructive quality evaluation of fruits and vegetables *Hortic Rev.* **20** 1–120

[225] Clarke C J and Mazucco R A 1997 Application of magnetic resonance imaging to preand post-harvest studies of fruits and vegetables *Postharvest Biol. Technol.* **11** 1–21

[226] Létal J and Hájek M 2003 'texture' analysis of MR images of apples during ripening and storage *LWT- Food Sci Technol.* **36** 19–27

[227] Campos I, Masot R, Alcañiz M, Gil L, Soto J and Vivancos J L *et al* 2010 Accurate concentration determination of anions nitrate, nitrite and chloride in minced meat using a voltammetric electronic tongue *Sens. Actuators* B **149** 71–8

[228] Bertram H C, Andersen R H and Andersen H J 2014 Development in myofibrillar water distribution of two pork qualities during 10-month freezer storage *Meat Sci.* **75** 128–33

[229] Gruwel M L, Latta P, Matwiy B and Tomanek B 2010 Characterization of food stuffs using magnetic resonance elastography *Food Res. Int.* **43** 2087–92

[230] Badii F and Howell N K 2006 Fish gelatin: Structure, gelling properties and interaction with egg albumen proteins *Food Hydrocoll.* **20** 630–40

[231] Venkatesh M S and Raghavan G S V 2004 An overview of microwave processing and dielectric properties of agri-food materials *Biosystems Eng.* **88** 1–18

[232] Clerjon S and Damez J-L 2009 Microwave sensing for an objective evaluation of meat ageing *J. Food Eng.* **94** 379–89

[233] Clerjon S, Daudin J-D and Damez J-L 2003 Water activity and dielectric properties of gels in the frequency range 200 MHz–6 GHz *Food Chem.* **82** 87–97

[234] Nyfors E 2000 Industrial microwave sensors—a review *Subsurf. Sens. Technol.* **1** 23–43

[235] Nelson S O and Trabelsi S 2008 Dielectric spectroscopy measurements on fruit, meat, and grain *Trans. ASABE* **51** 1829–34

[236] Parasoglou P, Parrott E P J, Zeitler J A, Rasburn J, Powell H and Gladden L F *et al* 2010 Quantitative water content measurements in food wafers using terahertz radiation *Terahertz Sci. Technol.* **3** 172–82

[237] Kim G-J, Kim J-I, Jeon S-G, Kim J, Park K-K and Oh C-H 2012 Enhanced continuous-wave terahertz imaging with a horn antenna for food inspection *J Infrared Milli Terahz Waves* 33

[238] Gowen A A, O'Sullivan C and O'Donnell C P 2012 Terahertz time domain spectroscopy and imaging: Emerging techniques for food process monitoring and quality control *Trends Food Sci. Technol.* **25** 40–6

[239] Ok G, Kim H J, Chun H S and Choi S-W 2014 Foreign-body detection in dry food using continuous sub-terahertz wave imaging *Food Control* **42** 284–9

[240] Qin J, Xie L and Ying Y 2015 Determination of tetracycline hydrochloride by terahertz spectroscopy with PLSR model *Food Chem.* **170** 415–22

[241] Kim G-J, Kim J-I, Jeon S-G, Kim J, Park K-K and Oh C-H 2012 Enhanced continuous-wave terahertz imaging with a horn antenna for food inspection *J. Infrared Milli Terahz Waves* **33** 657–64

[242] Jepsen P U, Møller U and Merbold H 2007 Investigation of aqueous alcohol and sugar solutions with reflection terahertz time-domain spectroscopy *Opt. Express* **22** 14717–37

[243] Jepsen P U, Møller U and Merbold H 2007 Investigation of aqueous alcohol and sugar solutions with reflection terahertz time-domain spectroscopy *Opt. Express* **15** 14717–37

[244] Jordens C and Koch M 2008 Detection of foreign bodies in chocolate with pulsed terahertz spectroscopy *Opt. Eng.* **47** 037003

[245] Hua Y F and Zhang H J 2012 Qualitative and quantitative detection of pesticides with terahertz time-domain spectroscopy *IEEE Trans. Microwave Theory Tech.* **58** 2064–70

[246] Redo-Sanchez A and Zhang X C 2011 Assessment of terahertz spectroscopy to detect antibiotic residues in food and feed matrices *Analyst* **136** 1733–8

[247] Li J S 2010 Optical parameters of vegetable oil studied by terahertz time-domain spectroscopy *Appl. Spectrosc.* **64** 231–4

[248] Banerjee D, Spiegel W, Thomson M D, Schabel S and Roskos H G 2008 Quantitative moisture content detection in food wafers. Infrared, millimeter, and terahertz waves *Opt. Express* **16** 9060–6

[249] Tian X, Wang J and Cui S 2013 Analysis of pork adulteration in minced mutton using electronic nose of metal oxide sensors *J. Food Eng.* **119** 744–9

[250] Morsy M K, Zór K, Kostesha N, Alstrøm T S, Heiskanen A and El-Tanahi H *et al* 2016 Development and validation of a colorimetric sensor array for fish spoilage monitoring *Food Control* **60** 346–52

[251] Ryu G-S, Park K H, Park W-T, Kim Y-H and Noh Y-Y 2015 High-performance diketopyrrolopyrrole-based organic field-effect transistors for flexible gas sensors *Org. Electron.* **23** 76–81

[252] Kang Q, Zhu X, Ma X, Kong L, Xu W and Shen D 2015 Response of an electrodeless quartz crystal microbalance in gaseous phase and monitoring adsorption of iodine vapor on zeolitic-imidazolate framework-8 film *Sens. Actuators* B **220** 472–80

[253] Oller-Moreno S, Singla-Buxarrais G, Jiménez-Soto J M, Pardo A, Garrido-Delgado R and Arce L *et al* 2015 Sliding window multi-curve resolution: application to gas chromatography–ion mobility spectrometry *Sens. Actuators* B **217** 13–21

[254] Peris M and Escuder-Gilabert L 2013 On-line monitoring of food fermentation processes using electronic noses and electronic tongues: a review *Anal. Chim. Acta* **804** 29–36

[255] Vlasov Y, Legin A and Rudnitskaya A 2002 Electronic tongues and their analytical application *Anal. Bioanal. Chem.* 36–146

[256] Prieto N, Oliveri P, Leardi R, Gay M, Apetrei C and Rodriguez-Méndez M L *et al* 2013 Application of a GA–PLS strategy for variable reduction of electronic tongue signals *Sens. Actuators* B **183** 52–7

[257] Wang D, Wang X, Liu T and Liu Y 2012 Prediction of total viable counts on chilled pork using an electronic nose combined with support vector machine *Meat Sci.* **90** 373–7

[258] Labrador R H, Masot R, Alcañiz M, Baigts D, Soto J and Martínez-Mañez R *et al* 2010 Prediction of NaCl, nitrate and nitrite contents in minced meat by using a voltammetric electronic tongue and an impedimetric sensor *Food Chem.* **122** 864–70

[259] Phat C, Moon B and Lee C 2016 Evaluation of umami taste in mushroom extracts by chemical analysis, sensory evaluation, and an electronic tongue system *Food Chem.* **192** 1068–77

[260] Alimelli A, Pennazza G, Santonico M and Paolesse R 2007 Fish freshness detection by a computer screen photoassisted based gas sensor array *Anal. Chim. Acta* **582** 320–8

[261] Cheng J-H, Sun D-W, Pu H and Zhu Z 2015 Development of hyperspectral imaging coupled with chemometric analysis to monitor K value for evaluation of chemical spoilage in fish fillets *Food Chem.* **185** 245–53

[262] Baldeón E O, Alcañiz M, Masot R, Fuentes E M, Barat J M and Grau R 2015 Voltammetry pulse array developed to determine the antioxidant activity of camu–camu (Myrciaria dubia (H.B.K.) McVaug) and tumbo (Passiflora mollisima (Kunth) L.H. Bailey) juices employing voltammetric electronic tongues *Food Control* **54** 181–7

[263] Qiu S, Wang J and Gao L 2015 Qualification and quantisation of processed strawberry juice based on electronic nose and tongue *LWT – Food Sci. Technol.* **60** 115–23

[264] Dias L G, Sequeira C and Veloso A C A 2014 Sousa MEBC, Peres AM. Evaluation of healthy and sensory indexes of sweetened beverages using an electronic tongue *Anal. Chim. Acta* **848** 32–42

[265] Veloso A C A, Dias L G, Rodrigues N, Pereira J A and Peres A M Sensory intensity assessment of olive oils using an electronic tongue *Talanta*

[266] Simoes Costa A M, Costa Sobral M M, Delgadillo I, Cerdeira A and Rudnitskaya A 2015 Astringency quantification in wine: comparison of the electronic tongue and FT-MIR spectroscopy *Sens. Actuators* B **207** 1095–103

[267] Rakow N and Suslick K 2000 A colorimetric sensor array for odour visualization *Nature* **406** 710–3

[268] Suslick K S, Rakow N A and Sen A 2004 Colorimetric sensor arrays for molecular recognition *Tetrahedron* **60** 11133–8

[269] Huang X, Zou X, Zhao J, Shi J, Zhang X and Li Z *et al* 2014 Sensing the quality parameters of Chinese traditional Yao-meat by using a colorimetric sensor combined with genetic algorithm partial least squares regression *Meat Sci.* **98** 203–10

[270] Soga T, Jimbo Y, Suzuki K and Citterio D 2013 Inkjet-printed paper-based colorimetric sensor array for the discrimination of volatile primary amines *Anal. Chem.* **85** 8973–8

[271] Chen Q, Li H, Ouyang Q and Zhao J 2014 Identification of spoilage bacteria using a simple colorimetric sensor array *Sens. Actuators* B **205** 1–8

[272] Huang X, Zhou F and Jiang F 2011 Evaluation of pork freshness using olfaction visualization detection technique *Nongye Jixie Xuebao/Trans. Chin. Soc. Agricul. Mach.* **42** 142–524

[273] Zaragozá P, Fuentes A, Ruiz-Rico M, Vivancos J-L, Fernández-Segovia I and Ros-Lis J V *et al* 2015 Development of a colorimetric sensor array for squid spoilage assessment *Food Chem.* **175** 315–21

[274] Huang X, Xin J and Zhao J 2011 A novel technique for rapid evaluation of fish freshness using colorimetric sensor array *J. Food Eng.* **105** 632–7

[275] Zhang C, Bailey D P and Suslick K S 2006 Colorimetric sensor arrays for the analysis of beers: a feasibility study *J. Agric. Food Chem.* **54** 4925–31

[276] Qin H, Huo D, Zhang L, Yang L, Zhang S and Yang M *et al* 2012 Colorimetric artificial nose for identification of Chinese liquor with different geographic origins *Food Res. Int.* **45** 45–51

[277] Chen Q, Liu A, Zhao J, Ouyang Q, Sun Z and Huang L 2013 Monitoring vinegar acetic fermentation using a colorimetric sensor array *Sens. Actuators* B **183** 608–16

[278] Xiaobo Z, Jian Z and Jiewen Z 2008 Identification of four vinegars by olfaction visualization technology *Trans. Chin. Soc. Agricul. Eng.* **24** 165–8

[279] Xiaobo Z, Jiewen Z and Shouyi W 2002 Identification of two types of vinegar using a gas sensor array and genetic program *Trans. Chin. Soc. Agricul. Eng.* **18** 226–30

[280] Suslick B, Feng L and Suslick K 2010 Discrimination of complex mixtures by a colorimetric sensor array coffee aromas *Anal. Chem.* **82** 2067–73

[281] Musto C J and Suslick K S 2010 Differential sensing of sugars by colorimetric arrays *Curr. Opin. Chem. Biol. Model Syst./Biomol. Syn. Mod.* **14** 758–66

[282] Huo D, Wu Y, Yang M, Fa H, Luo X and Hou C 2014 Discrimination of Chinese green tea according to varieties and grade levels using artificial nose and tongue based on colorimetric sensor arrays *Food Chem.* **145** 639–45

[283] Chen Q, Liu A, Zhao J and Ouyang Q 2013 Classification of tea category using a portable electronic nose based on an odor imaging sensor array *J. Pharm. Biomed. Anal.* **84** 77–83

IOP Publishing

Physics in Food Manufacturing
Case studies in fundamental and applied research
Megan J Povey

Chapter 11

Lyotropic liquid crystalline phases for the formulation of future functional foods

Amin Sadeghpour, Filippo Ceola, Richard Smith-Uchotski, Arwen I I Tyler and Michael Rappolt

11.1 Introduction to lyotropic liquid crystalline nanoparticles (mesosomes)

Recent advances in nanomaterials manufacturing are inherited from soft matter nanotechnology and novel approaches in formulating smart and generic solutions in various production fields [1]. The structure-induced functionalities of lyotropic liquid crystalline (LLC) phases have made a significant contribution in biomedical applications such as in the development of novel contrast enhancers for magnetic resonance imaging (MRI) [2, 3]. Within the food and pharmaceutical sciences, LLC-based nanomaterials have shown great potential in the development of new delivery systems [4–6]. The self-assembly from surfactant-like molecules leads to formation of nanoscale hierarchical structures with unique features including [6]: (i) high surface per volume ratio allowing high cargo load; (ii) controlled release behaviour originated from their special and tunable nanometre sized channel networks; (iii) capability to load bioactive molecules with different physico-chemical properties, e.g. hydrophobic and hydrophilic compounds; (iv) effective encapsulation capabilities which improve stability and bioavailability of loaded compounds and reduce their toxicity, and (v) very good biocompatibility and high biomembrane absorption properties resulting in enhanced transdermal penetration.

Various types of molecules under different conditions self-assemble into different LLC phases. Inverse bicontinuous cubic (V_2) phases spontaneously form when monoacylglycerols such as monoolein or phytantriol are mixed with excess water at room temperature. Inverse hexagonal (H_2), ordered inverse micellar phases (such as the one based on spacegroup Fd3m) and the inverse micellar (L_2) phases can be formed upon addition of simple oils, e.g. tetradecane or oleic acid, to glycerol monooleate, however pure or mixtures of lipids can also be used to drive the

...0-7503-2596-7ch11

formation of these phases. The inverse hexagonal phase consists of hexagonally cylindrical inverse micelles with rod-like water channels surrounded in a continuous oil phase, whereas in the bicontinuous cubic phases there are two intertwined continuous water channel networks separated by lipid bilayers (figure 11.1). The bicontinuous cubic phases are quite stable over a wide temperature range (only demonstrating contracted lattice spacing by about 15% at 65 °C) and can be formulated to be insensitive to even extreme variations in pH and salinity, which are normally met within the gastrointestinal tract. The aqueous dispersions of LLCs as dispersed lipid nanoparticles were first reported in 1981 by Lindström *et al* in a fat digestion study [7]. Later in 1996, Landh and Larsson patented the preparation of particles comprising different types of liquid crystalline phases in their interiors [8].

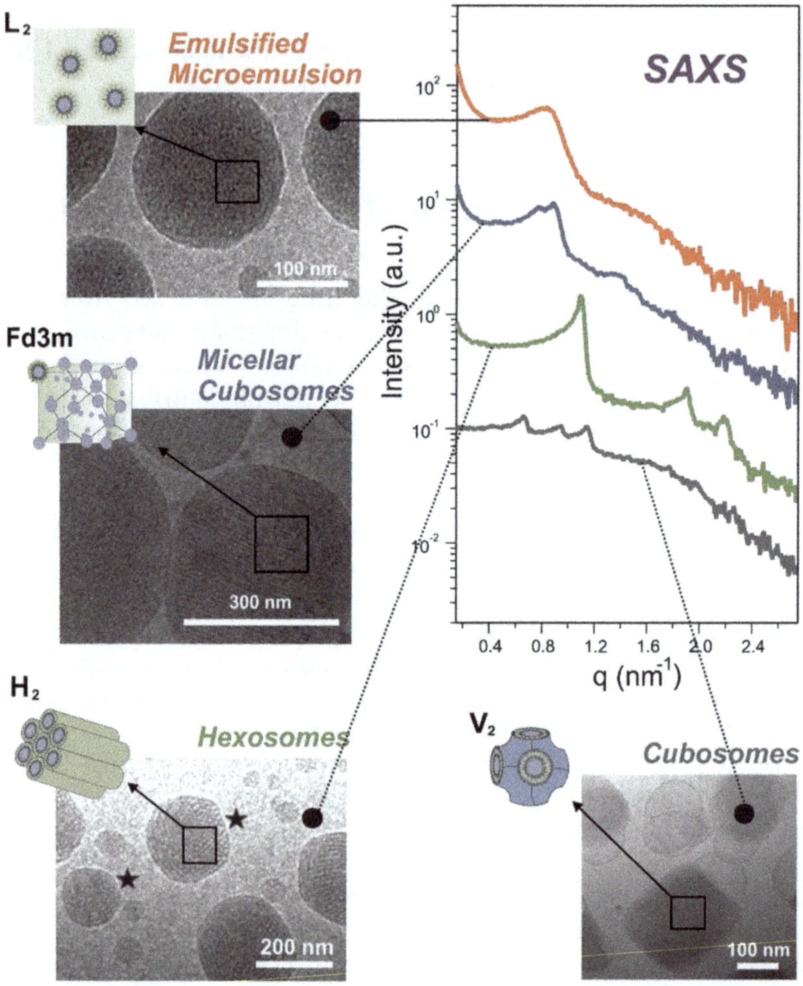

Figure 11.1. The cryo-TEM images of mesosomes from four different monoglyceride-based LLC phases [12] and their corresponding small angle x-ray scattering (SAXS) patterns.

These stabilised particles with internally self-assembled mesophases are sometimes referred to as *ISAsomes* or *Mesosomes* [9–11].

11.2 Stabilisation and functionalisation of mesosomes

Research and innovations based on LLC bicontinuous cubic phase particles (*cubosomes*) and the inverted hexagonal phase (*hexosomes*) have been developed considerably in recent years [13–15]. Different stabilizers have been used in formulation of mesosomes to enhance physico-chemical characteristics as well as their biocompatibility and food/pharmacological safety (see table 11.1). The most common stabiliser widely used over the last two decades is the FDA approved block copolymer Pluronic® F127. Investigations are ongoing to introduce new alternatives for dispersing cubosomes and hexosomes. Solid nanoparticles, proteins and polymer aggregates have been successfully applied in stabilisation of LLCs. Clay minerals, silica nanoparticles, nanocellulose, ethoxylated phytosterol and β-casein are examples of such a class of materials. It is worth mentioning that application of solid nanoparticles as stabilisers at the oil–water interface (also called Pickering stabilisation) [16] serves higher stability and surfactant-free possibility in formulation of mesosomes, and hence receives greater interest in mesosome research [10, 17, 18]. Nonetheless, screening for highly functioning stabilizers which fulfil the requirements of biosafety and at the same time offer high stability against extreme pH conditions and high concentrations of salt and other bioactive molecules in the gastric tract (e.g. bile salts) remains an important task.

Aside from nanostructural tunability of internal phases, the LLC mesosome allows further controlled release of loaded molecules via controlling their size, surface properties, surface packing density and the addition of hydrogels. In addition, different underlying mesophases are demonstrated to be more appropriate for different bioactive compounds and/or delivery via various administration routes [33, 34]. Moreover, the mesosomes' effective encapsulating properties serve to protect the loaded compounds from biodegradation, inactivation or unwanted interactions. Eventually, the release from LLC- based mesosomes can be triggered by structural transitions induced by temperature or pH. The latter offers the unique

Table 11.1. Various types of stabilizers applied for the formulation of mesosomes.

Type of stabilizer	Stabilised internal phase
Clay mineral particles [19, 20]	L_2, Fd3m, H_2, V_2[a]
Silica particles [17, 18, 21]	L_2, Fd3m, H_2 and V_2 (only partial stabilisation)
Copolymers [22–27]	L_2, Fd3m, H_2, V_2
Proteins [26, 28, 29]	H_2, V_2
Nanotubes [30]	L_2, Fd3m, H_2, V_2
Biopolymers [31, 32]	V_2

[a] L_2 (inverse micellar phase); Fd3m (inverse micellar face-centred cubic phase); H_2 (inverse hexagonal phase); V_2 (inverse cubic phase).

opportunity to actively modulate the structures and steer their functionalities as smart delivery systems [34].

11.3 Advanced nanostructural analysis of cubosomes

Curved bilayer-based biomembranes [35–37] come in the form of bicontinuous cubic phases [38]. Three types of bicontinuous cubic membrane phases have been unequivocally identified [35]: the gyroid of Schoen [39] (space group Ia3d), the Schwarz' [40] double-diamond (space group Pn3m) and primitive cubic phases (space group Im3m). In these cubic phases, the bilayers are draped around an infinite periodic minimal surface (displaying zero mean curvature at any point). That is, the position of this minimal surface coincides with the mid-plane (methyl trough region) of the curved bilayers (see dark blue electron density maps in figure 11.2), creating two distinct water-channel networks one being interwoven with the other. Scattering data was firstly described theoretically in 1994 by Clerc and Dubois-Violette [41],

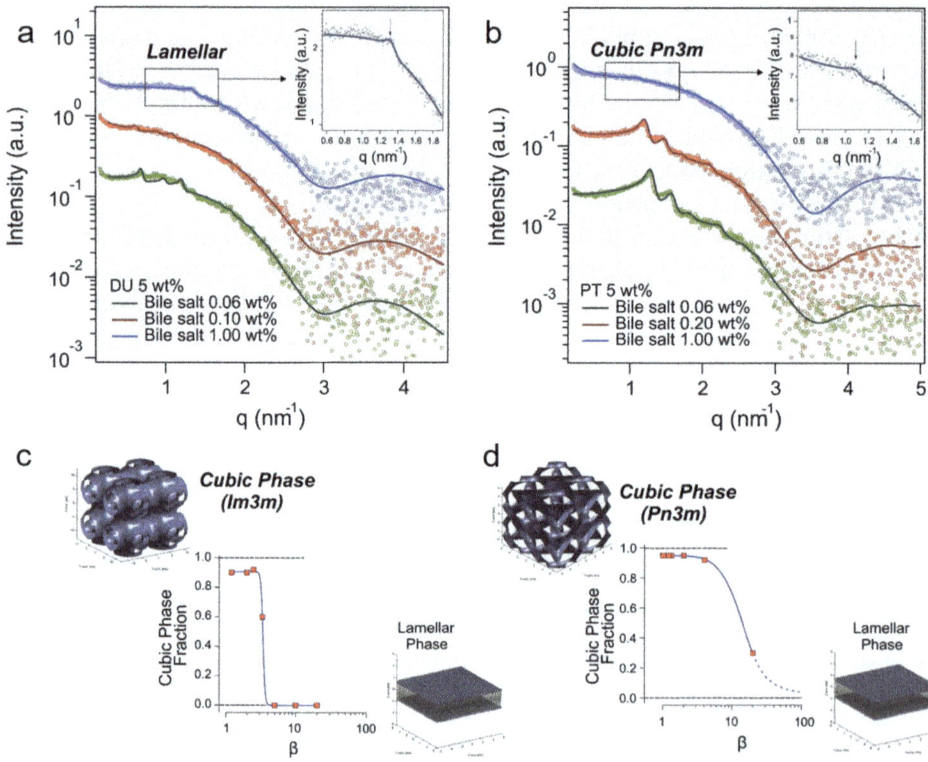

Figure 11.2. SAXS patterns for (a) monoglyceride- and (b) phytantriol-based emulsions stabilized by various concentrations of bile salts. The solid lines represent the best global fits to the data. The insets in panels (a) and (b) show traces of multilamellar vesicles and of the Pn3m phase, respectively. The bile-salt-driven transformation of cubic to lamellar phases via the coexistence regime is depicted by the computed fractions of cubic phases as a function of bile-salt to lipid ratio β for the monoglyceride and phytantriol samples in (c) and (d), respectively. Reprinted with permission from [44]. Copyright 2018 American Chemical Society.

modelling the bilayer with a simple step-function (for a more recent discussion on the bicontinuous cubic phase analysis, see references [42, 43]).

Studying a wide compositional range of bile-salt lipid dispersions, in which lamellar and bicontinuous cubic phases co-exist [44] (figure 11.2), we applied for the first time an extended global small-angle-x-ray scattering analysis method, allowing us to combine the scattering intensities from lamellar and cubic phases. The total scattering intensity can be written as

$$I_t(q) = f\ I_L(q) + (1 - f)I_Q, \tag{11.1}$$

where f represents the phase fraction of lamellar phase, $I_L(q)$ is the scattering intensity of the lamellar and $I_Q(q)$ is the scattering intensity of the cubic phase. The intensities of each single cubic phase reflection can be calculated according to a model described by Garstecki and Holyst [45, 46]:

$$I_{hkl}^{(mod)}(L) = \mathcal{M}_{hkl}\left[\frac{F_{hkl}^S}{a^2}\frac{\sin(\alpha_{hkl}\pi(h^2 + k^2 + l^2)^{1/2}L/a)}{\alpha_{hkl}\pi(h^2 + k^2 + l^2)^{1/2}}\right]^2, \tag{11.2}$$

in which the $\mathcal{M}_{hkl}$ is the multiplicity for each reflection, L/a is the dimensionless thickness of the bilayer separating the two continuous phases, a is the lattice parameter, α_{hkl} are a correction factor and $\frac{F_{hkl}^S}{a^2}$ are the normalized structure factors. The total intensity contribution taking into account the experimental broadening of the reflections is then formulated as:

$$I_{hkl}^{(exp/mod)}(q) = \frac{1}{\sigma_r(2\pi)^{1/2}}\sum\nolimits_{hkl} exp\left[\frac{-(q_{hkl} - q)^2}{2\sigma_r{}^2}\right]I_{hkl}^{(mod)}, \tag{11.3}$$

where σ_r is related to the standard deviation of the recorded Bragg reflections. Finally, the refined form factor values of the cubic phases, F_{hkl}, obtained from the global fits (equation (11.3)), were used to reconstruct the 3D electron density maps by applying standard Fourier transform procedures.

11.4 Delivery of natural polyphenolic compounds

Polyphenols are plant bioactive materials with demonstrated health benefits and applications in the development of food supplements. Flavonoids are polyphenols with two phenyl rings and one heterocyclic group that have shown several reported health benefits mainly originated from their radical scavenging property. For instance, strong intracellular antioxidant activity is demonstrated for anthocyanins from bilberry and blueberry [47]. Epigallocatechin-3-gallate (EGCG), an abundant polyphenol in green tea, prevents tumour formation and growth while tea polyphenols are shown to suppress colon cancer cells activity in *in vitro* studies [48]. Similarly, lemon verbena from herbal tea has high antioxidant activity. More information on the health benefits and possible restrictions of plant based polyphenols is provided in table 11.2. Despite the demonstrated health benefits of polyphenols in *in vitro* studies, their functions and bioactivities seems to remain very

Table 11.2. Health benefits and restrictions of selected polyphenols.

Type of polyphenol	Health benefits	Health risks	Restrictions	Natural sources
Naringenin	Antioxidant, anti-inflammatory and chemoprotective activity [49]	Active chelator of metallic ions [49]	Low absorption (6%) and rapid elimination [50]	Citrus fruits
Hesperetin	Antioxidant, anti-inflammatory, blood lipid and cholesterol lowering activity [50]		Low absorption (4%) and rapid elimination [50]	Citrus fruits
Genistein	Anticancer activity [51]		Low water solubility, low absorption (9%) [52]	Pomegranate seeds [53]
Daidzein	Antioxidant activity		30% bioavailability measured in urine [52]	Pomegranate seeds [53]
Luteolin	Anticancer, anti-hypertension and anti-inflammatory activity [54]		Poor water solubility [54] low oral absorption [55]	Green pepper, celery, broccoli, and parsley
Apigenin	Anti-inflammatory, antioxidant and anticarcinogenic activity [56]		Poor water solubility and unsatisfactory cutaneous permeability [56] Low absorption (1.2%) [57]	Parsley and onion [56]
Quercetin	Anticancer activity (breast, lung and prostate cancer) [58]	Interactions with soy proteins [59]	Limited bioavailability, reduced antioxidant activity through interactions with proteins [60]	Berries
Kaempferol	Antioxidant, anti-inflammatory, antimicrobial, anticancer, cardioprotective, neuroprotective, antidiabetic activity [61]		Poor water-solubility and low bioavailability (2%) [62]	Broccoli, apples
Tiliroside	Antioxidant activity [63], Inhibiting of neuroinflammation [64], anti-hyperglycemic effects [65]		Inhibitory effects on carbohydrate digestion [66]	Potentilla species [65]

Epigallocatechin-3-gallat (EGCG)	Neuroprotective, stroke risk lowering [67] and anticancer (colon cancer) activity [48]	Complex formation with proteins [68, 69]	Low bioavailability (0.2%–2%) [70]	Green tea
Cyanidin	Anticancer (skin and colon cancer [71]) and anti-diabetic activity [72]		Low bioavailability (<1%) and unstable *in vivo* [73, 74]	Wild mulberry/ raspberries

limited in *in vivo* models. Particularly in humans, very low bioavailability or bioactivity have been reported. The most important factors imposing such low biofunctionality in polyphenols include: (i) their interactions with proteins or enzymes throughout the gastrointestinal (GI) tract prior to absorption; (ii) poor solubility in aqueous media; and (iii) interactions with the other food molecules. To address these challenges, a smart formulation strategy with the necessary nanoscale design needs to be set up in manufacturing the new generation of polyphenol-based food supplements.

11.5 Influence of vitamin K_1 on the phase diagram of monolinolein

The aim of this study was to control the morphology of LLC phases of monolinolein (MLO; Dimodan® U from DuPont) by adding different concentrations vitamin K_1. Vitamin K_1 is a fat soluble vitamin that has essential roles in bone mineralisation, blood clotting and in the production of the GAS6 protein [75, 76]. It is involved in the formation of γ-carboxyglutamic acid residues formed through the vitamin K_1 dependent carboxylase reaction [76]. The Gla domains are found in bone related proteins such as osteocalcin, calcification inhibiting matrix Gla protein, perostin and the Gla rich proteins as well as blood clotting related proteins [77]. SAXS was employed to discern the underlying structures of the lyotropic phases in the absence and increasing concentrations of vitamin K_1. The temperature-composition phase diagram was established under excess of water conditions (see table 11.3).

The phase transitions ($V_2 \rightarrow H_2 \rightarrow L_2$) upon addition of vitamin K_1 is due to the induction of increasing negative interfacial curvature. Vitamin K_1 is a highly hydrophobic molecule with an octanol/water partition coefficient of 4.85 [78].

Table 11.3. The lattice parameter *a* (nm) as a function of *δ* and temperature. The Pn3m phase regions is colour-coded yellow, the H_2 phase in red and the L_2 phase in blue.

	The Lattice Parameters (nm)						
T (°C)	δ62	δ70	δ75	δ80	δ83	δ89	δ100
75	3.47	3.90	3.70	3.85	4.45	5.19	6.26
70	3.71	4.22	3.95	4.17	4.81	5.21	6.26
65	3.85	4.42	4.19	4.57	5.32	5.29	7.68
60	4.15	4.57	4.56	4.73	5.36	5.37	7.82
55	4.45	4.89	4.97	4.89	5.43	5.45	7.61
50	4.75	5.03	5.08	5.04	5.51	5.54	7.92
45	4.92	5.04	5.18	5.54	5.56	5.64	8.43
40	5.09	5.21	5.71	5.57	5.66	5.74	8.71
35	5.42	5.31	5.68	5.64	5.75	5.84	9.20
30	5.42	5.63	5.71	5.70	5.84	5.93	9.46
25	5.65	5.39	5.76	5.78	5.93	6.01	9.60
20	5.78	5.71	5.82	5.85	6.05	6.13	9.69
15	5.78	5.75	5.91	5.95	6.18	6.27	9.79

Note: δ = MLO conc./(MLO conc. + vitamin K_1 conc.).

Accordingly, it associates with the hydrophobic lipid chains of monolinolein. Their presence brings about increases in the lipid chain volume area relative to the head group size [79]. This reduces the effective head group areas in contact with the water phase resulting in phase transition sequence of increasing negative interfacial curvature. Increasing the temperature acts in a similar fashion. Heating the lipids increases the chain disorder and chain splay of the lipids [79]. Increasing the vitamin concentration and temperature favours the negative interfacial curvature resulting finally in the formation of the L_2 phase.

11.6 Polyunsaturated fatty acid—loaded mesosomes

The processing of food with nutritional additives including essential fatty acids, minerals such as calcium and iron, dietary fibre and vitamins satisfies a strongly growing market. This project's goal was to design and test a novel particulate delivery system with improved encapsulation and release of omega-3 fatty acids. A novel encapsulation concept basing on the formulation of food-grade nanostructured cubosomes (sub-micron sized) has been recently investigated in our lab, for an optimal food-compound release of polyunsaturated fatty acids (PUFAs). As model compounds we investigated the loading of cubosomes with eicosapentaenoic acid (EPA), a naturally occurring substance that is usually well tolerated at doses <1.5 g/day and has been extensively used for the prevention of cardiovascular diseases [80]. More recently, PUFAs have been also recognized to have anticancer activity [81]. There is also evidence suggesting improved efficacy and/or tolerability of conventional cancer chemotherapy when administered with omega-3. However, the administration of n-3-PUFAs can be optimized in order to improve the efficacy of delivery [82] and to moderate the unacceptable side-effect of diarrhoea and to eradicate the fishy taste through proper encapsulation. The objective of this work was to study the phase behaviour of eicosapentaenoic acid/monoglycerides mixture in excess of water and the investigation of a food-grade stabilizer to create edible nanoparticles for nutraceutical delivery.

In figure 11.3 the phase behaviour of eicosapentaenoic acid (EPA)/MLO mixtures in excess of water has been investigated. As expected, the phase diagram shows that EPA molecules induce increasing negative curvature at the lipid/water interfaces leading to the formation of bicontinuous cubic towards inverse micellar, following the order $V_2 \rightarrow H_2 \rightarrow Fd3m \rightarrow L_2$. It should be noted that the Fd3m phase was only observed in the dispersed particles and not in bulk, as shown in figure 11.3. This is mainly due to the high unsaturation degree of the EPA molecule and its relatively small head group, both leading to increasing negative spontaneous molecular curvatures of the mixtures.

Having a phase diagram of the EPA/monoglycerides system at hand, allowed choosing what seemed to be the most promising loading concentrations for the design of dispersed particles. $\delta = 98$ was chosen in order to investigate the formulation of a cubosome-based dispersion, $\delta = 85$ was related to the hexagonal phase, thus the formation of hexosomes were expected, and $\delta = 73$ was studied with the intention of producing micellar cubosomes. The latter dispersion has found great

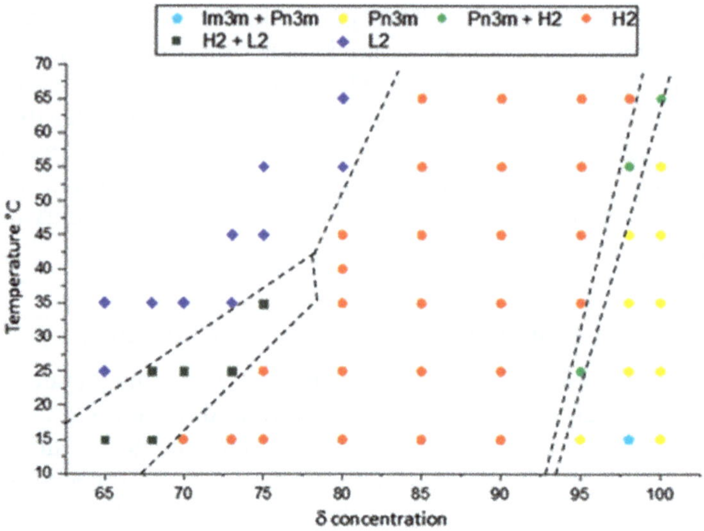

Figure 11.3. δ–*T* phase diagram of bulk MLO (Dimodan® U)/EPA mixtures in excess of water. Note, δ = MLO conc./(MLO conc. + EPA conc.).

Table 11.4. Pluronic® F127 and sucrose stearate stabilized particles (δ = 98) are compared in terms of lattice parameter, type of mesophase and water channel diameter (D_w), at three different temperatures.

Sample	Temperature [°C]	Lattice parameter, *a* [nm]	Phase	D_w[a] [nm]
Sucrose stearate, δ = 98	5	18.59	*Im3m*	8.2
	25	16.40	*Im3m*	6.9
	37	14.50	*Im3m*	5.7
Pluronic® F127, δ = 98	5	13.61	*Im3m*	5.2
	25	12.62	*Im3m*	4.6
	37	11.91	*Im3m*	4.2

[a] D_W was calculated according to $D_w/2\ (I_{m3m}) = 0.305a - l$, with *a* being the lattice parameter and *l* the lipid chain length [87].

attention [83–85] and it is particularly interesting because it is the first colloidal form present during lipid digestion. The possibility to tune particles' internal structure allows creating particles with different properties and function. For example, hexosomes show a slower hydrophobic compound release compared to cubosomes and a three times better bioavailability of the drug cinnarizine [82]. Differently, water channel size of cubosomes can be swollen up to 21.8 nm (lattice parameter of 47.0 nm) by the incorporation of charged lipids [86]. This allows loading and delivery of much larger compounds compared to hexosomes or micellar cubosomes.

With regards to this work, both Pluronic® F127 and sucrose stearate stabilized particles show an Im3m internal structure at δ = 98. However, the lattice parameter was quite different in the two cases: the triblock co-polymer forms a thick corona all around the particles but does not penetrate the water channels because of its large size (MW 12 600 gr mol^{-1}) (see table 11.4).

Despite promising results in terms of particles' internal nanostructure conservation, sucrose stearate did not meet the expectations in terms of stabilization capacity. The $\delta = 98$ sample is stable for 15 days from preparation, while $\delta = 85$ for 11 days and $\delta = 73$ for 9 days only. Clearly, the EPA content affects time stability in a negative way. Even though stabilization of EPA-loaded nanoparticles using sucrose stearate leads to a limited shelf life, application for this product can still be interesting to fortify food products with very short shelf life such as fresh milk and fresh fruit juice. Further work in this direction can lead to application in a much broader variety of fresh products such as soups and beverages in general. Studies on the bioavailability of delivered nutraceuticals are also to be conducted but the premises are very good as the nanoparticles' internal assembly mimics the structure that would normally form during fat digestion and simplify their absorption.

11.7 Conclusions and outlook

The applications of LLCs in encapsulation and the delivery of pharmaceuticals have been widely considered in recent years. Camptothecin [15], protopanaxadiol [88], dacarbazine [89] and fluorouracil [90] are the examples of nutraceuticals encapsulated into cubosomes in order to improve their delivery and hence their therapeutic behaviour. A comprehensive list of various bioactive molecules loaded into mesosomes has been reported recently in a review by Chong *et al* [4]. Nevertheless, the opportunities for enhancing the nutritional value of foods by addition of flavonoids are largely unexploited. In particular, the interactions of flavonoids with lipid self-assemblies, formulation of polyphenol-rich mesosomes and eventually engineering of nanostructured food supplements with proper taste, stability and digestibility behaviour is scarcely considered. In this respect, the challenges for highly stable and effective encapsulation of flavonoids through formulation of fully food-grade mesosomes which are able to protect bioactives throughout the different digestion stages and guarantee efficient delivery to the small intestine, still remains an essential task for the development of future functional food.

LLC mesosomes are potential nanomaterials to overcome the challenges concerning the restricted bioavailability, and thus, limited health benefits of current flavonoid-based food supplements. The flavonoids interactions with, and their release behaviour from various types of mesosomes have yet to be understood and further studies are required in order to provide intelligent formulations of mesosomes loaded with flavonoids and other bioactives as future smart food supplements.

Acknowledgments

The authors would like to thank Professor Megan Povey for her kind invitation to contribute to this innovating conference book on physics in food manufacturing.

References

[1] Sagalowicz L, Moccand C, Davidek T, Ghanbari R, Martiel I, Negrini R, Mezzenga R, Leser M E, Blank I and Michel M 2016 Lipid self-assembled structures for reactivity control in food *Phil. Trans. R. Soc. A - Mathem. Phy. Eng. Sci.* **374** 20150136

[2] Bye N, Hutt O E, Hinton T M, Acharya D P, Waddington L J, Moffat B A, Wright D K, Wang H X, Mulet X and Muir B W 2014 Nitroxide-loaded hexosomes provide MRI contrast *in vivo* *Langmuir* **30** 8898–906

[3] Muir B W *et al* 2012 Metal-free and MRI visible theranostic lyotropic liquid crystal nitroxide-based nanoparticles *Biomaterials* **33** 2723–33

[4] Chong J Y T, Mulet X, Boyd B J and Drummond C J 2015 *Advances in Planar Lipid Bilayers and Liposomes* ed A Iglič *et al* (New York: Academic) pp 131–87

[5] Kwon T K, Hong S K and Kim J C 2012 *In vitro* skin permeation of cubosomes containing triclosan *J. Ind. Eng. Chem.* **18** 563–7

[6] Rizwan S B, Boyd B J, Rades T and Hook S 2010 Bicontinuous cubic liquid crystals as sustained delivery systems for peptides and proteins *Exp. Opin. Drug Deliv.* **7** 1133–44

[7] Lindstrom M, Ljusbergwahren H, Larsson K and Borgstrom B 1981 Aqueous lipid phases of relevance to intestinal fat digestion and absorption *Lipids* **16** 749–54

[8] Landh T and Larsson K 1996 Particles, Method of Preparing Said Particles and Uses Thereof *US patent* US5531925A

[9] Iglesias G R, Pirolt F, Sadeghpour A, Tomšič M and Glatter O 2013 Lipid transfer in oil-in-water isasome emulsions: influence of arrested dynamics of the emulsion droplets entrapped in a hydrogel *Langmuir* **29** 15496–502

[10] Sadeghpour A, Pirolt F, Iglesias G R and Glatter O 2014 Lipid transfer between submicrometer sized pickering ISA some emulsions and the influence of added hydrogel *Langmuir* **30** 2639–47

[11] Tomšič M, Guillot S, Sagalowicz L, Leser M E and Glatter O 2009 Internally self-assembled thermoreversible gelling emulsions: ISAsomes in methylcellulose, κ-carrageenan, and mixed hydrogels *Langmuir* **25** 9525–34

[12] Yaghmur A, Østergaard J, Larsen S W, Jensen H, Larsen C and Rappolt M 2013 *Liposomes, Lipid Bilayers and Model Membranes: From Basic Research to Technology* ed G Pabst *et al* (Baco Raton, FL: Taylor and Francis) pp 341–60

[13] Baskaran R, Madheswaran T, Sundaramoorthy P, Kim H M and Yoo B K 2014 Entrapment of curcumin into monoolein-based liquid crystalline nanoparticle dispersion for enhancement of stability and anticancer activity *Int. J. Nanomed.* **9** 3119–30

[14] Sun W, Vallooran J J, Zabara A and Mezzenga R 2014 Controlling enzymatic activity and kinetics in swollen mesophases by physical nano-confinement *Nanoscale* **6** 6853–9

[15] Caltagirone C *et al* 2014 Cancer-cell-targeted theranostic cubosomes *Langmuir* **30** 6228–36

[16] Pickering S U 1907 Emulsions *J. Chem. Soc. Trans.* **91** 2001–21

[17] Sadeghpour A, Pirolt F and Glatter O 2013 Submicrometer-sized pickering emulsions stabilized by silica nanoparticles with adsorbed oleic acid *Langmuir* **29** 6004–12

[18] Muller F, Degousee T, Degrouard J, Brulet A and Salonen A 2015 Probing structure in submicronic aqueous assemblies of emulsified microemulsions and charged spherical colloids using SANS and cryo-TEM *J. Colloid Interf. Sci.* **446** 114–21

[19] Guillot S, Bergaya F, de Azevedo C, Warmont F and Tranchant J-F 2009 Internally structured pickering emulsions stabilized by clay mineral particles *J. Colloid Interf. Sci.* **333** 563–9

[20] Salonen A, Muller F and Glatter O 2008 Dispersions of internally liquid crystalline systems stabilized by charged disklike particles as pickering emulsions: basic properties and time-resolved behavior *Langmuir* **24** 5306–14

[21] Bhatt A B, Barnes T J and Prestidge C A 2015 Silica nanoparticle stabilization of liquid crystalline lipid dispersions: impact on enzymatic digestion and drug solubilization *Curr. Drug Deliv.* **12** 47–55

[22] Nakano M, Teshigawara T, Sugita A, Leesajakul W, Taniguchi A, Kamo T, Matsuoka H and Handa T 2002 Dispersions of liquid crystalline phases of the monoolein/oleic acid/ pluronic F127 system *Langmuir* **18** 9283–8

[23] Murgia S *et al* 2015 Cubosome formulations stabilized by a dansyl-conjugated block copolymer for possible nanomedicine applications *Colloids Surf.,* B **129** 87–94

[24] Chong J Y T, Mulet X, Keddie D J, Waddington L, Mudie S T, Boyd B J and Drummond C J 2015 Novel steric stabilizers for lyotropic liquid crystalline nanoparticles: PEGylated-phytanyl copolymers *Langmuir* **31** 2615–29

[25] Tilley A J, Drummond C J and Boyd B J 2013 Disposition and association of the steric stabilizer Pluronic® F127 in lyotropic liquid crystalline nanostructured particle dispersions *J. Colloid Interf. Sci.* **392** 288–96

[26] Sagalowicz L, Guillot S, Acquistapace S, Schmitt B, Maurer M, Yaghmur A, de Campo L, Rouvet M, Leser M and Glatter O 2013 Influence of vitamin E acetate and other lipids on the phase behavior of mesophases based on unsaturated monoglycerides *Langmuir* **29** 8222–32

[27] Chong J Y T, Mulet X, Waddington L J, Boyd B J and Drummond C J 2011 Steric stabilisation of self-assembled cubic lyotropic liquid crystalline nanoparticles: high throughput evaluation of triblock polyethylene oxide-polypropylene oxide-polyethylene oxide copolymers *Soft Matter* **7** 4768–77

[28] Zhai J L, Waddington L, Wooster T J, Aguilar M I and Boyd B J 2011 Revisiting beta-casein as a stabilizer for lipid liquid crystalline nanostructured particles *Langmuir* **27** 14757–66

[29] Angelova A, Angelov B, Papahadjopoulos-Sternberg B, Ollivon M and Bourgaux C 2005 Proteocubosomes: nanoporous vehicles with tertiary organized fluid interfaces *Langmuir* **21** 4138–43

[30] Patil-Sen Y, Sadeghpour A, Rappolt M and Kulkarni C V 2016 Facile preparation of internally self-assembled lipid particles stabilized by carbon nanotubes *J. Vis. Exp.* **108** e53489

[31] Uyama M, Nakano M, Yamashita J and Handa T 2009 Useful modified cellulose polymers as new emulsifiers of cubosomes *Langmuir* **25** 4336–8

[32] Spicer P T, Small W B, Lynch M L and Burns J L 2002 Dry powder precursors of cubic liquid crystalline nanoparticles (cubosomes) *J. Nanopart. Res.* **4** 297–311

[33] Yaghmur A, Larsen S W, Schmitt M, Østergaard J, Larsen C, Jensen H, Urtti A and Rappolt M 2011 *In situ* characterization of lipidic bupivacaine-loaded formulations *Soft Matter.* **7** 8291–5

[34] Guo C, Wang J, Cao F, Lee R J and Zhai G 2010 Lyotropic liquid crystal systems in drug delivery *Drug Disc. Today* **15** 1032–40

[35] Landh T 1995 From entangled membranes to eclectic morphologies: cubic membranes as subcellular space organizers *FEBS Lett.* **369** 13–7

[36] Deng Y, Almsherqi Z A, Ng M M L and Kohlwein S D 2010 Do viruses subvert cholesterol homeostasis to induce host cubic membranes? *Trends Cell Biol.* **20** 371–9

[37] Hyde S, Andersson S, Larsson K, Blum Z, Landh T, Lidin S and Ninham B W 1997 *The Language of Shape* (Amsterdam: Elsevier)

[38] Hyde S T 1996 Bicontinuous structures in lyotropic liquid crystals and crystalline hyperbolic surfaces *Curr. Opin. Solid State Mat. Sci.* **1** 653–62
[39] Schoen A H 1970 Infinite periodic minimal surfaces without self-intersections *NASA technical note* **TN D-5541**
[40] Schwarz H A 1890 *Gesammelte Mathematische Abhandlungen* vol 1 (Berlin: Springer)
[41] Clerc M and Dubois-Violette E 1994 X-ray scattering by bicontinuous cubic phases *J. Phys. II France* **4** 275–86
[42] Garstecki P and Hołyst R 2002 Scattering patterns of self-assembled cubic phases. 1. The Model *Langmuir* **18** 2519–28
[43] Garstecki P and Hołyst R 2002 Scattering patterns of self-assembled cubic phases. 2. Analysis of the Experimental Spectra *Langmuir* **18** 2529–37
[44] Sadeghpour A, Rappolt M, Misra S and Kulkarni C V 2018 Bile salts caught in the act: from emulsification to nanostructural reorganization of lipid self-assemblies *Langmuir* **34** 13626–37
[45] Garstecki P and Holyst R 2002 Scattering patterns of self-assembled cubic phases. 1. The model *Langmuir* **18** 2519–28
[46] Garstecki P and Holyst R 2002 Scattering patterns of self-assembled cubic phases. 2. Analysis of the experimental spectra *Langmuir* **18** 2529–37
[47] Bornsek S M, Ziberna L, Polak T, Vanzo A, Ulrih N P, Abram V, Tramer F and Passamonti S 2012 Bilberry and blueberry anthocyanins act as powerful intracellular antioxidants in mammalian cells *Food Chem.* **134** 1878–84
[48] Yang C S, Wang X, Lu G and Picinich S C 2009 Cancer prevention by tea: animal studies, molecular mechanisms and human relevance *Nat. Rev. Cancer* **9** 429–39
[49] Cavia-Saiz M, Busto M D, Pilar-Izquierdo M C, Ortega N, Perez-Mateos M and Muñiz P 2010 Antioxidant properties, radical scavenging activity and biomolecule protection capacity of flavonoid naringenin and its glycoside naringin: a comparative study *J. Sci. Food Agric.* **90** 1238–44
[50] Kanaze F I, Bounartzi M I, Georgarakis M and Niopas I 2006 Pharmacokinetics of the citrus flavanone aglycones hesperetin and naringenin after single oral administration in human subjects *Eur. J. Clin. Nutr.* **61** 472–7
[51] Sarkar F H, Li Y, Wang Z and Padhye S 2010 Lesson learned from nature for the development of novel anti-cancer agents: implication of isoflavone, curcumin, and their synthetic analogs *Curr. Pharmaceut. Des.* **16** 1801–12
[52] Setchell K D R *et al* 2003 Comparing the pharmacokinetics of daidzein and genistein with the use of 13C-labeled tracers in premenopausal women *Am. J. Clin. Nutr.* **77** 411–9
[53] Aviram M, Dornfeld L, Rosenblat M, Volkova N, Kaplan M, Colemann R, Hayek T, Presser D and Fuhrman B 2000 Pomegranate juice consumption reduces oxidative stress, atherogenic modifications to LDL, and platelet aggregation: studies in humans and in atherosclerotic apolipoprotein E-deficient mice *Am. J. Clin. Nutr.* **71** 1062–76
[54] Qiu J F, Gao X, Wang B L, Wei X W, Gou M L, Men K, Liu X Y, Guo G, Qian Z Y and Huang M J 2013 Preparation and characterization of monomethoxy poly(ethylene glycol)-poly(epsilon-caprolactone) micelles for the solubilization and *in vivo* delivery of luteolin *Int. J. Nanomed.* **8** 3061–9
[55] Zhou P, Li L-P, Luo S-Q, Jiang H-D and Zeng S 2008 Intestinal absorption of luteolin from peanut hull extract is more efficient than that from individual pure luteolin *J. Agric. Food Chem.* **56** 296–300

[56] Shen L-N, Zhang Y-T, Wang Q, Xu L and Feng N-P 2014 Enhanced *in vitro* and *in vivo* skin deposition of apigenin delivered using ethosomes *Int. J. Pharm.* **460** 280–8

[57] Patel D, Shukla S and Gupta S 2007 Apigenin and cancer chemoprevention: Progress, potential and promise (Review) *Int. J. Oncol.* **30** 233–45

[58] Nichenametla S N, Taruscio T G, Barney D L and Exon J H 2006 A review of the effects and mechanisms of polyphenolics in cancer *Crit. Rev. Food Sci. Nutr.* **46** 161–83

[59] Rawel H M, Czajka D, Rohn S and Kroll J 2002 Interactions of different phenolic acids and flavonoids with soy proteins *Int. J. Biol. Macromol.* **30** 137–50

[60] Arts M, Haenen G, Voss H P and Bast A 2001 Masking of antioxidant capacity by the interaction of flavonoids with protein *Food Chem. Toxicol.* **39** 787–91

[61] Calderon-Montano J M, Burgos-Moron E, Perez-Guerrero C and Lopez-Lazaro M 2011 A review on the dietary flavonoid kaempferol *Mini Rev. Med. Chem.* **11** 298–344

[62] Tzeng C-W, Yen F-L, Wu T-H, Ko H-H, Lee C-W, Tzeng W-S and Lin C-C 2011 Enhancement of dissolution and antioxidant activity of kaempferol using a nanoparticle engineering process *J. Agric. Food Chem.* **59** 5073–80

[63] Sala A, Recio M C, Schinella G R, Máñez S, Giner R M, Cerdá-Nicolás M and Ríos J-L 2003 Assessment of the anti-inflammatory activity and free radical scavenger activity of tiliroside *Eur. J. Pharmacol.* **461** 53–61

[64] Velagapudi R, Aderogba M and Olajide O A 2014 Tiliroside, a dietary glycosidic flavonoid, inhibits TRAF-6/NF-κB/p38-mediated neuroinflammation in activated BV2 microglia *Biochimica et Biophysica Acta (BBA) – Gen. Subj.* **1840** 3311–9

[65] Qin N, Li C-B, Jin M-N, Shi L-H, Duan H-Q and Niu W-Y 2011 Synthesis and biological activity of novel tiliroside derivants *Eur. J. Med. Chem.* **46** 5189–95

[66] Goto T, Horita M, Nagai H, Nagatomo A, Nishida N, Matsuura Y and Nagaoka S 2012 Tiliroside, a glycosidic flavonoid, inhibits carbohydrate digestion and glucose absorption in the gastrointestinal tract *Mol. Nutr. Food Res.* **56** 435–45

[67] Gundimeda U, McNeill T H, Barseghian B A, Tzeng W S, Rayudu D V, Cadenas E and Gopalakrishna R 2015 Polyphenols from green tea prevent antineuritogenic action of Nogo-A via 67-kDa laminin receptor and hydrogen peroxide *J. Neurochem.* **132** 70–84

[68] Charlton A J, Baxter N J, Khan M L, Moir A J G, Haslam E, Davies A P and Williamson M P 2002 Polyphenol/Peptide binding and precipitation *J. Agric. Food Chem.* **50** 1593–601

[69] Zanchi D, Poulain C, Konarev P, Tribet C and Svergun D I 2008 Colloidal stability of tannins: astringency, wine tasting and beyond *J. Phys. Condes. Matter* **20** 494224

[70] Nakagawa K, Okuda S and Miyazawa T 1997 Dose-dependent incorporation of tea catechins, (-)-epigallocatechin-3-gallate and (-)-epigallocatechin, into human plasma *Biosci. Biotechnol., Biochem.* **61** 1981–5

[71] Kang S-Y, Seeram N P, Nair M G and Bourquin L D 2003 Tart cherry anthocyanins inhibit tumor development in ApcMin mice and reduce proliferation of human colon cancer cells *Cancer Lett.* **194** 13–9

[72] Jayaprakasam B, Vareed S K, Olson L K and Nair M G 2005 Insulin secretion by bioactive anthocyanins and anthocyanidins present in fruits *J. Agric. Food Chem.* **53** 28–31

[73] Tsuda T, Horio F and Osawa T 1999 Absorption and metabolism of cyanidin 3-O-β-D-glucoside in rats *FEBS Lett.* **449** 179–82

[74] Ernst I M A, Wagner A E, Lipinski S, Skrbek S, Ruefer C E, Desel C and Rimbach G 2010 Cellular uptake, stability, visualization by 'Naturstoff reagent A', and multidrug resistance

protein 1 gene-regulatory activity of cyanidin in human keratinocytes *Pharmacol. Res.* **61** 253–8
[75] Hafizi S and Dahlbäck B 2006 Gas6 and protein S *FEBS J.* **273** 5231–44
[76] John W S 2013 *Handbook of Vitamins* 5th edn (Boca Raton, FL: CRC Press) pp 89–124
[77] Mann K G 1999 Biochemistry and physiology of blood coagulation *Thromb. Haemostasis* **82** 165–74
[78] Sangster J 1989 Octanol water partition coefficients of simple organic compounds *J. Phys. Chem. Ref. Data* **18** 1111–229
[79] Rappolt M 2006 *Advances in Planar Lipid Bilayers and Liposomes* ed A L Liu (New York: Academic) pp 253–83
[80] Cockbain A J, Toogood G J and Hull M A 2012 Omega-3 polyunsaturated fatty acids for the treatment and prevention of colorectal cancer *Gut* **61** 135–49
[81] Cockbain A J, Volpato M, Race A D, Munarini A, Fazio C, Belluzzi A, Loadman P M, Toogood G J and Hull M A 2014 Anticolorectal cancer activity of the omega-3 polyunsaturated fatty acid eicosapentaenoic acid *Gut* **63** 1760–8
[82] Boyd B J, Khoo S-M, Whittaker D V, Davey G and Porter C J H 2007 A lipid-based liquid crystalline matrix that provides sustained release and enhanced oral bioavailability for a model poorly water soluble drug in rats *Int. J. Pharm.* **340** 52–60
[83] Serieye S, Meducine F, Milosevic I, Fu L and Guillot S 2017 Interface tuning and stabilization of monoglyceride mesophase dispersions: food emulsifiers and mixtures efficiency *J. Colloid Interface Sci.* 26–34
[84] Guillot S, Salentinig S, Chemelli A, Sagalowicz L, Leser M E and Glatter O 2009 Influence of the stabilizer concentration on the internal liquid crystalline order and the size of oil-loaded monolinolein-based dispersion *Langmuir* **26** 6222–9
[85] Guillot S, Moitzi C, Salentinig S, Sagalowicz L, Leser M E and Glatter O 2006 Direct and indirect thermal transition from hexosomes to emulsified micro-emulions in oil-loaded monoglyceride-based particles *Colloids Surf.* 78–84
[86] Tyler A, Barringa M G, Parsons E S, McCarthy N L C, Ces O, Law R V, Seddon J M and Brooks N J 2015 Electrostatic swelling of bicontinuous cub lipid phases *Soft matter* **11** 3279–86
[87] Briggs J, Chung H and Caffrey M 1996 The temperature-composition phase diagram and mesophase structure characterization of the monoolein/water system *J. Phys. II* **6** 723–51
[88] Jin X, Zhang Z H, Sun E, Tan X B, Li S L, Cheng X D, You M and Jia X B 2013 Enhanced oral absorption of 20(S)-protopanaxadiol by self-assembled liquid crystalline nanoparticles containing piperine: *in vitro* and *in vivo* studies *Int. J. Nanomed.* **8** 641–52
[89] Bei D, Marszalek J and Youan B B C 2009 Formulation of dacarbazine-loaded cubosomes – Part I: influence of formulation variables *Aaps Pharmscitech* **10** 1032–9
[90] Nasr M, Ghorab M K and Abdelazem A 2015 *In vitro* and *in vivo* evaluation of cubosomes containing 5-fluorouracil for liver targeting *Acta Pharmaceut. Sin.* B **5** 79–88

IOP Publishing

Physics in Food Manufacturing
Case studies in fundamental and applied research
Megan J Povey

Chapter 12

3D printing: its future in food processing

Mat Francis

In this chapter the production of food materials via 3D printing (additive manufacturing (AM)) will be examined with a particular focus on extrusion and powder-based methods. In 2019, the global spend on 3D printers and materials was nearly US $10 billion. Due to its low cost and ability to rapidly prototype, 3D printing has been making inroads into aerospace and automotive engineering, industries which traditionally have used subtractive manufacturing. The food 3D printing market is expected to reach US $525 million by 2023. This chapter will examine the viability of applying 3D printing to the manufacture of food. The three main techniques used to produce food additively will be outlined as well as explanations of the advantages and some of the issues surrounding each method. The limitations of this technology as well as the processing challenges of integrating AM with other unit operations will also be discussed and the advantages of the production of food materials by AM will be presented.

12.1 Additive manufacturing

AM is the term used to describe the physico-spatial manufacture of objects by progressive addition, either by linear or layer deposition. This manufacturing method differs from subtractive or non-AM which is largely performed by advanced CNC (computer numerical control) machines which employ a suite of tools such as lathes, drills and bores in three dimensions to reduce an object to the required size and shape [1]. AM or rapid prototyping was first developed in 1981 by Hideo Kodama at the Nagoya Municipal Industrial Research Institute. This method used a mixture of polyester, styrene and an initiator [2]. A movable stage was immersed in the mixture which was selectively polymerised using UV illumination of the mixture as the stage was moved, constructing the model layer by layer. Different configurations were tested and transparent models were successfully created. A provisional patent was applied for but the application for the full patent was never completed due to funding issues.

doi:10.1088/978-0-7503-2596-7ch12

In 1984, Charles Hull developed a similar technique whereby custom parts could be produced through AM which he termed stereolithography or SLA [3]. This method uses UV irradiation focussed on the surface of a photoreactive polymer mixture. The patent for this process was awarded in 1986 and the same year Charles Hull started his own company, 3D systems, in Valencia California. 3D systems went on to release the first commercial additive manufacture device based on SLA in 1988 with the release of the SLA-1 [4]. The same year selective laser sintering or SLS was invented by Carl Deckard, an undergraduate at the University of Texas [5]. SLS uses a laser to sinter powdered particles together from a powder bed to create 3D structures layer by layer. After the top layer is sintered, the print is moved down one powder layer thickness and a fresh layer of powder is rolled over the top. The powdered bed is heated to keep the powder close to the melting point. A schematic of the SLS process is presented in figure 12.1. The polymer powders used in SLS are generally polyamides, polyaryletherketones, polystyrenes and thermoplastic elastomers [6, 7].

The next method of additive manufacture developed was fused deposition modelling or FDM (also known as fused filament fabrication or FFF). This method melts and extrudes thermoplastic filaments to create objects line by line [8]. Polymer is pushed into the heated nozzle using a stepper motor, the nozzle can be heated to different temperatures depending on the polymer being printed. The melted filament is then laid down line by line. The deposition is controlled with an XYZ gantry and the placing of the filament is determined by a computer design which slices the digital 3D model into individual layers. Upon cooling, the polymer solidifies forming a continuous structure. Due to the method of production there can be 'layering' if the layers do not fuse together. The gantry, extruded filament and the printing nozzle are shown in figure 12.2. In 1992, Scott Crump was awarded the patent for FDM and the first FDM commercial machine, called the 3D modeller, was released by his company Stratasys. This machine allowed the printing of waxes and polymers and cost around $182 000.

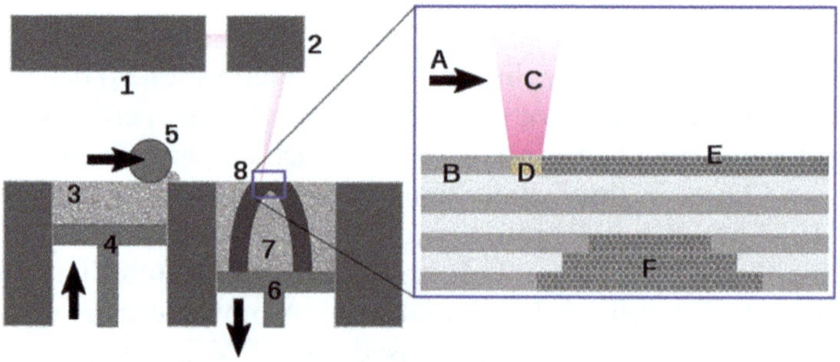

Figure 12.1. Selective laser sintering process (image courtesy of Gringer). 1 laser; 2 scanner system; 3 powder delivery system; 4 powder delivery piston; 5 roller; 6 fabrication piston; 7 fabrication powder bed; 8 object being fabricated (see inset); A laser scanning direction; B sintered powder particles (brown state); C laser beam; D laser sintering; E pre-placed powder bed (green state); F un-sintered material in previous layers.

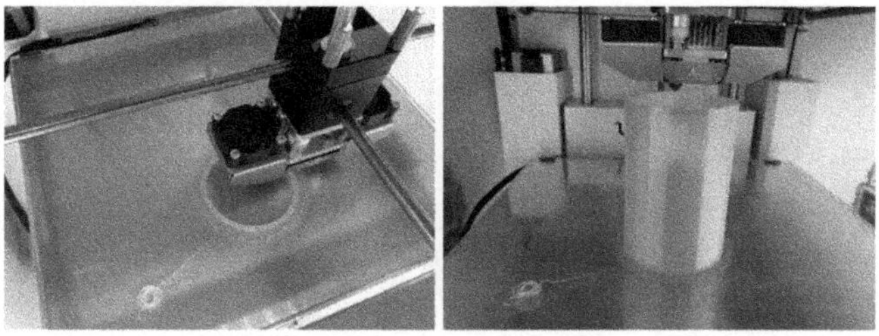

Figure 12.2. The FFF process, the print is built layer by layer from the bottom upwards.

In 1993, MIT developed a powder binding AM process based on custom ink jet technology and termed it '3D printing' to differentiate it from standard ink jet printing [9]. In this process a structure is built from a powder bed through the selected application of a binding agent which consists of adhesive polymers. The entire structure is subsequently heat-treated and the unbound powder is removed. During development, it was found that the cohesiveness of the powder bed needed to be increased through the addition of moisture so it could withstand the ballistic force of the binder droplets [10]. Due to the number of 3D printing technologies, this process is now termed binder jetting. An alternative to this is material jetting whereby a photopolymer resin is heated to achieve the optimal viscosity for printing. The print head subsequently travels over the build plate and deposits droplets of resin while a UV source on the print head cures the material, which essentially makes this method a combination of SLA and a jetting technique. Material jetting is popular for the production of printed electronics and medical devices [11, 12].

In 1995, the Fraunhofer Institute ILT in Aachen Germany developed the selective laser melting (SLM) process which uses a high intensity laser to bind metal powder particles together. During this process, the particles reach a completely molten state [13]. This technique is occasionally considered as a subset of the SLS technique since it uses a similar mechanism to bind plastic particles. However, there are important differences such as the temperatures used and the intensity of the laser beam used in SLM which makes the process harder to control and can cause warping in the resulting structures [14]. There is an alternative to SLM called electron beam melting (EBM) which uses an electron beam instead of a laser to sinter the metal particles. In EBM, electrons are accelerated to 60 keV under vacuum and focused onto the metal powder using electromagnetic lenses [15]. EBM allows for more power to be used than SLM but is material limited to metal powders and the build must be performed under helium conditions to limit the risk of electrostatic charging and smoke events [16, 17].

Since 2010, there have been several advances in the existing additive manufacture technologies. Digital Light Printing or DLP is a variation of SLA which was developed in 2012 [18]. In this method a digital light projector illuminates the photosensitive resin one complete layer of the deposited print at a time instead of at a single point. This makes DLP much faster than other methods of AM which is limited to print layers line by line. This technique also has reduced the problem of layering and increased homogeneity by using a continuous elevation of the build plate [11].

Unveiled as a commercial product in 2016, nanoparticle jetting (NPJ) is a combination of SLS and binder jetting. In the NPJ process irregularly shaped metal and plastic nanoparticles (build and support materials, respectively) are suspended in a liquid medium and then printed in a similar fashion as binder jetting onto a heated surface. The heat removes the continuous phase and deposits the dispersed plastic and metal constituents. After the supports are removed the metal constituents are sintered creating the final component. The advantages of NPJ are an increased resolution of the physical construction without a reduction in build time [19]. This process has been used to print microscale, low cost electronics [20, 21]. Nanoparticles have also been used in binder jetting whereby a solution of nanoparticles is jetted onto a bed of powdered metal in place of the standard adhesive binder (figure 12.3). Once in the interstices between the metal powder particles, the jetted nanoparticles are sintered (at a temperature lower than the sintering temperature of the metal particles). This method has been demonstrated to improve the structural integrity of the printed parts, compared with standard binder jetting. Depending on the concentration of nanoparticles in suspension there can be an up to a 150% increase in the flexural strength of the un-sintered parts [22].

In 2015 ISO/ASTM 52900 was introduced to standardise the terminology used in additive manufacture [23]. As of 2018, FDM was the most widely used technique, followed by SLS and SLA [24]. FDM has the advantage of being low cost, low maintenance and is able to be used with a wide range of materials. The drawbacks of FDM however include the long build time, potential delamination with changing temperatures and the visual evidence of layering on prints [25]. Prints built using powder-based techniques like SLS and SLM can be stronger and of higher resolution than those printed using other techniques, however, there are issues with handling and recycling the powders used in these processes [26]. Recently, there has been a great deal of research in the use of multi-process techniques (the combination of 3D printing technologies with existing manufacturing techniques

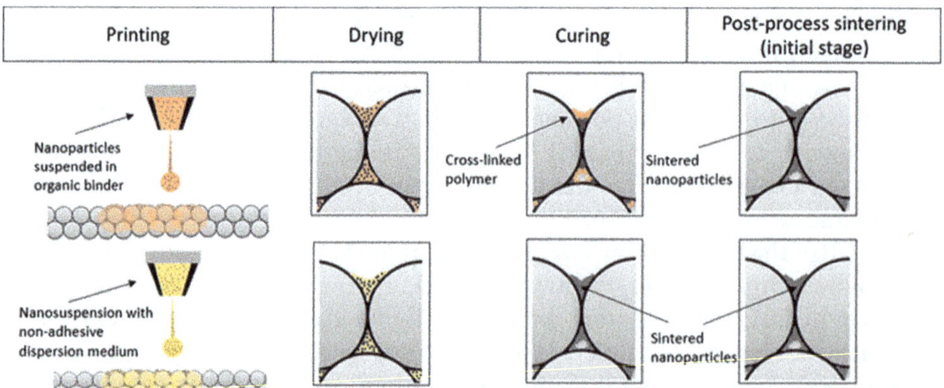

Figure 12.3. Schematic of nanoparticles used in the binder jetting process compared with standard binder jetting (reproduced from [22]).

or the integration of different 3D printing technologies [27]) and the ability to print multifunctional components, such as those with multiple materials, varying densities or ones which contain meta materials to reduce vibrations [28–30].

The three biggest manufacturing sectors that use AM are aerospace, consumer goods (including medical devices) and industrial goods [24]. Additive manufacture in the health sector has developed devices with increased bio-compatibility [31] and has also been used for printing replacement organs [32, 33]. AM has also been used in the construction industry with a house being constructed in 2014 entirely using 3D printing [34]. This demonstrates that 3D printing is capable of printing objects ranging in scale from nm to m. In general, components made using additive manufacture can have a longer lifetime, can be made with less energy and have a lower frequency of repair [35].

12.2 The production of food using AM

AM was first applied to food printing in 2007, when Cornell University released the FAB@Home Model 1, an open source design 3D printer which used FFF technology to print food using liquid food products as the feed material. Fab@Home is capable of producing everything from chocolate sculptures [36] to the scaffolds used in bone tissue engineering [37]. A similar printer, the Replicating Rapid 3D printer or 'RepRap', was the world's first open source 3D printer and it was designed to be able to print most of its own parts [38]. The RepRap printer also uses FFF technology to print objects at a very low cost [39]. The RepRap has been used to print processed cheese [40] and chocolate [41]. FAB@Home and RepRap are credited to be responsible for the marked increase of 'print at home' devices and the increasing popularity of 3D printing since their introduction [42–44].

Paste extruders, such as the Discovery paste extruder, are popular add-ons to FFF printers that can be used to adapt them from polymer printing to food printing. The issue with these extruders is that they have slow print speeds of around 20 mm s^{-1} [45]. FFF has been used to print a range of foods including pizza (both the dough and the tomato sauce were extruded), bread sticks in the shape of spoons, cat's paw-shaped dumplings and edible vases printed from spinach, beetroot, carrot and flour. Examples of some extruded foods using FFF are shown in figure 12.4. FFF has also been used to decorate the tops of baked goods with customised designs. The Foodini printer, which utilises FFF, can use pureed foods (or viscous

Figure 12.4. FFF printed foods: gummi frogs and chocolate figures (images courtesy of Cory Doctorow and Karlis Dambrans, Flickr).

liquids) as the printing medium. The benefit of these devices is that they can print nutritious food by using pureed vegetables as the printing material which potentially could increase the dietary intake of vegetables.

In addition to vegetables, meat has also been printed directly using FFF. In a study by Lipton *et al* [46], turkey and scallop meat were pureed into pastes and transglutaminase (a food grade enzyme used to bind proteins together) was added to the materials immediately prior to printing. This ensured that the meat pastes had the necessary rheological properties for extrusion but would reform protein networks after being printed. The extruded meat samples were subsequently cooked using a sous-vide. Upon cooking even though the samples shrank and warped, the material was judged to have the appropriate taste and texture by chefs. Surimi gels, made from silver carp flesh, were also successfully printed by altering the sodium chloride content of the gel mixture to cause it to harden following printing. A concentration of 1.5 g of NaCl per 100 g of gel was found to be an appropriate concentration for forming solid food prints using these fish gels [47].

Traditional meat production is responsible for more CO_2 emissions than any other food material and meat consumption is projected to increase over the next 20 years [48]. Meat production introduces additional pollution into the environment in the form of bacterial contamination of water and soil [49, 50]. In addition to environmental effects, excessive consumption of red and traditionally processed meats has been linked to an increased incidence of stroke, heart disease and cancer [51–53]. Due to this, alternative sources of meat are being investigated. One alternative to farmed meat is cultured meat where instead of traditional animal husbandry, which is subtractive, the meat is grown from the cellular level (additive). To form cultured meat, stem cells are combined with porous collagen spheres and placed inside a bioreactor. The bioreactor contains a growth medium which enables the stem cells to differentiate into myofibres (muscle tissue). It is currently possible to 3D print accurate and functional biological structures in this manner as this technology has been used in the medical sector to print tissues, organs and scaffolds for cells to grow on [32, 54]. For this to be applied at an industrial scale, however, an appropriate scaffold, i.e. the internal structure for the myofibres to grow, needs to be developed [55]. Modern Meadow, an American based start-up, is researching the viability of 3D printing stem cells onto agarose rods to create cultured meat structures which, if successful, would enable this process to be scaled up for meat production [56].

Other alternatives to farmed meat are analogues which consist of vegetable, fungal or insect-based proteins [57, 58]. Techniques like extrusion cooking and mixing/coagulation are commonly used to form meat analogues from plant and fungal protein sources [58, 59]. 3D printing is being examined as a method to produce meat replacements as the technique could print accurate structures and therefore make more realistic replacements. Imitation beef and chicken have been successfully printed using a combination of FFF and proteins isolated from insect and algal sources [60]. The textural properties of extruded meat analogues based on plant proteins have been found to be dependent on moisture concentration so control of the feed material characteristics are essential to create an analogue with the appropriate texture [61]. Concentrates of oat protein and faba bean protein have been used to extrusion print food structures in the form of a lattice. These materials were found to not have sufficiently high viscosity to retain their shape after printing and post processing in the form of freezing and baking was used to improve structural rigidity [62].

In South America, Asia and Africa insects are currently consumed as a protein source with over 1900 different species being consumed [63]. In America and Europe, insects and algae are currently used as fish or animal feed or as a food supplement instead of as a protein replacement in their own right [58, 64, 65]. Western countries have not traditionally consumed insects and there is a significant barrier to their acceptance as a food material with minimal processing [66]. The solution has been seen to present insect protein as part of more familiar food materials or within formulations [67]. By combining 3D printing and insect protein, there is the potential to create novel foods which might be more appealing to consumers. An example of this is the use of powdered mealworms, which are one of the best insect protein sources, to fortify flour. Mealworm powder fortified wheat flour has been made into a dough and successfully extruded using a Delta 2040 3D printer to form cylindrical shapes, which were frozen and then baked [63]. One issue found during the printing of the dough was that the enrichment of the wheat flour with high levels of insect protein resulted in a softer dough and over extrusion of the print, where too much material is extruded, and the resulting print is inaccurate and of low quality. Insect protein enriched flour has also been used to print butterfly and honeycomb shapes, showing that accuracy can be achieved with fortified flour doughs if the concentration of insect protein is maintained at appropriate levels [60]. Insect protein powders have also been added to soft cheese, chocolate and icing with spices like cinnamon and ginger added to achieve the correct consistency for 3D printing and to achieve structures with long term stability [68].

SLS has been used to successfully print powder-based food materials. The main materials studied for SLS food printing are sugar crystals or sugar rich powders such as Nesquik [69]. In food SLS printing a bed of powder is selectively sintered using either hot air or a laser [70]. The CandyFab project, which uses SLS technology, has printed large objects out of sugar including chains, toroidal sculptures, dodecahedrons and a Möbius strip. Examples of these powder-based 3D printed foods are shown in figure 12.5. In addition to sugar based materials, SLS has been used to print structures with different spice powders including cylinders, pyramids and pentagons using cinnamon, curry or paprika powders [71]. The main issue with SLS food printing is that it has limited food applications as food powders must be able to bind together through laser or hot air sintering and form a strong enough structure to maintain this shape.

The binder jetting process can be used to selectively deposit food particles suspended in a food safe binder upon a build plate. This method can be used to form solid food structures from these edible particles upon recrystallisation. One study used a binder of water, ethanol, xanthan gum and Tween 20 to print and recrystallise milled cellulose. This mixture created a brittle structure but after glucomannon was added to the binder mixture, stable 3D structures were

Figure 12.5. Sugar sculptures printed using SLS (images courtesy of Erik Charlton and Windell Oskay, Flickr).

formed [72]. In a separate study, alcohol has been used as a binder for sugar to binder jet print candy structures [62]. Binder jetting is thought to be the most versatile of the food printing technologies as it can be used to reconstitute a range of different powders and create a wide range of food products by altering the properties of the binder, yet it can achieve this with a small footprint. Rhee *et al* developed a 3D food printer which had 357 nozzles, providing a wide range of different materials which could be printed. The printing powders were created by cryogenically milling the different food materials. These powders were then mixed with water, printed and solidified using low heat. This technology has been examined by NASA as a solution for food production on long term space missions. The Systems and Materials Research Corporation is developing a 3D food printer that can print from a powdered store of proteins, carbohydrates, and micronutrients. The powders could come from a wide variety of sources, including non-traditional foods like insects and algae. This technology would potentially replace the meal-ready-to-eat (MRE) pouches that currently form the basis of astronaut's diets.

12.3 Properties of food materials used in 3D printing

The physical requirements of the food materials used as the feed material for 3D printing will differ depending on which printing technology is used. For FFF the most important physical properties are the rheological properties of the extruded material and the ultimate mechanical strength. Some foods are well suited to printing as upon heating they can be extruded (e.g. chocolate, cheese and icing mixtures). There are some issues when extrusion printing with these materials as the cooled food objects following printing will have altered characteristics to the initial materials. Le Tohic *et al* reported that cheese that has been melted and cooled through extrusion printing had a significant reduction in hardness and an altered microstructure, particularly of the fat droplets, even when compared with melted and cooled cheese [40]. Varying the cooling rate of structures extrusion printed with melted chocolate was also found to alter the ultimate rigidity of the structure [73].

Some materials will not have the required rheological properties, even upon heating, to be extrusion printed so will require additives to get the required extrusion of material during printing and the necessary rigidity and texture of the final product. Chocolate is a difficult material to print with as different levels of cocoa butter will alter the rheological properties of the material [74]. Magnesium stearate was added to dark chocolate to increase the flowability of the material but it also had the effect of altering the melting properties of the chocolate [75]. Understanding how the product crystallises on cooling and the glass transition temperature of the material being printed is essential for creating self-supporting structures [71, 76].

Cereal, meat, vegetable, dairy and fruit based materials all require additives to be printed successfully. The most common class of additives in food 3D printing are hydrocolloids. Hydrocolloids will modify the rheological properties and the texture of pureed food to allow it to be successfully printed. Instead of melting a material and creating a solid product upon cooling, these substances can be added to materials to allow them to be printed at lower temperatures and will form a gel network within the final object to stabilise it. Common hydrocolloid additives for food 3D printing include starch, pectin, gelatin, nanocellulose, alginate and carrageenan [77]. Dankar *et al* demonstrated that the concentration of agar and alginate can be varied to either reduce or increase the viscosity of the extruded material [78]. Mixtures of xanthum gum and gelatin at different concentrations were able to achieve a range of firmness and textures including self-supporting loose gels, gels akin to the texture of risotto and gels with the firmness of mushrooms [79]. Potato starch was added in different concentrations to mashed potato to determine the optimum concentration to achieve the required rigidity to form stable structures. Mashed potato with 2% potato starch was found to have the best extrudability and ultimate mechanical strength while mashed potato without starch lost cohesion over time [76].

The printing speed and the force applied to the material to achieve extrusion are important physical factors for FFF printing. The speed a material can be printed at will depend on its composition. 15% potato starch was added to lemon juice to create a novel food material, however due to the soft nature of the resulting gel, the print speed had to be limited to 35 mm s^{-1} to ensure a stable print [80]. Additionally, some materials will break up upon printing if too much force is applied, while others will require a certain amount of force so as not to obstruct the print head. When printing chocolate icing by extrusion, a slow speed but high pressure enables a smooth print with fewer air blockages [81]. When printing a banana and pectin mixture, Derossi *et al* reported that faster print speeds led to a reduced porosity in their final structure and an increased thickness as the filaments merged together [63].

In SLS, the definition of the print is directly related to the intensity and wavelength of the laser and the interaction between the laser and the food powder [82]. The fabricated parts also rely on the speed at which the laser traverses the powder bed with slower speeds generating higher density components [83]. For SLS the most important physical parameters of the powder are the melting temperature, the particle size and flowability [71]. There are limited powders that are appropriate for SLS as the powder must be agglomerate free and should be free flowing [76]. It has also been demonstrated that the crystal nature of the powder is important. Amorphous powders with elevated glass transition temperatures yield weaker, more brittle structures. These can be strengthened with the addition of another material with lower melting points. However, amorphous crystals do have the advantage of not causing product contraction upon cooling as is seen for crystalline materials [11].

The physical requirements for the suspended phase in binder jetting are very similar to those of the powder used in SLS but binder jetting has the added requirement that the continuous phase has the appropriate surface tension and viscosity [76]. Additionally, the saturation level of the binder (the amount of binder deposited through the print head) was found to affect the strength and surface quality more than the particle size of the powders or the binder properties [84]. The direction/orientation and speed at which the binder is applied was also found to have an effect on the accuracy of parts [10]. Following the application of binder, most prints formed using binder jetting are heat treated to form the final object. A study by Miyanaji *et al* examined the effect of heating temperatures after the powder had been bound together. It was found that higher temperatures reduced the porosity upon sintering and caused an increase in part shrinkage [85]. This demonstrates that of all the techniques, binder jetting is the most complicated to achieve repeatable, reliable prints with multiple materials.

12.4 Current challenges of AM of food

In addition to optimising each technique for each food material, one major challenge of using food 3D printing as a processing operation is the speed and reliability of the AM process itself [27]. Even with optimal settings and materials, it is difficult to repeatedly form consistent prints. As the technology matures, better understanding about how to control the final structures, sources of variation and the materials that can be combined in food 3D printing are being developed [73, 76, 80, 86]. As it is currently a slow operation, integrating this technique with other unit operations like baking or packaging will create bottlenecks in processing and manufacture. Food 3D printing has been increasing in speed over the past decade and the advent of multiple print nozzles, faster print speeds and the development of new technologies in traditional AM have all increased production rates, which will offer improved application in industrial or commercial settings. The effect of integration with other operations and in particular the effect of post processing on the printed structure and flavour are becoming better understood [87]. It is now well established that post processing methods like freezing and freeze-drying can be used to maintain the structure and texture of printed foods through cooking [62].

Another issue with the AM process is that the energy consumption of these processes is not well quantified. A study by Kellens *et al* estimated the energy cost of AM processes as 1 to 2 orders of magnitude higher than traditional machining or injection moulding and discussed the issue that there is little documentation of the production costs of the feed materials [88]. Additionally, barriers to AM becoming established as a sustainable process include the need for more attention to be focussed on the development and recycling of the materials used in the process and an improved energy use and quality of the prints [89]. As the technology matures, optimised digital 3D designs aim to reduce the amount of waste material during processing [90]. Overall, given it is an additive process, 3D printing is seen as a technology that will be capable of supporting sustainable production [91] but more thought needs to be given as to where it fits in the production chain and how it can be integrated with other processes.

Consumer acceptance of food 3D printing is one of the major hurdles facing this technology. Processed food was once viewed as convenient and affordable but over time consumers have come to prefer foods with lower levels of processing [92]. Ultra-processed foods, foods which have a high degree of processing, have come under scrutiny recently as being potentially harmful to human health, being linked to cancer and obesity [93–95]. Clean labelling in foods has recently gained momentum as consumers want to know the ingredients and their provenance. Due to this, food companies have begun to reformulate their foods with clearer labels and fewer ingredients where possible [96]. 3D printed food is generally viewed as highly processed and the formulations used for printing require numerous additives to get the appropriate rheological properties so education about the safety of these additives, their purpose and the process itself will be essential. Additionally, when introducing new 3D printed products the degree of novelty has to be carefully controlled as some products can be considered too novel by consumers [97]. This can lead to consumers being concerned that 3D printed foods are not 'real' and are only 'food-like' [98]. It has been shown that these attitudes can be overcome through education about 3D printed foods and how they are made [99]. Acceptance of these foods will increase as consumers are exposed to more 3D printed food and are influenced by social pressures. It has been demonstrated that consumers will follow emerging consumption patterns [100].

There is currently a lack of regulatory framework for the production of 3D printed foods [81]. Food production is required to meet stringent standards to ensure food safety and to reduce the risk to consumers. Given that 3D printed foods can be made from a range of additives, regulations need to be brought in to limit ingredients and concentrations. There is also a food safety concern with food 3D printing as its very nature will allow food to be easily created in a range of environments, including those which are not food safe. Additionally, the process of additively manufacturing some food involves the heating and cooling of organic materials (which generally have high water content). This has led to a high degree of bacterial and yeast contamination during the production of food printed in this way [101]. The contamination has been attributed to the contact between the food material and the different physical parts of the 3D printer. Therefore, hygienic design principles will need to be employed when developing the next generation of food 3D printers. 3D printed food has also been demonstrated to have a short shelf life with many foods only being stable for two hours [87, 102]. One solution for this is to store foods in their raw form and print the products on demand [76, 103].

12.5 Advantages of additive manufactured food

The main potential advantage of 3D printed food is the ability to create personalised foods. It has been demonstrated that by creating personalised food for consumers you are increasing their knowledge of food choices and nutrition [104]. Modern increased portion sizes have been linked with increased energy intake and obesity [105, 106], thus, 3D printing will give consumers the ability to control portion size which has become of increasing importance due to the rise of obesity in the western world. In a study by Cohen *et al*, some of the advantages that a non-food professional could expect from home-based 3D printed food were discussed. The major benefits are the time saved through the automation of the food production process and the increased knowledge that the computer controlled food printer will be able to provide consumers real time, for example, information on portion size and the nutritional content of the food they print [79].

One of the major current applications of food 3D printing is creating textured food materials from puree for the elderly and patients with dysphagia. Dysphagia, the weakening or loss of muscle control during swallowing, affects 60% of the residents in nursing homes [77]. Dysphagia can lead to feelings of isolation, depression and anxiety during mealtimes. 84% of patients think that eating should be an enjoyable experience but only 45% of patients with dysphagia find it so [107]. A common solution is to alter the consistency of the food to make it easier to swallow [108]. However, there is a potential for patients with dysphagia to develop a dislike of pureed foods due to the appearance, monotonous flavour and consistency of the food [109]. Food 3D printing allows for textural control of the food materials that are printed due to the resolution at which materials can be printed and the wide range of different additives that can be added to pureed foods to alter their structure and flavour. PERFORMANCE—PERsonalised FOod using Rapid MAnufacturing for the Nutrition of elderly ConsumErs is a European Union project that is developing new 3D printers to print food specifically for the elderly [76, 102]. 3D printers are increasingly being used in hospitals and schools for medicine, patient care and education [110–112].

Currently in the UK, one in three children is overweight or obese [113]. Additionally, 80% of children do not eat recommended levels of fruit and vegetables. It is difficult to encourage children to eat vegetables as often they will have a bitter taste and textural differences that contrasts sharply with the sweet and savoury food they primarily consume [114]. 3D printing is currently being used in a wide range of different settings in schools including design, spatial awareness, engineering and the environment so children are becoming familiar with the technology [115]. Food 3D printing is a potential method to create novel food shapes for children to encourage them to eat more fruit and vegetables [116]. Enabling children to design their own food structures which they subsequently consume will not only increase their vegetable intake but educate them on food processing and 3D printing. As 3D food printing technology develops, it has been argued that allowing children to develop their own designs is important for their personal development and for the technology at large [117]. There is some evidence that 3D printed food has been seen to encourage both healthy consumption and an interest in technology in younger children [111].

Recent developments in new product development (NPD) of food has focussed on the creation of functional foods, i.e. foods which offer health benefits in addition to meeting basic nutritional needs. The increasing market for functional food over the last two decades has been attributed to the increasing population and current health trends [118]. The main ingredients that add functionality to foods are fibre, probiotics, fatty acids and micronutrients such as flavonoids [119]. These are generally added to traditional foods through fortification, but 3D printing offers an opportunity to develop novel foods with added health benefits. An example of these ingredients being incorporated in 3D printed foods is the development of a non-dairy cereal snack that contains probiotics [120]. It has been demonstrated that novel foods will be more acceptable to certain sectors of the population but acceptance levels can change depending on the formulation [121] so it will be essential to promote the health benefits of the products.

One third of food produced each year is wasted with end consumers being the greatest generators of food waste [122]. Food waste is mainly produced in the home due to a lack of skills for reusing leftovers [123]. By printing in the home precise quantities of food there will no longer be an issue of leftover food. As 3D printed food is an additive process as opposed to the traditional subtractive methods of cooking, printed food will use less material and produce less waste than traditional subtractive processing [124]. Additionally, 3D printed food can be made from a range of waste materials so potentially any uneaten food could be used to print new foods, removing the skill issue currently hampering leftover reuse. Currently, there are examples of fruit juices and cereal bars that are already made from waste materials so consumer acceptance of eating food waste in principle should not be an issue [125].

12.6 The future of additive manufacture in food processing

The rate of development of food 3D printing has been extremely rapid. From the first food printers developed in 2007, just over a decade ago which could only print a few materials, the printers of today are able to print more materials, faster and with higher accuracy. Due to the advantages of food 3D printing and the ability of people being able to personalise their food it is not a technology that is likely to go away. The most likely early adopters will be schools and nursing homes as 3D printed food is an excellent educational tool and can be used for beneficial health outcomes. The commercial food sector has also adopted the technology as there is already 3D printed food in restaurants in Spain and America. Additionally, Byflow has opened restaurants in London and the Netherlands where the entire menu is comprised of 3D printed food. 3D food printing technology is rapidly changing and as it is better understood, the process itself will be easier to control and the potential applications expanded.

There are many challenges still facing the additive manufacture of food such as integration of these techniques with other unit operations, which can operate at higher speeds, and a lack of regulatory frameworks. As the speed of 3D printing increases and the technology develops, the integration of this process with other operations will become more viable. As more products are created through food 3D printing and brought to market, regulations and standards will have to be brought in to cover these materials. Whether 3D printing will ever be able to produce customised food at the rate at which similar techniques operate in the food industry (e.g. extrusion cooking) or if it will remain a predominantly customised food drive technology remains to be seen.

References

[1] Mourtzis D, Milas N and Athinaios N 2018 Towards machine shop 4.0: a general machine model for CNC machine-tools through OPC-UA *Procedia CIRP* **78** 301–6

[2] Kodama H 1981 Automatic method for fabricating a three-dimensional plastic model with photo-hardening polymer *Rev. Sci. Instrum.* **52** 1770–3

[3] Ambrosi A and Pumera M 2016 3D-printing technologies for electrochemical applications *Chem. Soc. Rev.* **45** 2740–55

[4] Hull C W 2015 The birth of 3D printing IRI achievement award address *Res. Technol. Manag.* **58** 25–9

[5] Deckard C and Beaman J J 1988 Process and control issues in selective laser sintering *ASME Prod. Eng. Div. PED.* **33** 191–7

[6] Schmidt M, Pohle D and Rechtenwald T 2007 Selective laser sintering of PEEK *CIRP Ann. – Manuf. Technol.* **56** 205–8

[7] Fish S, Booth J C and Kubiak S T *et al* 2015 Design and subsystem development of a high temperature selective laser sintering machine for enhanced process monitoring and control *Addit. Manuf.* **5** 60–7

[8] Ahn D, Kweon J H, Kwon S, Song J and Lee S 2009 Representation of surface roughness in fused deposition modeling *J. Mater. Process. Technol.* **209** 5593–600

[9] Sachs E, Cima M and Cornie J *et al* 1993 Three-dimensional printing the physics and implications *Ann. CRIP* **42** 257–60

[10] Miyanaji H, Momenzadeh N and Yang L 2018 Effect of printing speed on quality of printed parts in binder jetting process *Addit. Manuf.* **20** 1–10

[11] Stansbury J W and Idacavage M J 2016 3D printing with polymers: challenges among expanding options and opportunities *Dent. Mater.* **32** 54–64

[12] Mishra S, Barton K L, Alleyne A G, Ferreira P M and Rogers J A 2010 High-speed and drop-on-demand printing with a pulsed electrohydrodynamic jet *J. Micromech. Microeng.* **20** 095026

[13] Yadroitsev I, Bertrand P and Smurov I 2007 Parametric analysis of the selective laser melting process *Appl. Surf. Sci.* **253** 8064–9

[14] Kruth J P, Froyen L, Van Vaerenbergh J, Mercelis P, Rombouts M and Lauwers B 2004 Selective laser melting of iron-based powder *J. Mater. Process. Technol.* **149** 616–22

[15] Murr L E, Gaytan S M and Ramirez D A *et al* 2012 Metal fabrication by additive manufacturing using laser and electron beam melting technologies *J. Mater. Sci. Technol.* **28** 1–14

[16] Körner C 2016 Additive manufacturing of metallic components by selective electron beam melting – a review *Int. Mater. Rev.* **61** 361–77

[17] Ho C M B, Ng S H and Yoon Y J 2015 A review on 3D printed bioimplants *Int. J. Precis. Eng. Manuf.* **16** 1035–46

[18] Zheng X, Deotte J and Alonso M P *et al* 2012 Design and optimization of a light-emitting diode projection micro-stereolithography three-dimensional manufacturing system *Rev. Sci. Instrum.* **83** 125001

[19] Dilberoglu U M, Gharehpapagh B, Yaman U and Dolen M 2017 The role of additive manufacturing in the era of industry 4.0 *Procedia Manuf.* **11** 545–54

[20] Ko S H, Chung J, Hotz N, Nam K H and Grigoropoulos C P 2010 Metal nanoparticle direct inkjet printing for low-temperature 3D micro metal structure fabrication *J. Micromechan. Microeng.* **20** 125010

[21] Jeong J A, Lee J, Kim H, Kim H K and Na S I 2010 Ink-jet printed transparent electrode using nano-size indium tin oxide particles for organic photovoltaics *Sol. Energy Mater. Sol. Cells* **94** 1840–4

[22] Bai Y and Williams C B 2018 The effect of inkjetted nanoparticles on metal part properties in binder jetting additive manufacturing *Nanotechnology* **29** 395706

[23] Attar H, Bermingham M J, Ehtemam-Haghighi S, Dehghan-Manshadi A, Kent D and Dargusch M S 2019 Evaluation of the mechanical and wear properties of titanium produced by three different additive manufacturing methods for biomedical application *Mater. Sci. Eng.* A **760** 339–45

[24] Moreau C 2018 *The State of 3D Printing 2018* (Villejuif, France: Sculpteo)

[25] Huang S H, Liu P, Mokasdar A and Hou L 2013 Additive manufacturing and its societal impact: a literature review *Int. J. Adv. Manuf. Technol.* **67** 1191–203

[26] Gao W, Zhang Y and Ramanujan D *et al* 2015 The status, challenges, and future of additive manufacturing in engineering *CAD Comput. Aided Des.* **69** 65–89

[27] Guo N and Leu M C 2013 Additive manufacturing: Technology, applications and research needs *Front Mech. Eng.* **8** 215–43

[28] MacDonald E and Wicker R 2016 Multiprocess 3D printing for increasing component functionality *Science* **80** 353(6307)

[29] Tofail S A M, Koumoulos E P, Bandyopadhyay A, Bose S, O'Donoghue L and Charitidis C 2018 Additive manufacturing: scientific and technological challenges, market uptake and opportunities *Mater. Today* **21** 22–37

[30] Gornet T 2016 *History of Additive Manufacturing Non-SL Systems Introduction of Low-cost 3D Printers* (Fort Collins, Co: Wohlers Associates Inc.) pp 5–18

[31] Javaid M and Haleem A 2018 Additive manufacturing applications in medical cases: a literature based review *Alexandria J. Med.* **54** 411–22

[32] Melchels F P W, Domingos M A N, Klein T J, Malda J, Bartolo P J and Hutmacher D W 2012 Additive manufacturing of tissues and organs *Prog. Polym. Sci.* **37** 1079–104

[33] Mironov V, Kasyanov V and Markwald R R 2011 Organ printing: from bioprinter to organ biofabrication line *Curr. Opin. Biotechnol.* **22** 667–73

[34] Hager I, Golonka A and Putanowicz R 2016 3D printing of buildings and building components as the future of sustainable construction? *Procedia Eng.* **151** 292–9

[35] DebRoy T, Wei H L and Zuback J S *et al* 2018 Additive manufacturing of metallic components—process, structure and properties *Prog. Mater. Sci.* **92** 112–224

[36] Periard D, Schaal N, Schaal M, Malone E and Lipson H 2007 Printing food *18th Solid Freeform Fabrication Symp. SFF 2007* 564–74

[37] Diogo G S, Gaspar V M, Serra I R, Fradique R and Correia I J 2014 Manufacture of β-TCP/alginate scaffolds through a Fab@home model for application in bone tissue engineering *Biofabrication* **6** 025001

[38] Kentzer J, Koch B, Thiim M, Jones R W and Villumsen E 2011 An open source hardware-based mechatronics project: the replicating rapid 3-D printer *2011 4th Int. Conf. Mechatronics* 1–8

[39] Bradshaw S, Bowyer A and Haufe P 2010 The intellectual property implications of low-cost 3D printing *SCRIPTed.* **7** 5–31

[40] Le Tohic C, O'Sullivan J J and Drapala K P *et al* 2018 Effect of 3D printing on the structure and textural properties of processed cheese *J. Food Eng.* **220** 56–64

[41] Alimanova M, Zholdygarayev A and Kozhamzharova D 2017 Overview of a low-cost self-made 3D food printer *13th Int. Conf. on Electronics, Computer and Computation (ICECCO)*

[42] Lipton J I, Cohen D and Heinz M *et al* 2009 Fab @ home model 2: towards ubiquitous personal fabrication devices *Solid Freeform Fabrication Symposium*

[43] Petersen E and Pearce J 2017 Emergence of home manufacturing in the developed world: return on investment for open-source 3-D printers *Technologies* **5** 7

[44] Kocisko M, Teliskova M, Torok J and Petrus J 2017 Postprocess options for home 3D printers *Procedia Eng.* **196** 1065–71

[45] Pusch K, Hinton T J and Feinberg A W 2018 Large volume syringe pump extruder for desktop 3D printers *HardwareX* **3** 49–61

[46] Lipton J, Arnold D and Nigl F *et al* 2010 Mutli-material food printing with complex internal structure suitable for conventional post-processing *21st Annual Int. Solid Freeform Fabrication Symp.* 809–15

[47] Wang L, Zhang M, Bhandari B and Yang C 2018 Investigation on fish surimi gel as promising food material for 3D printing *J. Food Eng.* **220** 101–8

[48] Godfray H C J, Aveyard P and Garnett T *et al* 2018 Meat consumption, health, and the environment *Science* **361** eaam532

[49] Wang H, Xu J and Liu X *et al* 2018 Study on the pollution status and control measures for the livestock and poultry breeding industry in northeastern China *Environ. Sci. Pollut. Res.* **25** 4435–45

[50] Li L and Li M 2018 Relationship between livestock pollution and the environmental behaviour of farmers: a case study of Xiantao, China *Nat. Environ. Pollut. Technol.* **17** 269–76

[51] Ritchie H, Reay D S and Higgins P 2018 The impact of global dietary guidelines on climate change *Glob. Environ. Chang.* **49** 46–55

[52] Micha R, Wallace S K and Mozaffarian D 2010 Red and processed meat consumption and risk of incident coronary heart disease, stroke, and diabetes mellitus: a systematic review and meta-analysis *Circulation* **121** 2271–83

[53] Chen G C, Lv D B, Pang Z and Liu Q F 2013 Red and processed meat consumption and risk of stroke: a meta-analysis of prospective cohort studies *Eur. J. Clin. Nutr.* **67** 91–5

[54] An J, Teoh J E M, Suntornnond R and Chua C K 2015 Design and 3D printing of scaffolds and tissues *Engineering* **1** 261–8

[55] Bhat Z F and Fayaz H 2011 Prospectus of cultured meat – advancing meat alternatives *J. Food Sci. Technol.* **48** 125–40

[56] Portanguen S, Tournayre P, Sicard J, Astruc T and Mirade P S 2019 Toward the design of functional foods and biobased products by 3D printing: a review *Trends Food Sci. Technol.* **86** 188–98

[57] Smetana S, Mathys A, Knoch A and Heinz V 2015 Meat alternatives: life cycle assessment of most known meat substitutes *Int. J. Life Cycle Assess.* **20** 1254–67

[58] van der Weele C, Feindt P, Jan van der Goot A, van Mierlo B and van Boekel M 2019 Meat alternatives: an integrative comparison *Trends Food Sci. Technol.* **88** 505–12

[59] Grahl S, Palanisamy M, Strack M, Meier-Dinkel L, Toepfl S and Mörlein D 2018 Towards more sustainable meat alternatives: how technical parameters affect the sensory properties of extrusion products derived from soy and algae *J. Clean. Prod.* **198** 962–71

[60] Yang F, Zhang M and Bhandari B 2017 Recent development in 3D food printing *Crit. Rev. Food Sci. Nutr.* **57** 3145–53

[61] Chiang J H, Hardacre A K and Parker M E 2020 Extruded meat alternatives made from Maillard-reacted beef bone hydrolysate and plant proteins. Part I – Effect of moisture content *Int. J. Food Sci. Technol.* **55** 649–59

[62] Lille M, Nurmela A, Nordlund E, Metsä-Kortelainen S and Sozer N 2018 Applicability of protein and fiber-rich food materials in extrusion-based 3D printing *J. Food Eng.* **220** 20–7

[63] Severini C, Azzollini D, Albenzio M and Derossi A 2018 On printability, quality and nutritional properties of 3D printed cereal based snacks enriched with edible insects *Food Res. Int.* **106** 666–76

[64] Ścieszka S and Klewicka E 2018 Algae in food: a general review *Crit. Rev. Food Sci. Nutr.* **0** 1–10

[65] Makkar H P S, Tran G, Heuzé V and Ankers P 2014 State-of-the-art on use of insects as animal feed *Anim. Feed Sci. Technol.* **197** 1–33

[66] Gere A, Radványi D and Héberger K 2019 Which insect species can best be proposed for human consumption? *Innov. Food Sci. Emerg. Technol.* **52** 358–67

[67] Tan H S G, van den Berg E and Stieger M 2016 The influence of product preparation, familiarity and individual traits on the consumer acceptance of insects as food *Food Qual. Prefer.* **52** 222–31

[68] Soares S and Forkes A 2014 *Proc. of the 16th Int. Conf. on Engineering and Product Design Education* 426–31

[69] Wegrzyn T F, Golding M and Archer R H 2012 Food Layered Manufacture: A new process for constructing solid foods *Trends Food Sci. Technol.* **27** 66–72

[70] Sun J, Peng Z, Yan L, Fuh J and Hong G S 2015 3D food printing—An innovative way of mass customization in food fabrication *Int. J. Bioprinting* **1** 27–38

[71] Godoi F C, Prakash S and Bhandari B R 2016 3D printing technologies applied for food design: status and prospects *J. Food Eng.* **179** 44–54

[72] Holland S, Tuck C and Foster T 2018 Selective recrystallization of cellulose composite powders and microstructure creation through 3D binder jetting *Carbohydr. Polym.* **200** 229–38

[73] Lanaro M, Forrestal D P and Scheurer S *et al* 2017 3D printing complex chocolate objects: Platform design, optimization and evaluation *J. Food Eng.* **215** 13–22

[74] Afoakwa E O, Paterson A and Fowler M 2007 Factors influencing rheological and textural qualities in chocolate – a review *Trends Food Sci. Technol.* **18** 290–8

[75] Mantihal S, Prakash S, Godoi F C and Bhandari B 2017 Optimization of chocolate 3D printing by correlating thermal and flow properties with 3D structure modeling *Innov. Food Sci. Emerg. Technol.* **44** 21–9

[76] Liu Z, Zhang M, Bhandari B and Wang Y 2017 3D printing: Printing precision and application in food sector *Trends Food Sci. Technol.* **69** 83–94

[77] Tan C, Toh W Y, Wong G and Li L 2018 Extrusion-based 3D food printing – materials and machines *Int. J. Bioprinting* **4** 0–13

[78] Dankar I, Haddarah A, El Omar F, Sepulcre F and Pujolà M 2018 Assessing the microstructural and rheological changes induced by food additives on potato puree *Food Chem.* **240** 304–13

[79] Cohen D, Lipton J, Cutler M, Coulter D, Vesco A and Lipson H 2009 Hydrocolloid printing: a novel platform for customised food production *Solid Freeform Fabrication Symp.* 807–17

[80] Yang F, Zhang M, Bhandari B and Liu Y 2018 Investigation on lemon juice gel as food material for 3D printing and optimization of printing parameters *LWT – Food Sci. Technol.* **87** 67–76

[81] Dankar I, Haddarah A, Omar F E L, Sepulcre F and Pujolà M 2018 3D printing technology: the new era for food customization and elaboration *Trends Food Sci. Technol.* **75** 231–42

[82] Kruth J P, Levy G, Klocke F and Childs T H C 2007 Consolidation phenomena in laser and powder-bed based layered manufacturing *CIRP Ann. – Manuf. Technol.* **56** 730–59

[83] Shirazi S F S, Gharehkhani S and Mehrali M *et al* 2015 A review on powder-based additive manufacturing for tissue engineering: selective laser sintering and inkjet 3D printing *Sci. Technol. Adv. Mater.* **16** 1–20

[84] Vaezi M and Chua C K 2011 Effects of layer thickness and binder saturation level parameters on 3D printing process *Int. J. Adv. Manuf. Technol.* **53** 275–84

[85] Miyanaji H, Zhang S, Lassell A, Zandinejad A and Yang L 2016 Process development of porcelain ceramic material with binder jetting process for dental applications *JOM* **68** 831–41

[86] Vancauwenberghe V, Delele M A and Vanbiervliet J *et al* 2018 Model-based design and validation of food texture of 3D printed pectin-based food simulants *J. Food Eng.* **231** 72–82

[87] Sun J, Zhou W, Yan L, Huang D and Lin L Y 2018 Extrusion-based food printing for digitalized food design and nutrition control *J. Food Eng.* **220** 1–11

[88] Kellens K, Mertens R, Paraskevas D, Dewulf W and Duflou J R 2017 Environmental impact of additive manufacturing processes: does AM contribute to a more sustainable way of part manufacturing? *Procedia CIRP* **61** 582–7

[89] Peng T, Kellens K, Tang R, Chen C and Chen G 2018 Sustainability of additive manufacturing: an overview on its energy demand and environmental impact *Addit. Manuf.* **21** 694–704

[90] Griffiths C A, Howarth J, De Almeida-Rowbotham G, Rees A and Kerton R 2016 A design of experiments approach for the optimisation of energy and waste during the production of parts manufactured by 3D printing *J. Clean. Prod.* **139** 74–85

[91] Ford S and Despeisse M 2016 Additive manufacturing and sustainability: an exploratory study of the advantages and challenges *J. Clean. Prod.* **137** 1573–87

[92] Sajdakowska M, Jankowski P, Gutkowska K, Guzek D, Żakowska-Biemans S and Ozimek I 2018 Consumer acceptance of innovations in food: A survey among Polish consumers *J. Consum. Behav.* **17** 253–67

[93] Monteiro C A, Cannon G and Levy R B *et al* 2019 Ultra-processed foods: what they are and how to identify them *Public Health Nutr.* **22** 936–41

[94] Monteiro C A, Moubarac J C, Levy R B, Canella D S, Da Costa Louzada M L and Cannon G 2018 Household availability of ultra-processed foods and obesity in nineteen European countries *Public Health Nutr.* **21** 18–26

[95] Fiolet T, Srour B and Sellem L *et al* 2018 Consumption of ultra-processed foods and cancer risk: results from NutriNet-Santé prospective cohort *Brit. Med. J.* **360** k322

[96] Van Gunst A, Roodenburg A J C and Steenhuis I H M 2018 Reformulation as an integrated approach of four disciplines: a qualitative study with food companies *Foods* **7** 64

[97] Stolzenbach S, Bredie W L P and Byrne D V 2013 Consumer concepts in new product development of local foods: traditional versus novel honeys *Food Res. Int.* **52** 144–52

[98] Lupton D and Turner B 2018 'I can't get past the fact that it is printed': consumer attitudes to 3D printed food *Food, Cult. Soc.* **21** 402–18

[99] Brunner T A, Delley M and Denkel C 2018 Consumers' attitudes and change of attitude toward 3D-printed food *Food Qual. Prefer.* **68** 389–96

[100] Higgs S, Liu J, Collins E I M and Thomas J M 2019 Using social norms to encourage healthier eating *Nutr. Bull.* **44** 43–52

[101] Severini C, Derossi A, Ricci I, Caporizzi R and Fiore A 2018 Printing a blend of fruit and vegetables. New advances on critical variables and shelf life of 3D edible objects *J. Food Eng.* **220** 89–100

[102] Lipton J I, Cutler M, Nigl F, Cohen D and Lipson H 2015 Additive manufacturing for the food industry *Trends Food Sci. Technol.* **43** 114–23

[103] Porter K and Phipps J 2015 *3D Opportunity Serves it Up : Additive Manufacturing and Food* (New York: Deloitte University Press) pp 1–20

[104] German J B and Watzke H J 2004 Foods for health and delight *Compr. Rev. Food Sci. Food Saf.* **3** 145–51

[105] Rolls B J, Morris E L and Roe L S 2002 Portion size of food affects energy intake in normal-weight and overweight men and women *Am. J. Clin. Nutr.* **76** 1207–13

[106] Wansink B, Painter J E and North J 2005 Bottomless bowls: why visual cues of portion size may influence intake *Obes Res.* **13** 93–100

[107] Ekberg O, Hamdy S, Woisard V, Wuttge-Hannig A and Ortega P 2002 Social and psychological burden of dysphagia: its impact on diagnosis and treatment *Dysphagia* **17** 139–46

[108] Zargaraan A, Rastmanesh R, Fadavi G, Zayeri F and Mohammadifar M A 2013 Rheological aspects of dysphagia-oriented food products: a mini review *Food Sci. Hum. Wellness* **2** 173–8

[109] Kouzani A Z, Adams S and Whyte D J *et al* 2017 3D Printing of food for people with swallowing difficulties *KnE Eng.* **2** 23

[110] Huston C 2013 The impact of emerging technology on nursing care: warp speed ahead *OJIN: Online J. Issues Nurs.* **18** 1

[111] Hamilton C A and Alici G in het Panhuis M 2018 3D printing vegemite and marmite: redefining 'breadboards *J. Food Eng.* **220** 83–8

[112] Khaled S A, Burley J C, Alexander M R, Yang J and Roberts C J 2015 3D printing of five-in-one dose combination polypill with defined immediate and sustained release profiles *J. Control. Release* **217** 308–14

[113] Rudolf M, Perera R, Swanston D, Burberry J, Roberts K and Jebb S 2019 Observational analysis of disparities in obesity in children in the UK: has Leeds bucked the trend? *Pediatr. Obes.* 1–7

[114] Nekitsing C, Blundell-Birtill P, Cockroft J E and Hetherington M M 2018 Systematic review and meta-analysis of strategies to increase vegetable consumption in preschool children aged 2–5 years *Appetite* **127** 138–54

[115] Ford S and Minshall T 2019 Invited review article: Where and how 3D printing is used in teaching and education *Addit. Manuf.* **25** 131–50
[116] Derossi A, Caporizzi R, Azzollini D and Severini C 2018 Application of 3D printing for customized food. A case on the development of a fruit-based snack for children *J. Food Eng.* **220** 65–75
[117] Eisenberg M 2013 3D printing for children: What to build next? *Int. J. Child-Computer Interact.* **1** 7–13
[118] Gul K and Singh A K 2016 Jabeen R. Nutraceuticals and functional foods: the foods for the future world *Crit. Rev. Food Sci. Nutr.* **56** 2617–27
[119] Vicentini A, Liberatore L and Mastrocola D 2016 Functional foods: trends and development *Ital. J. Food Sci.* **28** 338–52
[120] Zhang L, Lou Y and Schutyser M A I 2018 3D printing of cereal-based food structures containing probiotics *Food Struct.* **18** 14–22
[121] Torri L, Tuccillo F, Bonelli S, Piraino S and Leone A 2020 The attitudes of Italian consumers towards jellyfish as novel food *Food Qual. Prefer.* **79** 103782
[122] Girotto F, Alibardi L and Cossu R 2015 Food waste generation and industrial uses: a review *Waste Manag.* **45** 32–41
[123] Stancu V, Haugaard P and Lähteenmäki L 2016 Determinants of consumer food waste behaviour: two routes to food waste *Appetite* **96** 7–17
[124] Jin Y, Du J and He Y 2017 Optimization of process planning for reducing material consumption in additive manufacturing *J. Manuf. Syst.* **44** 65–78
[125] Ferreira M S L, Santos M C P, Moro T M A, Basto G J, Andrade R M S and Gonçalves E C B A 2013 Formulation and characterization of functional foods based on fruit and vegetable residue flour *J. Food Sci. Technol.* **52** 822–30

Lightning Source UK Ltd.
Milton Keynes UK
UKHW051615250921
391114UK00003B/82

9 780750 325943